PROJECT ENERGY '93

Real Energy Technologies: Environmentally Responsible – Ready for Today

Conference Proceedings

Editors:

R.E. BILLINGS
E. DAYTON

International Academy of Science

Independence, Missouri
USA

International Academy of Science
26900 East Pink Hill Road
Independence, Missouri 64057-3284

Library of Congress Catalog Card Number: 93-061304

ISBN: 0-9631634-4-2

First Edition 1993

Printed in the United States of America.

FOREWORD

An unprecedented gathering of scientists and industry leaders from a wide range of energy fields met to present their latest research and applications of clean energy systems at the international conference, **Project Energy '93**.

Responding to the theme *"Real Energy Technologies: Environmentally Responsible -- Ready for Today"*, these representatives from all sides of the energy table discussed the newest energy-related technologies currently available for use on the one hand, and the latest innovations in energy systems already in use on the other. Among the presentations were reports on the latest developments in battery and energy storage systems, the historic Russian hydrogen airplane, and hydrogen fuel cell, electric, and natural gas vehicles.

The Conference itself was a demonstration in energy conservation, as parts of it were video conferenced from locations in New York, Washington, D.C., Virginia, and London. This demonstration showed the viability of recent innovations in using this technology, as well as pointing out that by distributing a technical conference electronically to multiple sites, the information disseminated becomes available to a much wider audience, while energy is conserved and environmental burdens reduced, since there is a reduced need for travel.

With the urgency of settling onto and into a technically viable, environmentally responsible, economically practical, sustainable energy system nipping at our heels, the International Academy of Science and the Project Energy '93 Organizing Committee and Sponsors present the Proceedings of this historic Conference.

Roger E. Billings
Eileen Dayton

ACKNOWLEDGMENTS

The Organizing Staff wishes to thank the following individuals for their support and assistance with Project Energy '93:

SPONSORS

Alternative Energy Research Organization
American Council for an Energy Efficient Economy
American Gas Association
American Hydrogen Association
American Institute of Chemical Engineers
American Society of Gas Engineers
American Solar Energy Society
American Wind Energy Association
Clean Energy Research Institute
Edison Electric Institute
Environmental Protection Agency
Fusion Power Associates
International Academy of Science
International Association for Hydrogen Energy
International Association for Energy Economics
Northeast Sustainable Energy Association
Solar Energy Industries Association

ORGANIZING COMMITTEE

T. Nejat Veziroglu
International Association for Hydrogen Energy

Robert Sechler
Society of Automotive Engineers

Rayford G. Anthony
American Institute of Chemical Engineers

Keith Rutledge
Solar Energy Expo & Rally

Roy Hamill
Sandia National Laboratory

Robert Wills
Skyline Engineering

Dorry Tooker
Brookhaven National Laboratory

Laural L. Heinze
Columbia Gas

Richard D. Stern
Environmental Protection Agency

Robert P. Wichert
Sacramento Municipal Utility District

Marv Clement
Pacific Northwest Laboratory

Nelson Hay
American Gas Association

Larry Sherwood
American Solar Energy Society

Richard L. Ittelilag
Missouri Public Service Division of UtiliCorp United

Michael Grabscheid
Northeast Sustainable Energy Association

INTERNATIONAL ACADEMY OF SCIENCE BOARD OF DIRECTORS

PROJECT ENERGY '93

CONTENTS

CONTENTS

CONTENTS

CONTENTS

CONTENTS

CONTENTS

CONTENTS

CHAPTER 1 -- OPENING SESSION

Compressed Digital Video Conferencing Demonstration
-- Introduction --

Roger E. Billings
President
International Academy of Science
Independence, Missouri, U.S.A.

Dr. Roger E. Billings
President
International Academy of Science

It is my sincere pleasure to welcome you to the beginning of Project Energy '93. This morning, scientists are gathering from all over the world and from throughout the United States to participate in a three-day discussion on various energy technologies -- their problems and their potentials for a better world.

At the beginning of this important conference, we would like to introduce a new technology which we believe, in the future years, will not only help conserve energy, but perhaps more importantly, will make the technologies discussed in these kinds of conferences available to wider and broader audiences. This is a technology which will be referred to today as Compressed Digital Video Technology. It is a technology which has been maturing for the past decade in the United States and around the world. This morning, by the miracle of video conferencing, we would like to open this conference by bringing you a message from several different parts of the world.

As we begin this demonstration, I would like to express my appreciation to U.S. Sprint and GPT, two important technology organizations who are collaborating with the International Academy of Science to make this presentation available.

At this point then, I would like to take us live to New York City where from the United Nations Dr. Carlos Nones Sucre will greet us.

• • • from New York • • •

Carlos Nones Sucre
Chief of Science Technology Branch
United Nations
New York, New York, USA

Dr. Carlos Nones Sucre
Chief of Science Technology Branch
United Nations

Thank you, Dr. Billings, for introducing me and for your invitation to speak, via teleconferencing, in the opening session of Project Energy '93 Conference organized by the International Academy of Science. It gives me great pleasure to welcome the hundreds of participants from the United States and a number of other countries now gathering in Kansas City, Missouri.

Your conference is particularly akin to the work of the United Nations in its utilization of the potential of compressed digital video technology, a form of communication which we follow with a special interest. Likewise, the subject matter of the Conference, namely the study of energy technologies that are environmentally sound and that are ready for commercial deployment, as well as the event's future-oriented approach, represents a welcome contribution to global ends.

As the United Nations prepares to commemorate its fiftieth anniversary, it is worth recalling the words of its Charter which reflect the ever actual goal of the international community of achieving "social progress and better standards of life in larger freedom".

Hence, we could see the United Nations global conferences on the Human Environment held in Stockholm in 1972; on New and Renewable Sources of Energy in Nairobi in 1980 and last year on Environment and Development -- also referred to as the Rio de Janeiro Earth Summit -- as important landmarks in the quest of the international community for a balanced and integrated approach to environmental development issues as well as for a new global partnership for sustainable development. The outcome of those conferences, and particularly Agenda 21 adopted by the Earth Summit, underscore the crucial importance that the choice energy sources and technologies have for the future of society in our planet.

Within that framework, may I add, the United Nations Secretariat was given a mandate by its General Assembly to promote and help to coordinate the analysis of technological developments and trends worldwide. For example, and International Association of Technology Assessment and Forecasting Institutions is being established this year under the auspices of the United Nations, with a goal of linking in their common endeavors -- especially through newly available electronic technology -- several hundred institutions involved in this field worldwide.

While the entire programme of Project Energy '93 will be followed by me and my colleagues as a source of new ideas, I wish to underline the particular importance that we attribute to the presentations to be made this morning on *The Coming Energy Revolution* (by Christopher Flavin of the Worldwatch Institute, Washington, D.C.) and on *The Architecture of Energy Systems for a Brighter Twenty-First Century* (by David Scott of the University of Victoria, Canada).

I look forward to Project Energy '94 next year in which at least four groups in separate conference sites will participate simultaneously in discussing and sharing scientific and technical information. I also wish to express the hope that by the United Nations fiftieth anniversary in 1995, something close to worldwide scientific teleconferencing will have been tried.

May I conclude in congratulating the participants and organizers, for making this worthy initiative possible. I sincerely wish that your deliberations will be successful and personally enriching to all.

Thank you.

• • • from Washington D.C. • • •

John R. Brodman
Acting Deputy Assistant Secretary, International Affairs,
Department of Energy
Washington, D.C. USA

Mr. John R. Brodman
Acting Deputy Assistant Secretary, International Affairs
Department of Energy

Good morning, Dr. Billings. On behalf of the United States Department of Energy I would like to welcome all participants to the Project Energy '93 Conference. The Secretary of Energy, Hazel O'Leary, regrets that she cannot be with you this morning, but she fully and completely recognizes the importance of your work.

The scientific community that all of you represent is the future. You are the twenty-first century. As time goes on, it becomes increasingly difficult to balance our competing goals of energy security, economic competitiveness, and environmental enhancement. That is why we have shifted our priorities to meet the needs of the changing world. Our new mission at the Department of Energy will result in significant funding increases to enhance energy efficiency and develop new and renewable energy resources; to support basic research and the goal of advanced science, math, technology, and engineering education; to support the transfer of new and developing technology to the private sector; and to work with industry and the scientific community to develop environmentally beneficial green technologies including new techniques for waste cleanup and environmental protection.

Every dollar saved for energy efficiency and commercialization of renewable technologies strengthens our economy, reduces pollution, and lowers our dependence on unreliable supplies of oil.

What you are doing in Kansas City over the next three days is extremely important. One measure of your success will be your ability to improve the lives of ordinary men. As I said before, you and your colleagues are the future.

With that I want to thank you for the opportunity to greet you this morning, and I extend to you our sincerest wishes for a successful conference. Thank you, Dr. Billings.

• • • from London • • •

Roger King
Director of Public Affairs
Society of Motor Manufacturers and Traders, Ltd.
London, U.K.

Mr. Roger King
Director of Public Affairs
The Society of Motor Manufacturers and Traders, Ltd.

Good morning, Dr. Billings. A very warm welcome to all at this quite excellent conference you have organized on what is starting actually the longest day. I think in London we are just a little bit further ahead on that longest day than you are, but you have allowed yourself extra daylight, I hope, to discuss what is a very vital issue in terms of the development of our planet.

Now, I have looked at what Sir Hal Miller was scheduled to talk to you about before he was unfortunately affected by an influenza outbreak. The title of his short presentation was, "To study the cooperation between government, industry, and the consumer in order to implement new and better energy technologies." I think as an entrance examination for a top class university, that is a question that would require a great deal of effort and knowledge to answer. But, of course, in terms of the use of energy, it has given us, as a civilization, enormous opportunities to increase our social well being and to enable primary, of course, mobility. That is what the SMMT stands for in Great Britain, The Society of Motor Manufactures and Traders, the trade association for our motor industry. We are as aware as anyone of the challenge that we face as an industry in meeting that basic desire of mankind for mobility and also eliminating as much as possible the impact that desire has upon man's surroundings.

How can we go about meeting the requirements of everyone for the mobility they desire? Indeed, it is a case of working between government, between industry, and consumer -- a partnership, if you like. At the very outset, what we will require as we push back the frontiers of technology is a steady, stable economic climate -- not one which goes in a series of booms and busts so that developments which cost a lot of money cannot be appreciated in terms of commercial payback. So we certainly need that in order to set about meaningful development, design, and research. We need government to provide us with better roads, a better public transport system, and, as I have said, sensible financial policies. Given those options, industry can of course set out to develop new ideas. There is no shortage of them. The question is whether the technologies involved can actually come to a commercial fruition. We are still some way from that.

Meanwhile, we have to, as an industry, refine ever further known technologies -- looking at the internal combustion engine, improving its efficiency; looking at lean burn technology; looking at diesel vehicles; looking at hybrid cars, a mixture of diesel and electric; and, of course, ultimately looking at alternative energy. For indeed, as we move into the next millennium, that is the only real way that we can go forward together.

So of all those things, by working hard towards developing alternative energy, by looking at the techniques applicable, by looking at how far we can push back the frontiers of research and development, your conference will certainly pave the way over its three days for us to learn from it as to how we can develop the technologies of tomorrow.

Our warmest congratulations to you all on this conference. Every best wish. From a sunny London, thank you for inviting us to participate in the opening sequence of your conference.

• • • from Virginia • • •

Christopher J. Rooney
President, Sprint Data Services Division
Sprint
Reston, Virginia, U.S.A.

Christopher J. Rooney
President, Sprint Data Services Division
Sprint, U.S.A.

I am honored to be with you all this morning, and I commend the Academy for this gathering. As you indicated, Dr. Billings, we have been pioneering this technology for about ten years, and we really appreciate the opportunity to be with you this morning and use it in this very important form of Project Energy '93. Because, as you have indicated, what this technology achieves, as we saw it ten years ago, was the ability it had to allow others to share technical information across wide expanses without travel and thereby preserve precious energy and preserve the environment. I am very pleased to hear about the '94 conference via video from four sites, because that is really the way the world is going.

I wanted to share just a few moments this morning with your audience there that in basic science, mathematics, engineering, language training, this technology is being used across the world today to create what we call multiple campuses, where you have limited resource in terms of professors. They can be teaching the same class in two, three, four, or even six universities across the world in a virtual campus environment.

We are using this same technology to retrain our work force here in the United States to allow multiple training sites to simultaneously be retrained in industry and in government. We are using the same technology in rural school systems, where there are limited instructors for certain topics, in order to bring things like Japanese training to summer schools in rural southwest Virginia and Kentucky.

So this is a technology that we are very, very supportive of in supporting science in general -- both domestically in the U.S. and international science. We think it is a science application that will also allow tremendous conservation of our environment and of our energy in the future. We are beginning to see the application of video teleconferencing in the home, which will help us further take the office to the home, thereby eliminating the need for people to commute to their offices within the next ten years -- eliminating the need for so much motor travel and taxi travel, allowing people to eliminate parking spaces or reduce office spaces and the need to heat and cool those office spaces.

Video teleconferencing, in the application of this technology, should allow us to have tremendous energy savings both here in the U.S. and internationally and allow us to protect those valuable environmental resources -- to bring that together in a way that will not cause us to lose any efficiency or any productivity. In many ways, we see it increasing efficiency and productivity.

We are honored that you all would use this technology. Our home, our corporate office, ironically, is there in Kansas City. Sprint has about 10,000 employees in the Kansas City area. So we welcome you to Kansas City. We certainly are honored to participate today. On behalf of U.S. Sprint we wish you a very, very successful conference. Thank you all.

Compressed Digital Video Conferencing Demonstration -- Conclusion --

Roger E. Billings
President
International Academy of Science
Independence, Missouri, U.S.A.

Dr. Roger E. Billings
President
International Academy of Science

The International Academy of Science is involved in the development of a video conferencing technology which we have named Windeo™. The Windeo™ technology is being specifically designed to give us the ability to conduct the kind of conference that we would like to conduct in 1994. We have made an estimate that if this technology were to be made available to the scientific community at large, it would be possible to expand the attendance at these technical conferences by almost an order of magnitude while at the same time, conserving over 6 million barrels of oil for jet fuel travel each year, resulting in a reduction of carbon dioxide by-product to the greenhouse effect of 141,000 tons per year.

If we expand the technology from just the scientific community into the conferences of the business community, we now see a potential reduction in jet fuel consumption of 60 million barrels a year, which relates to 4.4% of the total airline fuel consumption and a reduction of CO_2 pollution to the environment of 1.2 million tons of carbon dioxide each year.

This substantial reduction in the utilization of our planet's resources is very significant, but I still believe much more significant is the potential of increasing the attendance at these technical conferences.

The problem with international video conferencing traditionally has been the expense. When you transmit very large quantities of information over long distances, it requires communication capabilities with wide bandwidths, and these capabilities traditionally have been quite expensive. Recently, primarily through the miracle of fiberoptics and the advancements in the microprocessor, things are changing. What has traditionally been cost- prohibitive is now becoming the only decision that makes any sense from the point of view of budget constraints.

One of the things that has made a major impact on the bandwidth requirement, or the communication channel requirement, is a technology referred to as "Compressed Digital Video". As we have been listening to these speakers (and I am hoping that in a minute we can pull them back so you can look at this one more time), we have only been transmitting the picture that has changed. We send a whole frame at first and then, through the continuing cycles as the screen is updated, the computer looks at one picture, compares it with the next picture, and determines what in the picture changed. The changes are the only information that is being transmitted over these loops. As a result, we are talking about major reductions in the amount of communication capability required.

In Figure 1 you can see that we are talking typically an order of five to one or 20% of the information is all that needs to be sent in order to have a very acceptable video presentation, as you have seen today.

The Windeo™ technology is a software program coupled with a conference management and organization scheme that takes this compressed digital video one more step. In Figure 2 you see we have the screen divided into a dynamic window and a static window. In this case, which we hope to demonstrate in the conference next year, the static window would contain the speaker's viewgraphs, slides, or charts. Typically, that information can be transmitted one time, held in the local computer, and then displayed on the screen for as long as the speaker desires to discuss that information. Meanwhile, the figure of the person speaking is in the small window, thereby achieving a very substantial reduction in the amount of communication bandwidth required.

In the example which we have before you, the only information which is being updated live is the information in the very small window. As a result of that, we have another order of magnitude drop in the bandwidth requirement of the conferencing.

In Figure 3 you see what we would propose at the next conference. You will notice that I have been a little ambitious and added Tokyo to the form, and we hope we will be able to add other sights. One of the problems is, if we begin the conference on schedule in Kansas City, in London the closing session will end at around 10:00 pm. Unfortunately, in Tokyo the conference will take place starting late at night and go on until very early in the morning. I explained to my friends in Tokyo, however, that that is no different than flying around the world and getting up with jetlag. This way they can have jetlag without the jet.

At any rate, the potential of having these kinds of conferences take place, and even having a panel discussion with individuals from each of the different locations, is a very real possibility with today's technology.

Now going just one step further, in 1995 we would like to expand the Windeo™ technology to include what we call Desktop Video Conferencing. With the advancements in modem digital communications, the expansions of Internet and other communication technologies, and of course with the advancement of Compressed Digital Video itself and the Windeo™ techniques, we believe it will be possible to actually add a person to an ongoing, live video conference by having them dial up the conference center by modem and then participate with their own personal computer at their home. If their personal computer does not have a method of capturing a video signal (if they do not have a camera and a video acquisition device), they still will be able to participate as a recipient of the conference and will actually see this picture on their screen. They will see the speaker speaking. They will hear live the reaction of the audience and the questions, and they will be able to type in a question on their computer or even call an 800 number and ask a question by audio. So I think the potential of the video conferencing is very exciting, and we expect very great things in the future.

Figure 1

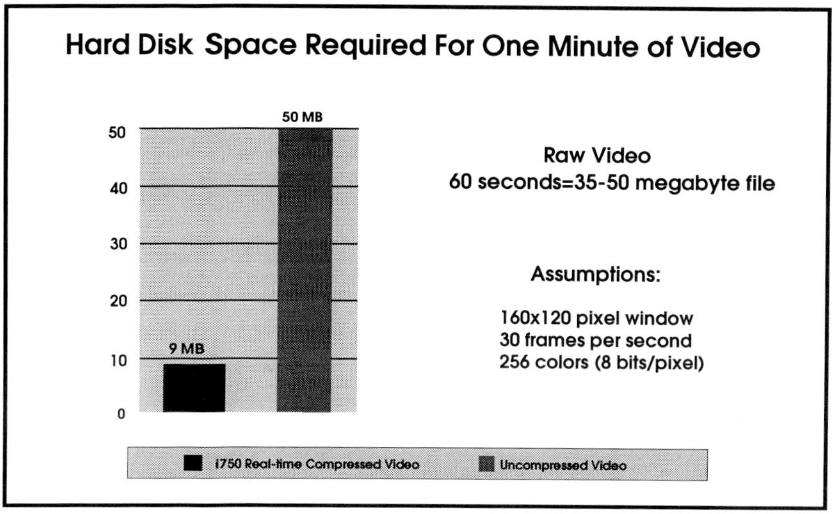

Figure 2

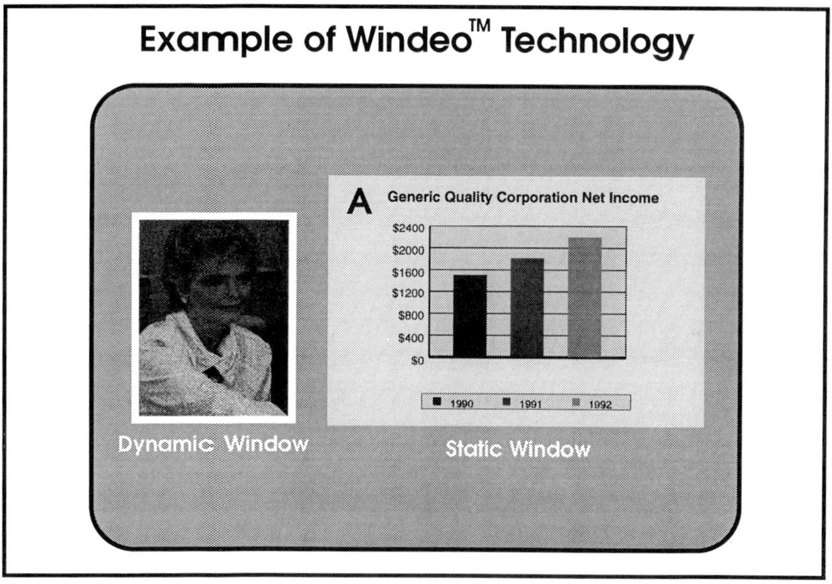

Figure 3

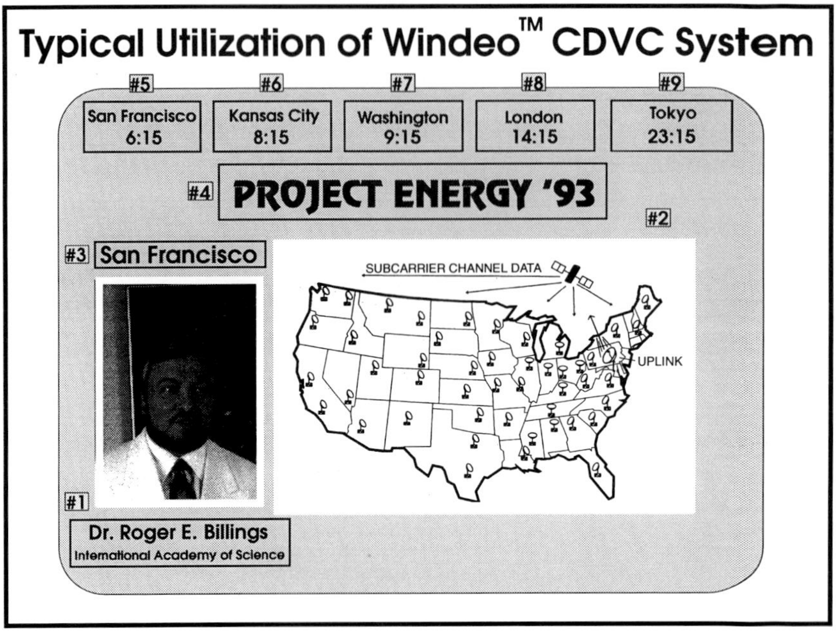

Typical Utilization of Windeo™ CDVC System
#5
San Francisco 6:15
#6
Kansas City 8:15
#7
Washington 9:15
#8
London 14:15
#9
Tokyo 23:15
#4
PROJECT ENERGY '93
#2
#3
San Francisco
SUBCARRIER CHANNEL DATA
UPLINK
#1
Dr. Roger E. Billings
International Academy of Science

The Coming Energy Revolution

Christopher Flavin
Vice-President for Research
Worldwatch Institute, U.S.A.
1776 Massachusetts Ave., NW
Washington, D.C. 20036-1904 U.S.A.

Background:

Mr. Flavin's research efforts focus on energy resource, technology, and policy issues. He has served as an advisor for the Clinton-Gore Campaign, the Washington-based Business Council for a Sustainable Energy Future, and the Government of Bhutan, and has participated in the United Nations Conference on New and Renewable Sources of Energy in Nairobi, and in the Earth Summit in Rio de Janeiro.

DE-CARBONIZING THE U.S. ENERGY SYSTEM
A COST-EFFECTIVE NATIONAL STRATEGY

President Clinton's commitment to return U.S. emissions of greenhouse gases to the 1990 level in the year 2000 offers both a challenge and an opportunity. The challenge is to meet the obligations of the Rio Climate Treaty of 1992, and lead worldwide efforts to stabilize the global atmosphere. The opportunity is to reorder our national energy priorities in ways that will strengthen the nation's economic competitiveness. Many federal energy programs remain oriented to the energy technologies and priorities of the past; the new Climate Action Plan provides a timely chance to redirect them.

A host of new, cost-effective energy supply technologies now reaching the market will allow both energy bills and carbon dioxide emissions to be cut. Indeed, growing evidence suggests that the kind of rapid technological change that has transformed the computer and telecommunications industries may begin to sweep the energy sector in the decade ahead. Failure to develop a strong new energy strategy will leave the United States poorly positioned to compete in the fast-moving global economy of the twenty-first century.

Today's energy supply system is dominated by fossil fuels--accounting for roughly 85 percent of the primary supply. Oil alone provides 40 percent of the total, half of which is now imported. The other 15 percent of the nation's energy comes mainly from hydro, biomass, and nuclear energy. In order to reduce carbon dioxide emissions, there are two basic energy supply strategies. The first is to improve the carbon-efficiency of the energy supply system by relying on cleaner fossil fuels, and on more advanced technologies. The second is to develop a range of carbon-free renewable energy sources, including new options such as solar, wind, and geothermal energy. In order to stabilize and then reduce U.S. carbon dioxide emissions, it will be necessary to pursue both strategies simultaneously.

NEW POWER OPTIONS

Natural gas is the cleanest of the fossil fuels, capable of substituting for all of the important current uses of oil and coal, including power generation and transportation. It virtually eliminates two of the more serious problems related to coal combustion--emission of sulfur dioxide and heavy metals. In addition, natural gas produces 30 percent less carbon dioxide per BTU compared to oil, and 43 percent less than coal. (See Figure 1.) More importantly, natural gas is well-suited for use in efficient, decentralized energy systems that allow even lower rates of emission.

Production and use of natural gas have risen steadily in recent years--up by more than 20 percent in the United States since 1986. Based on new resource estimates, it appears that gas use can be increased substantially during the next two decades, providing more energy for heating and cooling buildings and powering factories. In addition, power plants and automobiles represent nearly untapped markets for natural gas--now dominated respectively by coal (55 percent of power generation) and oil (95 percent of automotive fuels). If the natural gas market continues to function freely, and if the remaining barriers to gas demand are eliminated, gas may surpass oil as the country's leading energy source within a decade. Unlike oil, more than 95 percent of the gas is likely to come from North America.

Electric power generation accounts for one-third of U.S. carbon dioxide emissions, but new technologies will allow those figures to be reduced substantially. Gas-fired, combined-cycle power plants installed in the past few years are 35 percent more efficient than a typical coal plant. These plants allow a 60 percent reduction in carbon dioxide emissions compared to coal plants; more efficient gas turbines that will be available by the end of the decade allow a 70 percent reduction. (See Figure 2.) Combined-cycle plants running on gasified coal have also been built, but they allow only a 16 percent reduction in carbon dioxide emissions. Wind, solar, geothermal, and closed-loop biomass power plants allow a nearly 100 percent reduction in carbon dioxide emissions.

Natural gas-fired combined cycle and cogeneration plants represent more than half the new power plants under construction in the United States. However, the need to limit carbon dioxide emissions, together with the Clean Air Act requirements to reduce sulfur dioxide and nitrogen oxide emissions, suggest the need for more aggressive efforts to replace coal and oil with natural gas in electric power generation. More than 1400 small coal and oil-fired power plants will reach the age of 40 by the year 2000--most of them small, inefficient, and heavily polluting. Collectively, these plants represent 30 million tons of annual carbon emissions. Many of these can either be closed or cost-effectively "re-powered" as efficient natural gas plants.

A more advanced technology, the electrochemical fuel cell, will allow even sharper reductions in carbon dioxide emissions within the decade. Fuel cells are relatively small, modular devices that can convert natural gas to electricity, with about the same efficiency as an advanced combined-cycle system. However, fuel cells' smaller sizes allow them to be placed in a range of small industrial, commercial, and even residential buildings, which means that the waste heat is readily available for heating and cooling, and electric line losses are reduced. The resulting full-cycle system will permit a 70-80 percent reduction in carbon dioxide emissions compared to today's systems. Fuel cells are currently being produced and tested by a number of companies in the United States and Japan. The technology's workability has been proven, and widespread deployment is expected by the year 2000.

RENEWABLE ENERGY TECHNOLOGIES

Similar breakthroughs are occurring in a range of technologies for generating electricity without the use of fossil fuels. Utility-scale, grid-connected power stations running on wind, geothermal, biomass, and solar energy, have all been successfully deployed in recent years. California, for example, leads the world in wind power, with more than 16,000 turbines (1,700 MW) installed already--sufficient to meet the power needs of San Francisco.

These renewable energy resources are actually far more abundant than are fossil fuels, capable of meeting U.S. power needs many times over. It is estimated, for example, that solar power plants covering an area of 23,000 square miles could meet all U.S. power needs. Government assessments show that the wind resource of the Dakotas alone is adequate to meet 90 percent of U.S. power needs. And a Union of Concerned Scientists study shows that agricultural wastes and energy crops could meet between 10 and 50 percent of most Midwest states' power needs. Full-cycle analysis shows that some of the renewable energy technologies are less land-intensive than coal combustion.

As with fuel cells, a combination of technological advances, and the economics inherent in mass production of modular devices will continue to drive down the costs of renewable energy technologies in the years ahead. The cost of wind power has already fallen from 32 cents per kilowatt-hour in 1982 to 8 cents in the late eighties, and less than 6 cents for the new generation of wind turbines now being marketed. Among the advances that have made such gains possible are improved aerodynamics, lower cost blades, and better electronic controls. Analysts believe that mass production combined with low-cost financing could cut wind power costs to less than 4 cents per kilowatt-hour a decade from now. At that point, wind power plants are likely to be produced on a large scale, in much the way that tractors and appliances are today.

The most abundant renewable energy source is sunshine, which is now being harnessed commercially by photovoltaic cells. Some 50 megawatts of solar cell generating capacity was manufactured worldwide in 1992, most of it going to off-the-grid applications such as rural homes, water pumps, and hand-held calculators. However, large utility markets for solar cells beckon in the late nineties--providing supplementary power at the far reaches of power grids, and supplementary power on rooftops. These applications could provide hundreds of megawatts of market in the late nineties, bridging the way to far greater markets after the year 2000.

THE NEED FOR FEDERAL LEADERSHIP

Diversifying the nation's electric supply by reducing the use of coal, and increasing the use of gas and renewable energy, can reduce the carbon dioxide emissions in power generation by up to 30 percent in the year 2010 (which would reduce projected national emissions by 10 percent.) The key is to reduce the dominance of coal--from 56 percent to 30 percent--while introducing a combination of efficient gas-fired plants (30 percent of supply), and carbon-free renewable energy generators (30 percent) to provide the bulk of the remaining power needs. (These figures are based on an assumed growth in electricity demand of 1.5 percent per year, and a decline in the nuclear share from 22 to 10 percent.)

Such goals will be spurred by the increasingly competitive power industry, and the falling cost of new technologies. Combined-cycle natural gas and wind plants have both fallen to between 4 and 6 cents per kilowatt-hour in the past few years, providing low-cost options for reducing carbon dioxide emissions. (See Figures 3 and 4.) (A decade ago, utilities paid 8-10 cents per kilowatt-hour for new coal and nuclear plants.) However, these reductions will only be achieved if active federal efforts to commercialize new technologies are accompanied by state government commitments to use their ratemaking powers to encourage fuel diversification and low-emission power plants.

While more advanced generation technologies are ready to be deployed on a large scale, they still face formidable market barriers that are causing U.S. industry to fall behind the aggressive commercial efforts of Europe and Japan. The biggest barrier is that most electric utilities are not yet comfortable with renewable energy technologies, and are reluctant to commit to generating sources that have not yet been widely deployed. So far, the federal commitment to renewables has been largely rhetorical. Commercial development has not been effectively supported, and the federal energy technology budget continues to be dominated by less promising options such as nuclear fusion and new methods of coal combustion.

The first step in a genuine commitment to a new generation of technologies is to increase R&D expenditures so that they can continue to advance rapidly. (Fiscal 1995 will offer the Clinton administration its first real opportunity to present an energy R&D budget that is consistent with new economic and environmental realities, including the new Climate Action Plan.) In order to achieve identified technological goals, renewable energy R&D spending needs to be tripled during the next few years to a figure of roughly $600 million per year. Support for advanced natural gas technologies also needs a substantial increase--to a level of $200 million annually.

The next step is to launch an aggressive market development program that includes specific goals for commercializing renewable energy technologies and fuel cells during the rest of this decade. Examples might include installing 100 megawatts of grid-connected photovoltaics, 1,000 megawatts of fuel cells, and 5,000 megawatts of wind generators by the year 2000. Such efforts would need to include funds for co-financing renewable energy projects that cumulatively represent markets that are large enough to spur low-cost, assembly-line production. (Recently, over a dozen utilities formed an alliance to purchase large quantities of solar cells for use on their systems, and are now looking for government support.)

European governments already have established specific renewable energy goals and market development programs, and are planning to install thousands of megawatts of wind and solar plants during the remainder of the decade. The United States has the technological and resource base to make it the world leader in renewable energy, but that leadership will soon be squandered without a more active partnership between government and industry.

Although the electric power sector is the readiest source of supply-side carbon dioxide reductions, there are other sizable opportunities for fuel substitution and supply side efficiency. One is the potential to displace automotive gasoline and diesel fuel with natural gas and closed-cycle biomass fuels such as ethanol. If deployed using current technology, such fuels would allow only a 10-20 percent reduction in carbon emissions, but new, more-efficient technologies that are likely to be available by the end of the decade could allow reductions per vehicle-mile of 70 percent or more. Additional reductions can be achieved by widespread adoption of solar water heating in homes and solar process heating in industry.

ENERGY FOR THE TWENTY-FIRST CENTURY

In sum, if new energy supply technologies are combined with more efficient end-use technologies, it should not be difficult to turn the recent trend of rising carbon dioxide emissions around. A federal commitment to remove market barriers, combined with modest programs for commercializing selected technologies, will allow the U.S. to reduce emissions to less than the 1990 level in the year 2000, exceeding the President's goal. A real commitment to this path by the Clinton administration could lead to a self-sustaining, market-driven move to a low carbon energy economy in the following decade. Indeed, a 10 percent reduction in emissions by 2010 is now well within reach.

Commitments to improved efficiency and new supply technologies are in the end not only complementary but catalytic. As energy requirements are reduced substantially in the decades ahead, it will be increasingly easy to turn to renewable energy options. Even if their BTU cost is a bit higher, the cost of delivered energy services may well be lower. It would, for example, be difficult to fully fuel today's vehicle fleet with biofuels and natural gas. However, a new generation of 100 mile-per-gallon equivalent electric hybrid cars will easily be able to run on domestic reserves of those fuels.

In the long run, the path of efficient energy supply and end-use technologies will likely lead to a high-tech, twenty-first century energy economy that runs on gaseous hydrogen. Such a system would evolve gradually from the electric and natural gas infrastructure we have in place today. In fact, hydrogen fuel is so physically and chemically similar to methane that the transition is not likely to be difficult.

Fifty years from now, solar and wind power plants in remote areas will be able to convert water to hydrogen that is shipped by pipeline to cities and industries. The hydrogen fuel can then be used for home and industrial use, as well as for running vehicles. With widespread use of fuel cells and rooftop solar cells, a large fraction of the electric power is likely to be produced within buildings--some of it having been stored and carried in the form of gaseous hydrogen. Virtually all of the technologies to build such a system have already been developed, and the remaining challenges are ones of cost reduction and mass production.

Among the rewards of achieving an efficient hydrogen economy: carbon dioxide emissions that would be at least 90 percent lower than today's. But even more important than the reductions achieved in the United States is the fact if we pursue such a path, it will encourage other countries, including the developing nations, to pursue a similar path. In the long-run, the most efficient way for the United States to turn around the global trend of rising carbon dioxide emissions is to invest in making the new energy systems more economical at home, and then promoting their export to other lands.

Figure 1.
Carbon Content of Fossil Fuels

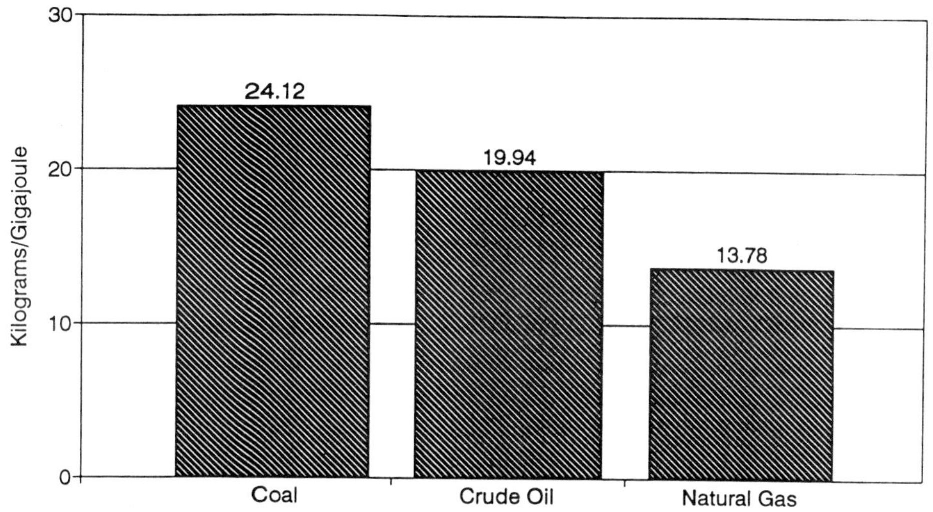

Figure 2.
CO2 Emissions by Generating Source

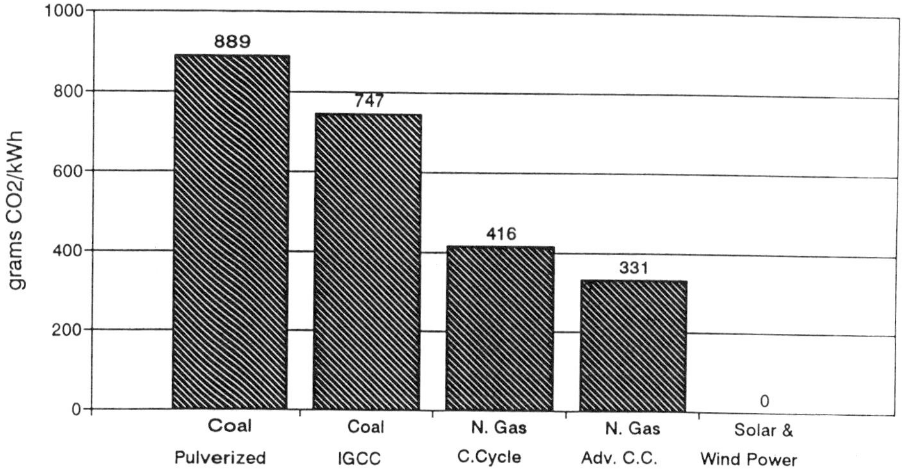

Figure 3. New Power Plants, Mid 1980's
Electricity Generating Costs

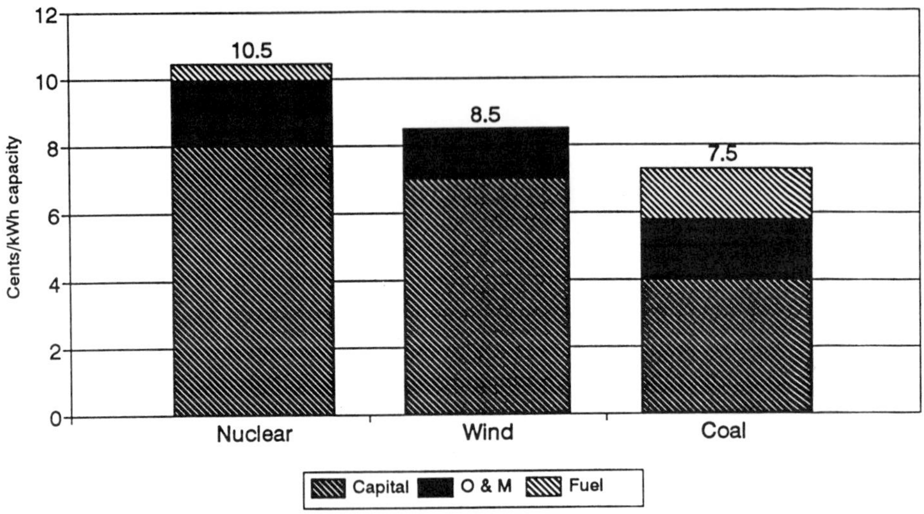

Figure 4. New Power Plants, Mid 1990's
Electricity Generating Costs

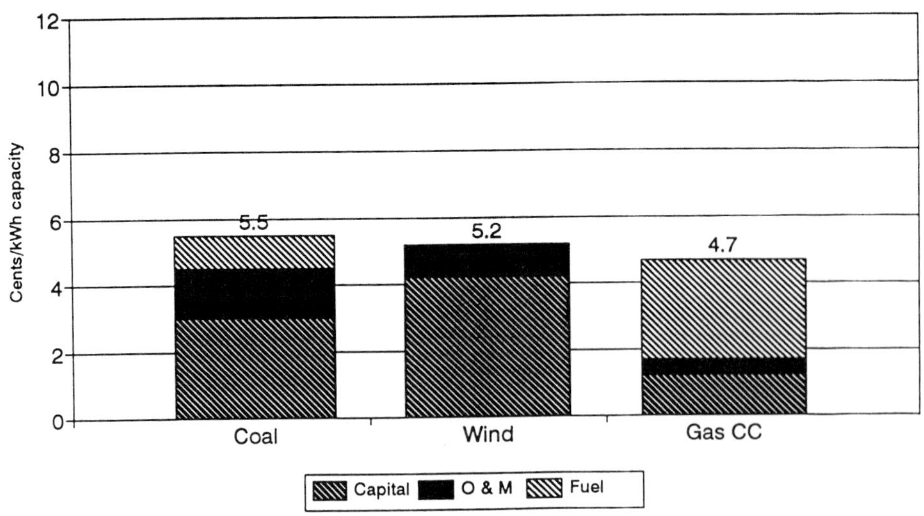

The Architecture of Energy Systems for a Brighter 21st Century

David S. Scott
Director, Institute for Integrated Energy Systems
University of Victoria
Victoria, Canada

Background:

Dr. Scott is Director of the Institute for Integrated Energy Systems (IESVic) and Professor of Mechanical Engineering at the University of Victoria. IESVic is engaged in the development of energy systems that emphasize low carbon fuels, particularly hydrogen and natural gas. Previously, Dr. Scott chaired the Canadian Advisory Group on Hydrogen Opportunities, was the founding Executive Director of the Institute for Hydrogen Systems, and was Professor and Chairman of Mechanical Engineering at the University of Toronto.

THE ARCHITECTURE OF ENERGY SYSTEMS for a BRIGHTER 21st CENTURY

David Sanborn Scott
Institute for Integrated Energy Systems
P.O. Box 3055
Victoria B.C. V8W 3P6
Canada

It is seldom I follow a speaker with whom I agree more than Mr. Christopher Flavin. Chris talked about the importance of the system. What I will discuss is the architecture of that system. I want to give you an idea of what the architecture is like, and then comment on what parts of that architecture are changing. The important issue is how we can use knowledge of the way the architecture is evolving to develop a methodology for teasing out dangers and opportunities. Part of that methodology is to first identify the barriers and attractors to this evolution. Looking at the barriers and attractors can highlight both corporate and national opportunities — overlapping opportunities with different flavours. But the real point is something different: An examination of the evolution makes it clear that there *are* pathways to a brighter 21st century.

These are not pathways of hunkering down, being poor, and accepting fewer energy services. The teeter-totter mythology around energy services is that there are two extremes: one is cleaning the environment and being poor and the other is living with a dirty place and being rich. That myth is nonsense. Except in some very short–term scenarios — like *retrofitting* dust collectors on a smoke-stack of an *established* industrial process — cleaner is almost always cheaper. History has shown cleaner is cheaper (that is, yields more wealth) over and over again.

So, the brighter 21st century of which I speak *will* provide environmental security. I use the word "security" because I am talking about climatic impact, economic growth, and improved, not diminished, energy services which together will lead to improvements in our quality of life.

Now, to put these ideas in context, I want to tell you about "smelling land." In 1974, my nine-year-old son and I sailed a small racing yacht from Cork on the south coast of Ireland, to the Scilly Isles off the western tip of England, and then east to Southhampton on the south coast of England. We sailed through a gale on the passage south to the Scillys and then turned east in a light breeze but with another gale in the forecast.

Day's end in a small boat at sea with uncertain weather brings a curious mixture of tranquility and foreboding. The skipper, Sammy Sampson, a friend of Edward Heath who had helmed Heath's yacht *Morning Cloud*, looked back at the Scillys and said, "You know, during a thick fog a famous British admiral drove his fleet hard aground on those islands." Then he spoke to a footnote to that history.

It seems there was a cabin boy that went to the admiral when they were still some distance off the islands and said, "Sir, I smell land. I think we should put about." Now, advice from cabin boys to admirals of the Royal Navy are not welcome. So, he was reprimanded and sent away. But, he must have gone on deck for another whiff, because soon he was back again saying, "Excuse me sir. I smell land. I think we should put about."

Now discipline allowed the Royal Navy to rule the seas — and thereby to claim so much of what was becoming the "new world." And discipline was why the cabin boy was swinging by his neck from a yard–arm, dead, when the fleet crunched ashore upon the land that he had been smelling.

Now, what intrigues me about this story is: What was that cabin boy smelling when he said he was smelling land?

He was smelling what we smell when we say we are smelling the sea. He was smelling life in the intertidal zone. That mixture of seaweed, kelp, crabs and mollusks, all of whom make their home between the high and low water marks of any tidal coast.

If we live our life on land, a whiff of damp seaweed is a whiff of the sea. But if we live our life at sea, and cabin boys were seldom allowed ashore, then a whiff of the intertidal zone is a whiff of land. Correct, but not conventional whiffery.

So to me "smelling land" has become a metaphor for seeing things the other way around. I think that is what this conference is about — seeing things the other way around. So if you have been wondering "why is this bald bloke telling us all this" now you know. Because, if there is one thing I'd like you to take away from my talk, it is that seeing things the other way around is valuable and needed.

Sometimes, "smelling land" is just fun. But sometimes it is really needed if problems are to be resolved or opportunities captured. But the parable also speaks to dangers for the smellers, if they dare speak of what they smell. We don't hang people from yard–arms anymore. But people do lose jobs. And they certainly place in jeopardy that warm feeling of being taken seriously. Smellers of land in the corporate and government ladders are shunned to staff positions. They are safe on the flanks, but are dangerous to corporate stability if they are left on a line of responsibility. In most government staff positions "smellers of land" are tolerated even less well. All of you know the dangers.

The parable of the cabin boy also speaks to the dangers to ships of state or corporations — if they discourage the whiffers or if they trivialize what the whiffers are saying. I believe that to understand the evolving energy system — to avoid the dangers and capture the opportunity (because frankly the system is rich in both) — smelling land is essential.

Before we go on to today's energy systems, we should acknowledge some "smellers of land" that are among us today. One is the man responsible for this conference: Roger Billings. At the banquet we will honour another: Willis is an example of a very rare breed; people who are *both* corporate leaders and smellers of land.

Now on to the system. Figure 1 represents how we in our Institute for Integrated Energy Systems view the architecture of the system. We place "energy services" at the left of the diagram. We are convinced that the system is driven by the services — people buy energy *services* they don't buy energy. Next, to treat the system holistically we must consider:

- Material extracted from the surroundings at each point in the chain.
- Material diverted within the surroundings at each point in the chain.
- Emissions to the environment at each point in the chain.

Material from the surroundings might be coal, oil or uranium. But we must also include the diversion of material flows caused by such things as dams for hydroelectric generation.

Emissions to the environment must not only consider *what* is emitted, but *where* in the system chain it is emitted, and in what *concentration*. This last idea can be illustrated by considering transportation as the "service" and methane (CH_4) as the "source." If methane is carried on-board, then the emitted carbon dioxide (CO_2) comes out a *mobile* tailpipe *diluted* by the nitrogen contained in the air that carried the oxygen to the engine for combustion. On the other hand, if the service-to-methane pathway involved steam methane reforming (SMR) to produce hydrogen — with hydrogen used as the on-board currency — then the CO_2 comes off at a *stationary* site. Moreover, depending upon the source of exogenous heat, either some or all of the CO_2 emitted from the SMR process is relatively pure and that provides another commercial opportunity. If anthropogenic CO_2 became as severe a climatic threat as some now think, the CO_2 might even be collected, liquefied and disposed in the Kaltwasser sphere of deep oceans. There are several opportunities for either using or disposing CO_2 if it comes from a stationary source like SMR hydrogen production. But the opportunities vanish if the CO_2

comes from a moving vehicle: It *does* matter where in the system CO_2 is emitted.

One implication of Figure 1 is that energy sources don't matter very much. Now *that* is a "white knuckle" statement, particularly for those of us who attend energy conferences and read articles describing mainstream thinking on energy issues. In part it is a "white knuckle" statement because we are often part of, aligned with, or receive financial support from, one or more *sourcing* industries. Sources matter to the industry. But what has shaped *culture* is the way the system delivers services.

For example, the late 20th century has been profoundly shaped by technologies which require the currency electricity: FAX machines and computers, telephones and televisions, all of which determine our perception of what we do and what others have done. It manifests itself in instant replays — of a touchdown pass at a football game or a "smart" missile hitting a bunker in Baghdad. One of the reasons these technologies have been so contagious — why they have penetrated every niche of our world — is that they all use electricity. That makes them *blind* to the source harvested to make the electricity. What shapes the late 20th century are images of Arafat and Shamir shaking hands, Challenger explosions, and financial transactions allowed by telecurrencies. The energy sources that power these images and transactions are much less important.

We often hear the claim that civilization has an insatiable demand for energy. I believe that is incorrect. Rather, civilization has an insatiable demand for energy *services*. Most energy services have a surprisingly low *fundamental* need for energy. Reduced to fundamentals, there is, in most cases, almost no limit to which we can shrink the ratio of energy used to energy service provided.

Figure 1 shows the architecture of the energy system. The five *functional* links of this architecture are constant. Also, the fundamental services — keeping warm, being able to communicate, or wanting to be healthy — are as true for hunting-gathering societies or medieval societies as they are for us today. What *does* evolve is the way we *realize* the three middle links [Service Technologies]-[Currencies]-[Transformer Technologies] and what we select from the right hand link [Sources]. These three central links and what we select from the right hand link *do* evolve. And for these four we should ask:

- What is the *RATE* of the evolution?
- What are the *BARRIERS & ATTRACTORS* to this evolution?

So now I ask you to imagine this constant five-link chain having another dimension perpendicular to the page. This dimension is TIME. So as this basic architecture moves forward in time the three middle links evolve. When the orderly evolution hits a snag in one part of the system, it has hit a "barrier." When evolution gets tugged ahead at one part of the system, it has encountered an "attractor."

Barriers can be *technical*, such as the difficulty of on-board storage of low-carbon fuels. They can be *cultural*, such as the public perception that we were running out of natural gas in the late '70s and early '80s. They can be *legislative*, such as the regulations that prevented natural gas being used to generate electricity.

Similarly, *attractors* can be technical, such as the fuel cell; or they can be cultural, like the growing public sensitivity to the need for clean atmospheres; or, they can be legislative, such as the recent Southern California and European clean air laws.

My point is that looking at the barriers and attractors to continued orderly evolution can illuminate business and national opportunities. Such "barrier/attractor analysis" will provide you with a construct — a procedure — for more effectively "smelling land."

Now let's leave for a moment the generalized architecture of Figure 1, to move on to Figure 2 which shows the energy system at some time in the past. We can see the consistency of the architecture and the consistency of *generic* services. The top row and the *menus* of the left-most and of the right-most columns are unchanging. What changes is the other stuff. Now it becomes fun. I had my own fun with Figure 2 — which I structured a little differently than Figure 1 because, this time, there *IS* a correlation along each row. If we go even further back in time than what is illustrated in Figure 2, and then select illumination as the service, we could set out a sequence something like: [Illumination]→[Lamps]→[Whale Oil]→[Sailing Ships]→[Whales]

This illustrates the fact that, after we have picked an example, let's say illumination which Figure 2 shows as lamp and town gas, we can always go back to some earlier time to think the "what was it like" through again. In this case we think of a different kind of lamp, one that used whale oil. This requires the currency to be whale oil rather than town gas. In turn, this forces the transformer technologies to be sailing ships. Of course, the energy source was whales. That was a time when we hunted animals rather than drilled rock for our oil.

Let's pause here. Because this naturally begs another statement — a statement that may cause more "white knuckles." Whales are a renewable energy source. And this flags a truth: there are more difficulties with renewable energy sources than many people acknowledge. These difficulties will become increasingly apparent with time, but for now we must live through the rhetoric that "renewables are the answer." I'm not saying "Please don't use renewables." But, as with everything we do, they must be thought through carefully. The fact that one source is labelled renewable and another source isn't, does not make the labels imply that one source is "good" and the other "bad."

Now we come to the question of the *RATE* of systemic evolution. Many of you will have seen different versions of Figure 3. The concept of displaying historical energy system data in this manner originated with the International Institute for Applied Systems Analysis (IIASA). What we do in Victoria is just update the data every year. This is the latest version. Figure 3 plots energy sources. But the evolution of sources can give us a good feeling for the cadence by which the overall system is changing. A while back I mentioned legislative and cultural barriers to the use of natural gas. You can see from Figure 3 how these barriers influenced systemic evolution. People said the "bridge to the future" was coal. You will remember the phrase. But nowadays reality is beginning to set in. People are recognizing that the availability of natural gas is immense. And simultaneously, they are recognizing that coal has grave difficulties.

Now that the legislated/cultural barriers to new markets for natural gas are being removed, and recognizing the elasticity represented by the curves for natural gas and coal in Figure 3, what do you expect for the future of coal and natural gas? It is something to think about.

By the way, I can't pass this point in the talk without saying something which may be my third "white knuckle" statement: *"Running out" is not a factor.* I won't go through all the whys! I will make only one comment. In our Institute Dr. Holger Rogner performed an analysis that considered *only* those fossil resources in the world which we *already have found.* Now, of these resources, if we used only those which we can recover at less than $10 a barrel of oil equivalent, we would *triple* the CO_2 content of our atmosphere. So the use of fossil fuels is *not* capped by availability but by the environment. Depletion is a real factor in public policy because it is a real factor in public perception. It is not a factor in reality. Of course the geography of *where* Earth's resources are located is a real issue in geopolitics and, in turn, is kindling for international conflict.

One more issue. If you look again at Figure 3, it should be clear that natural gas will be unable to reach the projected market penetration of 60% to 65% without massively entering transportation. So the question is: How will natural gas get into transportation? Asking that question and picking the right answer will make some people, and some nations, one colossal pile of money.

Christopher Flavin mentioned this idea of "integrated resource planning." IRP is now becoming a buzz phrase *within* utilities — and that is good because it is a first step. But the utilities must expand this thinking further. They must begin to think *across* utilities. For example, if you are running an electrical utility you must think about how the holistic system could be more efficient if utilities co-operated — through *technological* cross-links — with, for example, the gas utilities in the same jurisdiction. The second thing utilities must develop is effective downstream integration towards services. The job of an electrical utility must not stop at delivering electricity. If they are to be successful, utilities must make it their strategy to always think of their business as delivering services. If I were leading strategic planning for a natural gas company I would say: Make a bunch of signs that paraphrase "It's the economy, stupid!" and stick them up all round the place reading — "Sell Services, Not Gas! Stupid!"

Now let's leave rate-of-change to take a peek at Figure 4 which speaks to the patterns of the "deep future." This deep future is, say, fifty to a hundred years from now — probably closer to a hundred years. You can see that the system will provide the same generic services, it will simply provide them better. There will be more flexibility in how we communicate and store information. Everything will be more efficient. Technologies will exhibit longer life times, higher efficiencies and cleanliness. That will be the nature of emerging technologies. Fuel cells will come to dominate energy conversion technology — a hundred years from now they will have displaced most internal combustion engines. The relationship between the fuel cell and the internal combustion engine is analogous to the relationship between the microelectronics chip and the vacuum tube we had in our radios listening to "The Green Hornet" during the 1940's.

Figure 5 gives us another optic teasing out barriers and attractors to systemic evolution. The palette is a free-form representation of energy services and how they are sourced. Like any free-form representation we can quibble details. But there is value in the macro message.

The palette has two domains, one served by chemical fuels and one served by electricity. Fossil sources have access to the full palette. But the striking feature of today's system is that our new/old hopes for sustainable energy sources are effectively restricted to manufacturing electricity — and therefore are *TRAPPED IN AN ELECTRICITY SERVICE GHETTO.*

This highlights the question: How do we get solar, or wind, or tidal, or the next(s) to fly an airplane? The answer is hydrogen. The bridge for our new/old hopes to services which are now powered by chemical fuels is hydrogen. But as Figure 5 shows, this bridge is now incomplete. Technologies are missing. Institutional and physical infrastructures do not exist. Barriers need removing. Attractors need improving.

This bridge is where the winners will plant seeds. It is where people, corporations and nations will become rich.

Curiously, a head start in completing the planks and trusses for this bridge has little to do with the immediate unleashing of new/old hopes. Fuel cells were first deployed and liquid hydrogen storage greatly enhanced using hydrogen derived from natural gas to fly to the moon. Just as your computer or telephone is blind to the source that made the electricity, a fuel cell powered by liquid hydrogen is blind to the source that made the hydrogen. So methane-sourced hydrogen is a platform for technologies that will complete the bridge.

The service palette shows that the hydrogen bridge is needed before *any* of the new/old hopes can fulfil their promise. It is amazing that the need for this bridge is so rarely seen by those promoting their favourite new energy source.

The fact is, energy sources are not the problem. Nature gave us a plethora of sources. The trick is to harvest these sources in ways that allow us to manufacture currencies which can provide the services that are most needed.

Now let's put some of these ideas in context. The context of the sweep of planetary history.

Most of us in this room were alive when life flew off planet Earth to stand on another heavenly body. (But not all of us: I am delighted to see there are some young folks in the audience.) Landing on the moon was an extraordinary event. When I rank it with other events in the maturing of Earth, only a few are its equal. One may be when anaerobic bacterial life began about 3.6 billion years ago. A second may be when oxygen — the excrement of later life forms — first dominated atmospheric chemistry, holding it out of equilibrium from about 2.5 billion years ago until today. After the second I am not convinced there is a *planetary development* event that can rank with the significance of humans standing on the moon — not because it was a "giant step for mankind," which almost trivializes it, but because it marks the beginning of an epoch for the planet.

At the time, the moon landings fired and engaged all of civilization. Now that engagement is little more than a warm ember. But the legacies of the space program, and in particular landing on the moon, are many. Today I will speak of two:

1. THE VIEW FROM SPACE

The first photographs of an Earthrise from behind the moon altered our perspective of Earth in a visceral way. Of course, we still have our jealousies, wars and religious differences, but the view of Earth from space triggered an irreversible change in how we think about the planet. Seen as a whole, Earth became something we needed to protect.

Soon after, space activity *also* allowed many of the risks to Earth to be *quantified*. If it wasn't for satellites we would have learned much later, and with much less precision, about ozone holes — and mega icebergs floating north from Antarctica's Ross Ice Shelf.

From this perspective, and in the fullness of time, history books may observe that the space program saved planet Earth.

2. THE STAPLE CURRENCIES OF SPACE EXPLORATION

Electricity and the H_2-O_2 pair are the staple energy currencies of space technology. The mission that deployed the first H_2-O_2 technologies was driven by a grand objective, not by immediate economic payback or avoiding environmental danger. But now, these are precisely the technologies that must be brought down to Earth in order to protect our planet from the dangers the satellite data quantified.

We learned a lot about the coming hydrogen age in space. We learned about fuel cells, hydrogen liquefaction, safety, materials, and so forth. But we learned about them when cost was not as important as completing the mission.

Now we are in different times. The glamour and excitement have gone. We witness the fracturing of planned economies in Eastern Europe. Global war may have left us, but only to be replaced by widespread political flares and regional fighting. Then there is the environmental imperative, with some believing it is little more than cleaning up smoke and grime, but with a growing recognition that planetary environmental changes may place our civilization at risk.

Finally, we witness the economic decline of North America. It is not something I like to identify. But it is very real. And without a grand vision of how we can turn things around by engaging in some new *mission* — a mission as significant as building dikes to keep out the sea, or building cathedrals to glorify God, or flying to the moon — there seems little prospect of arresting this decline.

Perhaps, as the French might say, we are in a time of "la pause créatrice." A time for getting Earth's house-in-order, after the first steps that showed *space colonization is possible.* After their first deployment in space applications, hydrogen technologies must now be made more efficient and much less expensive. They must be made capable of colonizing Earth-bound society over the next 50-100 years. They must be deployed to roll back growing environmental stress. At the same time, they will improve the standard of living, the quality of life, and they *could* revitalize the economies of all industrial nations.

Earth-bound deployment will also ready these technologies for the grand age of space colonization. For we cannot colonize space with today's cost of space vehicles, or flying them, or without the hydrogen-electricity systems needed for space outposts. Our chance lies in first developing these technologies for widespread use on Earth. If that is done, it will also bring costs in line for space colonization.

So I see our sliver-of-time in three phases:

- The late 20th century; glorified by a cost-be-damned demonstration that we *can* send emissaries from Earth's biosphere to another heavenly body. *The currencies used for this mission were electricity and the* H_2-O_2 *pair.*
- The first half of the 21st century; characterized by bringing these technologies down-to-Earth where they will undergo massive cost reduction and then widespread deployment. *This will create a new wave of wealth creating industries and lead to improvements in quality of life and increased environmental security.*
- The last half of 21st century and early 22nd century; when we again turn our eyes to the heavens and space colonization. *It will be a time of Earthlet spores that use hydrogen technologies. Early forms of these technologies appeared during Earth's first steps into space during the* 20th *century. But they will have matured and been readied for space colonization by Earth-bound deployment during the early* 21st *century.*

Each of these steps needs the preceding step.

It is sad the industrial nations are not exploiting this legacy. We *could* exploit it — to create wealth for ourselves — to clean the place up for everyone on our planet — and to achieve a type of national and international leadership appropriate for the 21st century.

Now, as a Canadian I have a bit of a "little brother complex" in North America. But we Canadian "little brothers" are just as tied to America's future in North America as the cabin boy and admiral were tied to each other's future as they sailed towards the Isles of Scilly. And in that sense I'll be a cabin boy: It pains me to see you Americans squander your legacy — your space program legacy of technologies so needed as the basic technological building blocks for our world to pass into a brighter future. But we Canadians also contributed to your space program. So in truth both of us are squandering our opportunity to lead the world with appropriate technologies for the 21st century. Instead of using the experience gained in space program to developing hydrogen-fuelled technologies that could secure our future, *both* Canadian and American governments claw at the past — for example, fighting to keep North American car drivers paying the lowest price for gasoline in the world, or by denying we might be moving into an era of unprecedented climatic instabilities caused by using that gasoline.

Earlier today, Roger King said a very interesting thing. He reminded us that today is June 21, the longest day of the year. As he said that, I know how I wanted to end this talk. Because Roger reminded me of that other "longest day" — June 4, 1944 — the Normandy Invasion day of World War II. After World War II this country, the United States was, to my knowledge, the first nation in history to *massively* pour resources into defeated nations. It came to be called the Marshall Plan. It required a quick, unprecedented shift of national cultural objectives to suit the times — a shift from an almost monomaniacal need to win a war, to the balance and wisdom required to make the world a better place for all after having been on the winning side in a war. It seems to me that there is a parallel with our times. The question is: Having "won" the space race, how are we going to use the technologies developed for aerospace in a mission vitally appropriate to the times? In a mission that will make the world a better place for all?

The space program has uniquely placed the United States to lead the world in pioneering the hydrogen age. That could leave another kind of legacy for America. Not just a legacy of international economic recovery spread over five decades — as extraordinary an event of goodwill and success as that was. But a legacy for planet Earth that will last throughout the eons.

To do that we all must "smell land." Clearly we have the intellectual ability. But it may not be as easy to achieve the vision and cultural will to pull it off. In the 1940's your nation did it once. Perhaps this conference can be a trigger for America to do something similar, but even grander, once again.

I wish you Godspeed.

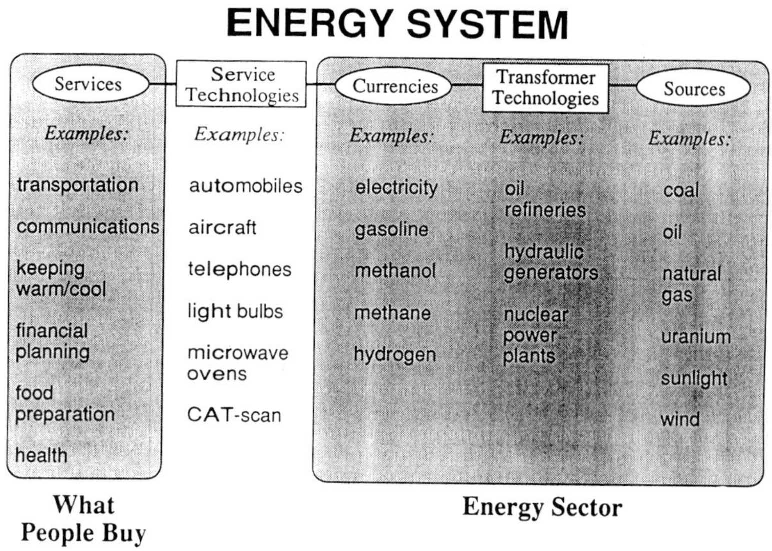

Source: David Sanborn Scott
Institute for Integrated Energy Systems
University of Victoria, Victoria, B.C.
CANADA

Figure 1: The energy system

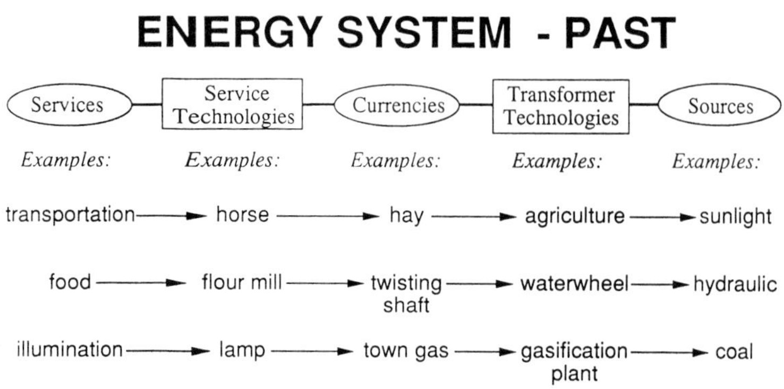

Source: David Sanborn Scott
Institute for Integrated Energy Systems
University of Victoria, Victoria, B.C.
CANADA

Figure 2: The energy system - past.

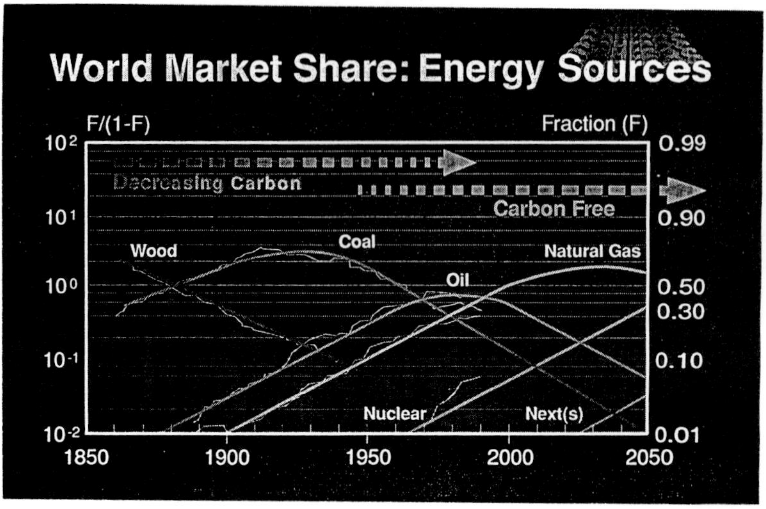

Figure 3: Energy source competition.

Source: David Sanborn Scott
Adapted from Marchetti and Nakicenovic,
IIASA Report RR-79-13, 1979

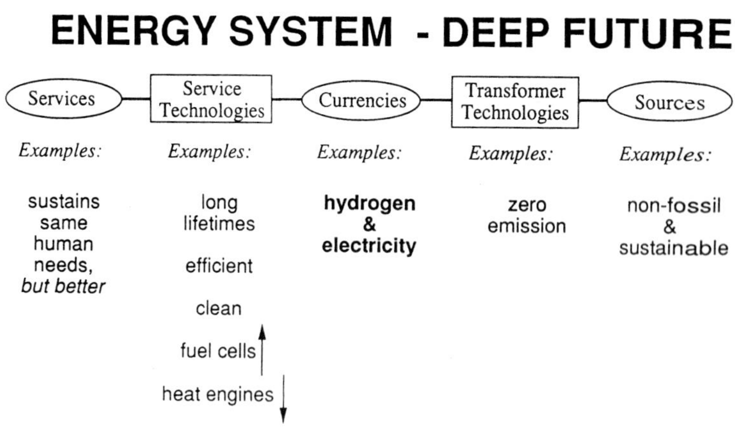

Figure 4: The energy system - deep future.

Source: David Sanborn Scott
Institute for Integrated Energy Systems
University of Victoria, Victoria, B.C.
CANADA

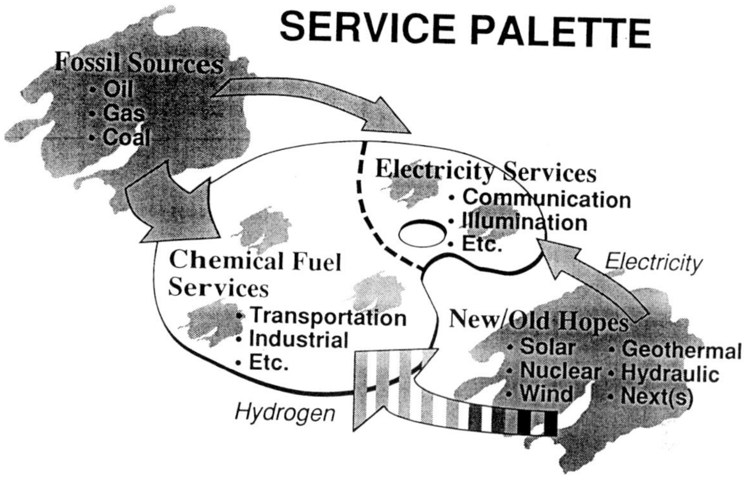

Figure 5: Service palette.

Source: David Sanborn Scott
Institute for Integrated Energy Systems
University of Victoria, Victoria, B.C.
CANADA

CHAPTER 2 -- HYDROGEN ENERGY

The Hydrogen Energy System

Nejat Veziroglu
President
International Association for Hydrogen Energy
Coral Gables, Florida, U.S.A.

Background:

Dr. Veziroglu is Director of the Clean Energy Research Institute at the University of Miami, where he has been a member of the Engineering faculty since 1962. He was one of the co-founders of the International Association for Hydrogen Energy (1975), is editor-in-chief of the *International Journal of Hydrogen Energy*, and is noted for his work in hydrogen and clean energy research.

The Hydrogen Energy System

Nejat Veziroglu
International Association for Hydrogen Energy
P.O. Box 248294
Coral Gables, FL 33124
U.S.A.

Ladies and gentlemen, welcome to the Project Energy '93, the second one in the series organized by the International Academy of Science. It is an honor to be speaking at both the Project Energy meetings, courtesy of visual aids this time. The motto of the Conference is a very appropriate one -- improving the quality of life, safeguarding the environment.

A year ago the historic Earth Summit in Rio, Brazil, brought home the importance of a healthy environment for improving the quality of life worldwide. It was attended by more than 100 heads of states, scores of ministers of the environment, ministers of energy, ministers of government and thousands of bureaucrats. They discussed the steps to take to solve the global problems such as the global warming, ozone layer depletion, population explosion, biodiversity, resources depletion and more in order to improve the living standards throughout the planet earth, the only one known to be hospitable to life.

Ladies and gentlemen, to solve the global problems and to achieve the goals set for the Project Energy '93, the most realistic solution, the most elegant solution is the Hydrogen Energy System. Hydrogen will not produce greenhouse gases, carbon dioxide and others. As a matter of fact, it will reverse the global warming and bring back the earth's temperature to preindustrial levels. Hydrogen does not produce ozone layer depleting chemicals. Hydrogen hydride refrigeration obviates the need for chloroflourocarbons and also results in quieter refrigerating and air conditioning. There will be no oil spills, no caustic discharges with hydrogen. Beaches, oceans and aquatic life will be protected from the use of hydrocarbons. Hydrogen will produce little or no acid rain ingredients, little or no pollution because of the advanced versatility of hydrogen. Because it can be converted to heat, mechanical power, electrical power without flame combustion, nitrogen oxide produced in the Hydrogen Energy System will be some 70% less than that of the present fossil fuel system. The Hydrogen Energy System, being a permanent and environmentally compatible energy system, will provide humankind with abundant clean energy, resulting in sustaining economic growth and higher living standards. As such, it will slow down and eventually stop the population growth. It will also protect biodiversity by conserving the environment. It will help slow down the energy-related resources depletion, because of hydrogen's high utilization efficiency. In summary -- hydrogen will improve the quality of life while safeguarding the environment.

I understand that this year, to this second conference in the series, several hundred high school science students have also been invited. This is a very worthy action, as the bright young science majors of today will build the sustainable future of tomorrow.

On behalf of the International Association for Hydrogen Energy, I would like to extend them a special welcome. The Association, or IAHE for short, recognizes the importance of informing the students about the great advantages of the Hydrogen Energy System. For this purpose it has established a student division as of January of this year. In the registration area there will be copies of the first issue of the student division newsletter appropriately named Hi. It also includes application forms for the division. There are also applications forms for IAHE.

We would like everybody, both students and graduates, to join and help build the energy system that our home, the planet earth, deserves, the energy system which will result in a sustainable future.

Finally, I would like to congratulate a hard-working, hydrogen energy builder, Roger Billings, and also his colleagues and staff, for organizing the Project Energy '93, and I wish every participant a very fruitful meeting.

Hydrogen Projects Running in Europe

Olof Tegström
Founder, Tebetron
Harnosand, Sweden

Background:

Mr. Tegström worked for ten years in engineering on electrovehicles before founding his own company, Tebetron, through which he has developed and implemented advanced residential energy systems in Sweden and Switzerland. He holds several patents on innovative automative machines and DC-speed controllers, and serves as a consultant to the U.N. as an alternative energy expert.

HYDROGEN PROJECTS RUNNING IN EUROPE

Olof Tegström
Tebetron
Johannesbergsg. 4
S-871 31 HÄRNÖSAND, Sweden

When we introduce any change to a world-wide system, we have to know the consequences of what we are doing. That is my motive in my involvement with application research. The following facts are enough to say -- we must change our energy system. As a result of our high living standard, the concentration of CO_2 in the atmosphere has increased more than 25% over the last 100 years, from 280 to 358 ppm. The UV radiation in the North Pole area has been reported as being 30-40% higher than it was last year, and we have measured 298 nm light frequencies -- shorter frequencies than ever before. These conditions are a result of the breakdown of the ozone layer. Climate change and damage to forest vegetation are other noticeable phenomena.

The hole in the ozone layer is also attributed to a new, unknown factor -- hydrogen as a by-product. It is released into the atmosphere in quantities of 17 million tons; that amount would have been sufficient to power 17 million busses. Sweden releases 200,000 tons of hydrogen per year, but the possibility of using that hydrogen for 1 million cars in Sweden has been stopped, according to the government, due to market disturbance. Today the atmospheric content of hydrogen is increasing .6% per year. Oil refineries and chemical industries are the biggest producers of this hydrogen.

Hydrogen is the cleanest alternative to fossil fuels and it can be produced on-site in small quantities without requiring world-wide transport. In Europe, the CO_2 tax will open up new opportunities for hydrogen to compete in that economy. Sweden has already put into effect a tax of 10 cents/liter gasoline as a first step towards implementing the CO_2 tax. Other countries and EC are moving in the same direction.

I work in a UN organization called APCTT, an organization for technology transfer to developing countries. My goal is to provide the latest hydrogen technology developed from within Europe. I would like now to give you the latest results from Europe.

Over the past four to five years, hydrogen projects have been running in Germany and Switzerland. These projects range in size from single family dwellings to villages and are all equipped with the accessories for a high standard of living.

In 1988, the German government announced a 7-billion-dollar development program for hydrogen and other alternative energy sources, like solar and wind. All the aforementioned projects have partly been financed by that money and have created many new jobs.

One of these projects utilizes solar energy collected from 50 square meters of solar cells mounted on the roof. This energy is subsequently used to break water molecules into hydrogen and oxygen. The generated hydrogen can then be used to power a specially converted Toyota. This Toyota is a hybrid vehicle that can be run on hydrogen, gasoline, or biogas with no loss in vehicle performance. Hydrogen has been used in this vehicle to provide more than 90% of the total fuel utilized. The solar-produced hydrogen is also used in several household appliances including a range and a washing machine.

Progress in other areas should also be mentioned. Due to mass production, electrolyzer prices have fallen by more than 70% over the past two years. The FIAT-affiliated company, IDROENERGY, manufactures these units and markets them to produce hydrogen for welding purposes. Electrolytically produced hydrogen is also used for sintering and melting in the glass and jewelry industries, where it is used in a new hydrogen application, often replacing oxygen and acetylene gases. With a production capability of twenty units per week, a price reduction of 74% has been realized for units producing between 1 and 7 Nm^3 hydrogen per hour. (These are the appropriately-sized units for private use in homes or villages.) Figure 1 gives more information on this product. In addition, Professor Grätzel in Lausanne University, Switzerland, has recently developed a new solar cell, offering a 90% reduction in cost when compared to traditional cells of today. (Figure 2.)

After ten years of research on a solar-parabolic reflector and a Stirling motor, Hydrogen Engineering Associates reported in a Senate Hearing held in March 1993 a hydrogen production price of 25.8 cents per Nm^3 or 80 cents per liter gasoline equivalent!

From an economics point of view, the larger the electrolyzer project, the lower the hydrogen cost. (See Figure 3.) An interesting story is also coming from Denmark where an electrolyzer from Chernobyl was recycled. Practically everything in the area was damaged in that catastrophe, but the 20-kW electrolyzer was salvaged and brought to Denmark. Today it operates with a 110-kW windmill and delivers all the energy to a village via hydrogen.

Small devices, like 4 HP motor generator sets and catalytic burners are on the market, and a recent agricultural hydrogen project is reported from Norway in which a hydrogen burner kills the parasites in the furrows without damaging the vegetation. In Germany (Lampenholdsheim) a steam generator capable of producing up to 100 MW of electricity utilizes hydrogen produced during periods of low demand by converting it to steam to help cover peak load requirements. This system utilizes technology developed for the Space Shuttle in its base construction.

A new market introduction of hydrogen busses, ranging in size from the Suzuki van to the 120-passenger Van Hool bus, has been reported from Belgium. These vehicles are marketed by Hydrogen Systems in Belgium, which is a member of the Begemann Group, with a stable turnover of $250 million. Hydrogen Systems, together with Windmaster, offers a system which includes a windmill used to produce the hydrogen required by the bus.

The two largest-ever hydrogen projects to date are now underway in Germany; they were first presented at the World Hydrogen Congress in 1992 in Paris as European-Canadian project EQHHPP and NHEG. Both projects will be completed around the year 1997 with 965 busses, 763 taxis, hydrogen airplanes, etc. In both cases, hydrogen is produced by hydropower, cooled down to a liquid (-253°C) and shipped per boat from Canada or Norway to Hamburg and Bremerhafen in Germany. (See Figure 4.) Several test vehicles, including busses and taxis, have been road tested for the past two years. Hydrogen shall be stored on these vehicles in its liquefied form. These two projects were originally initiated by EC and Germany. Today the number of participating companies has increased from 42 in 1991 to 84 in 1993. Economics behind the project include a filling station price (brutto) of 50 cents/Nm^3.

In 1995 we hope to see the first busses on the roads in Germany. Major cities around the world, where pollution problems are predominant, are showing considerable interest in this solution. We can congratulate the involved German companies with the emerging world market possibilities.

A major impact on the European market is being made by a new "climate convention" which began in mid-1992 in Frankfurt and has now been signed by over 300 European cities. By ecological planning for buildings and decreasing the number of vehicles in a city, CO_2 emissions shall be reduced 30% by the year 2005 and 50% by 2010.

The active involvement of these cities gives alternatives a major thrust toward commercialization and opens new market opportunities for Europe. It is hoped that the U.S.A. will follow this development.

Perhaps one day refueling stations with wind-produced hydrogen will be commonplace. Until then, the Hydrogen Organization in Europe is arranging research programs for interested companies. (Fax 46-611 15900 att. Tebetron).

IDROGENERGY s.r.l.
EXPORT OFFICE, VIGO IMPIANTI s.r.l.
Via A. Cacchi, 15, 16129 GENOVA
Tel. 010/561985 - Fax 5838888
P. IVA 02918940103

Nm^3/h	KW	IDROENERGY GAS GENERATOR PRICE LIST		
.33	1,7	Model 0.5 (220V single phase 50 Hz)	Lire	9.900,000
.66	3,5	Model 1.0 (220V single phase 50 Hz)	Lire	14.139,000
.66	3,5	Model 1.0 (380V 3 phase 50 Hz)	Lire	14.469,000
1,07	5,6	Model 1.6 (380V 3 phase 50 Hz)	Lire	17.663,000
2,13	11,2	Model 3.2 (380V 3 phase 50 Hz)	Lire	22.957,000
3,33	17,2	Model 5.0 (380V 3 phase 50 Hz)	Lire	32.760,000
6,66	35,0	Model 10.0 (380V 3 phase 50 Hz)	Lire	59.887,000

Figure 1

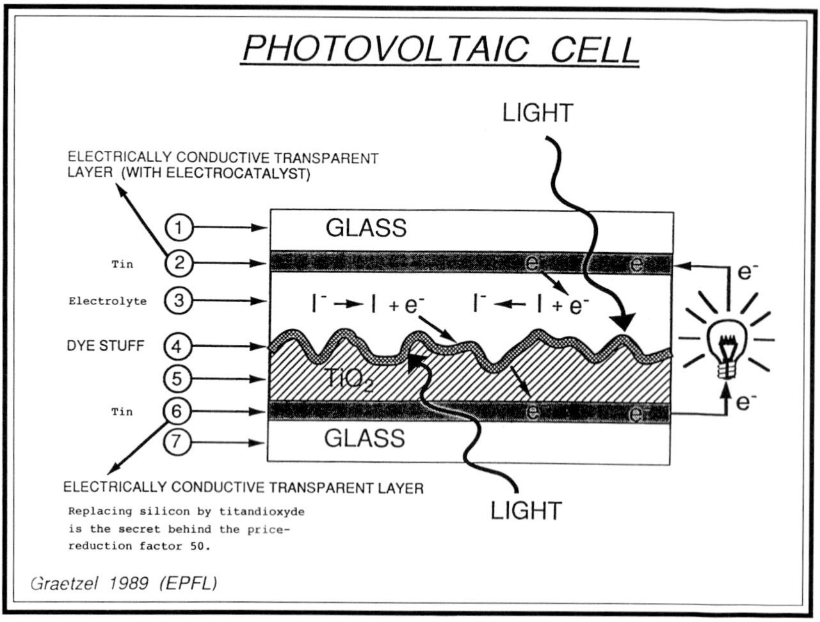

Figure 2

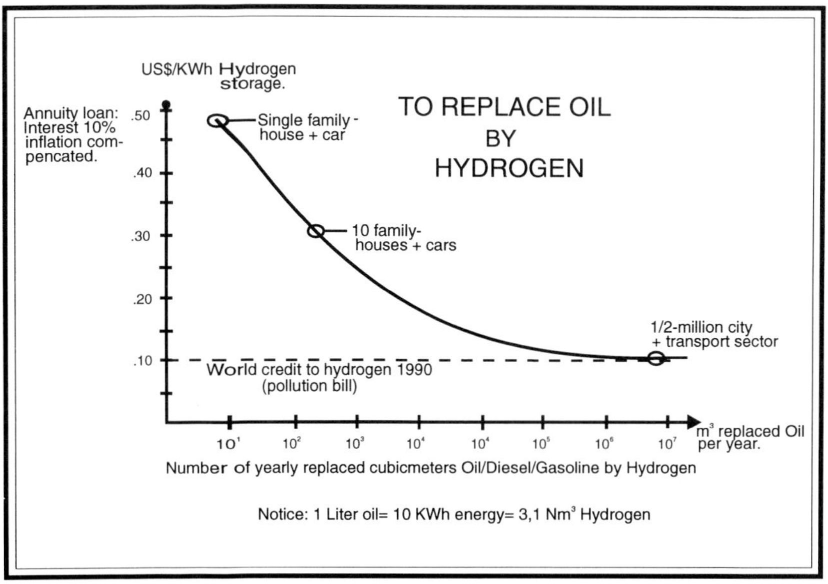

Figure 3

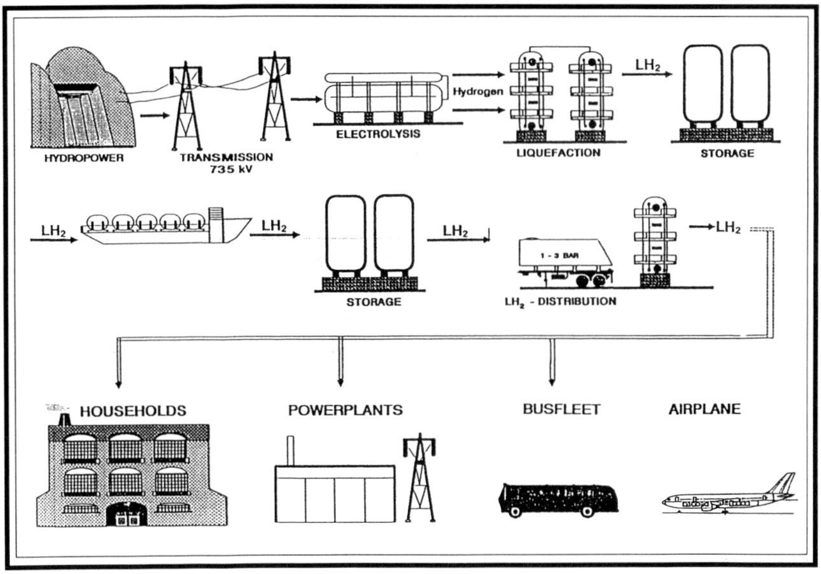

Figure 4

The Hydrogen World View

Roger E. Billings
President
International Academy of Science
Independence, Missouri, U.S.A.

Background:

Dr. Billings has, for twenty-seven years, pushed forward the field of hydrogen energy, developing numerous breakthrough technologies to help bring the world a viable alternative energy system.

THE HYDROGEN WORLD VIEW

Dr. Roger E. Billings
International Academy of Science
26900 East Pink Hill Road
Independence, MO 64057-3284

ABSTRACT

In the opinion of many experts, the element, hydrogen, will play a very important role in our energy future. Increased concerns about environmental pollution, depleting fossil fuel resources, and advances in utilization technologies are all making hydrogen a very attractive choice as a fuel and as an energy vector.

Hydrogen can be utilized in a diversity of energy applications. It is an excellent fuel for internal combustion engines, and can be safely stored in its powdered, metal hydride form. Substantial testing of hydrogen-fueled prototype vehicles has been undertaken which has demonstrated higher utilization efficiencies with hydrogen than with conventional fossil fuels.

Hydrogen has been utilized in a diversity of domestic applications. Conventional natural gas appliances can be readily converted to operate on hydrogen without major or expensive modifications to existing appliances.

Hydrogen can be very efficiently converted into electricity by a hydrogen fuel cell. Such a device can generate electric power for stationary applications or for electric vehicles. Hydrogen fuel cells are two to three times more efficient than conventional internal combustion engines.

HYDROGEN ENERGY

As fossil fuel reserves continue to diminish and the negative environmental impact of today's fossil fuel-based economy becomes increasingly apparent, more and more worldwide attention is being turned toward clean hydrogen energy. Presently, fossil fuels provide about 80% of total world energy demand (1); and if usage continues at the current rate, known reserves of oil and natural gas would only last for forty more years (2). Coal reserves, on the other hand, are much larger; but, of all the fossil fuels, coal is considered to be the most environmentally unfriendly alternative (3).

All fossil fuels emit various pollutants into the atmosphere -- carbon dioxide; nitric and nitrous oxides (NO_x); as well as sulfur dioxide (from sulfur containing fuels such as coal); and products of incomplete combustion, including carbon monoxide and unburned hydrocarbons. Various environmental problems, such as the greenhouse effect and acid rain, have been attributed to these pollutants, and certain human ailments, including respiratory diseases such as emphysema and lung cancer, are exacerbated by them.

Hydrogen, on the other hand, when combusted, produces no hydrocarbons, and no carbon or sulfur-containing by-products. In early hydrogen internal combustion engine conversions, significant quantities of NO_x were measured in vehicle emissions due to the high combustion chamber temperature. Specialized processes, however, such as water induction, have proven to eliminate virtually all of the nitrogen-containing by-products. Therefore, whether hydrogen is used to provide energy for combustion or chemical reaction (as in the case of the hydrogen fuel cell), the process is essentially, pollution-free, producing only pure water vapor as a product.

Hydrogen is the most abundant element in the universe. It is not normally found in its unreacted state in nature, however, and so it is almost always necessary to synthetically produce the hydrogen which shall be utilized. Hydrogen in made by splitting water into its component parts--hydrogen and oxygen--a process that utilizes energy. For this reason, hydrogen is not usually considered a source of energy, but rather is a form of stored energy. The energy to produce hydrogen from water can come from a wide variety of sources including solar, wind, hydroelectric, nuclear, natural gas reformation, and coal gasification. The produced hydrogen can then be stored, transported, and utilized in conventional energy consumption devices such as automobiles and gas appliances.

When hydrogen is oxidized, the result is the formation of water and the release of nearly as much energy as was required to synthesize the hydrogen in the first place. The reduction in the resulting energy liberated is referred to as the efficiency of the conversion cycle. In many cases, hydrogen's chemical properties make possible much higher utilization efficiencies than are possible with conventional fuels. In these cases, it is often possible to accomplish more useful work by converting the energy to hydrogen than by utilizing the primary fuel source.

Perhaps one of the most exciting advantages of hydrogen energy over conventional energy sources of today is that hydrogen offers a convenient method for capturing and storing energy supplied by renewable sources. Solar, wind, and geothermal power can be harnessed and collected for times of demand by using the renewable energy source and water to produce hydrogen, which can then be combusted or chemically combined with oxygen, producing the required energy--completely pollution-free.

PROPERTIES OF HYDROGEN FUEL

Hydrogen is quite different from many of the hydrocarbon fuels in common usage today. When compared to other common fuels on a per unit mass basis, hydrogen is found to have a considerably higher heating value. For example, hydrogen's heating value is 120 kJ/g compared to 50 kJ/g for natural gas and

45 kJ/g for gasoline (4). This makes hydrogen an especially attractive fuel for use in applications where fuel weight is of considerable importance, such as in aviation or space applications. Although hydrogen offers a much higher heating value per unit weight, hydrogen is also found to be substantially more voluminous than hydrocarbon fuels -- 3.4 m^3 of hydrogen contains the same amount of energy as 1 m^3 of methane. (See *Figure 1*.)

Another unusual chemical property of hydrogen as compared to hydrocarbon fuels is its flammability range. On the lean end, a mixture consisting of 4 percent hydrogen and 96 percent air is flammable. On the rich end, a mixture of 75 percent hydrogen and 25 percent air is still flammable as is any mixture within that range. By contrast, methane is flammable in concentrations of only 4 to 16 percent in air, and gasoline, in concentrations ranging between .8 and 6 percent.

Additionally, as shown in *Figure 1*, the flame speed for hydrogen in air is an order of magnitude greater than that of hydrocarbon fuels -- 265 cm/s compared to 34 cm/s for both natural gas and gasoline. The spontaneous ignition temperature for hydrogen is 571°C compared to 447°C for gasoline. The ignition energy for hydrogen is .02 millijoules whereas the ignition energy for methane and gasoline are .47 and 1.35 millijoules, respectively. Thus, the energy required to ignite hydrogen is an order of magnitude less than that required to ignite hydrocarbon fuels, but that energy must be supplied at a higher temperature than that required for most hydrocarbon fuels.

HYDROGEN PRODUCTION

Although hydrogen has been estimated to make up more than 75 percent of the mass of the universe, less than 1 ppm by volume exists in its free state in the atmosphere. On earth, hydrogen occurs primarily combined with oxygen in water, but it is also found in all organic matter, including living plants, petroleum, natural gas, coal, etc. Proposed methods for hydrogen production range from natural gas reformation to photosynthetic production utilizing blue green algae.

Natural Gas Reformation (Methane Steam-Reforming)

In the United States, most hydrogen is presently produced by natural gas reformation. In this process, methane mixed with steam is typically passed over a nickel oxide catalyst at an elevated temperature. The reforming reaction is highly endothermic and is usually performed in a fire tube reformer with the catalyst loaded in the tubes. The methane-steam reforming reaction is as follows:

$$CH_4 + H_2O \text{ ® } CO + 3H_2 \qquad (1)$$

This reaction is affected by the steam-to-carbon ratio, temperature, and pressure, as well as catalyst activity.

After the steam-reforming step, the product gas stream contains a considerable quantity of carbon monoxide which may be further reacted with additional steam, thereby increasing the hydrogen yield:

$$CO + H_2O \circledR CO_2 + H_2 \qquad (2)$$

This reaction is commonly known as the *CO shift,* or *water gas shift,* reaction. The conversion of CO by this reaction is slightly exothermic and favored by lower temperature. The lower practical operating temperature range for this reaction is between 180 and 200°C (9).

Coal Gasification

Coal gasification is the second most common method presently used for producing hydrogen. The chemistry of coal gasification is very similar to natural gas reformation. Steam is reacted with carbon in coal, thereby forming carbon monoxide and hydrogen. This low energy-density gas is subsequently sent through a CO shift reaction as used in the methane steam reforming process (10). The resulting products are hydrogen, carbon dioxide, and ash which is useful as a road base or cement additive. The basic reactions for coal gasification are as follows:

$$C + H_2O \circledR CO + H_2 \qquad (3)$$

$$CO + H_2 + H_2O \circledR 2H_2 + CO_2 \qquad (4)$$

In the coal gasification process, any sulfur present in the coal exits the gasifier as hydrogen sulfide. The hydrogen sulfide is removed by various processes such as a Holmes-Stretford unit where the sulfide is absorbed and regenerated. The resulting sulfur is filtered out as a cake (39 percent by weight) which is sold as a valuable feedstock.

Electrolysis

Environmentally, the best method for producing hydrogen would be one which is completely dependent on renewable resources. By utilizing water and electrical energy, hydrogen can be produced with no negative environmental effects using a process known as *electrolysis.*

When two electrodes are immersed in an aqueous ion-conducting electrolyte in the presence of a catalyst and electricity is applied across the cell, water is decomposed into its component parts -- hydrogen from the negative electrode, or cathode, and oxygen from the positively-charged anode.

Commercial electrolysis units have been available for many years. Various techniques and schemes are used to increase the efficiency of this conversion process. The higher the voltage, the faster the hydrogen and oxygen evolution, but the greater the quantity of wasted heat. The theoretical voltage for the decomposition of water at 25°C is 1.23 volts. Typical commercial electrolysis equipment have an efficiency ranging between 40 and 80 percent.

Early electrolysis systems typically used potassium hydroxide as the electrolyte. This equipment not only was very heavy and bulky, but also required extensive maintenance. Additional water had to continually be added for the process to continue, and electrolyte replacement was frequently required as impurities began to accumulate, causing buildup inside the cell.

For the past twenty years, the solid polymer electrolyte (SPE) has begun replacing potassium hydride in electrolysis systems. Electrodes coming into contact with the water-saturated membrane encounter an acidic environment equivalent to that of a ten weight percent sulfuric acid solution, although only pure water is actually circulated through the system. The membrane is proton-selective, allowing only hydrogen ions passage through -- on the anode side of the cell, the electric current pulls a hydrogen ion away from a water molecule and then transports it through the membrane. On the cathode, the hydrogen ion receives an electron, thereby evolving hydrogen.

Electrolysis systems based on the solid polymer electrolyte technology offer many advantages over the older technology. They are very compact and lightweight. Also, since the electrolyte is a solid membrane material and not a liquid, it doesn't wash away. In an SPE electrolyzer, a small amount of water is routinely drained from the cell, removing any impurities from within.

Because of their high efficiency (60 to 80 percent), SPE electrolyzers are ideal for converting renewable energy resources into hydrogen. For example, photovoltaic solar collectors can be connected to an electrolyzer to generate hydrogen during the daylight hours. This hydrogen can be stored and then very efficiently utilized in a diversity of applications. Furthermore, hydrogen produced by solar-powered electrolysis can be combined indiscriminately with hydrogen produced from other energy sources. In this way, it will be possible to use hydrogen produced from a variety of energy sources as the energy vector of future society.

Biological Production of Hydrogen

Hydrogen production by photosynthetic microorganisms is another method for capturing and accumulating the energy from sunlight. Blue-green algae, also known as cyanobacteria, and other anaerobic bacteria have been found to be capable of splitting water into hydrogen and oxygen by a light-driven process known as *biophotolysis* (11). Cyanobacteria typically possess three hydrogen-metabolizing enzymes: nitrogenase, membrane-bound uptake hydrogenase, and soluble hydrogenase. In general, the capacity to photoproduce hydrogen varies widely from one species to another depending on the levels of nitrogenase and hydrogenase in the organism. When incubated under anaerobic conditions in the presence of 1% carbon monoxide, as much as 1400 nmol H_2 mg dry wt^{-1} h^{-1} have been reported for *Anabaena* strain CA (12).

TRANSPORTING HYDROGEN

Today, small volumes of gaseous hydrogen are typically transported in high pressure containers via truck, train, and barge. Transmission of large quantities of gas using these methods, however, is impractical. The preferred method for transporting large volumes of hydrogen efficiently is through an underground pipeline system similar to existing natural gas networks. In fact, the existing natural gas distribution piping could be converted to hydrogen. Since this distribution system is generally a low pressure system, it is possible that existing natural gas pipelines would be adequate to deliver the required volume of hydrogen gas without danger of hydrogen embrittlement (13).

In addition, since hydrogen can be produced by any primary energy source and water, almost every major population center in the world could produce the hydrogen required for their own community. This would mean that locally-produced hydrogen could be transported through new transmission mains installed at the hydrogen plant and then distributed utilizing converted natural gas distribution systems.

ECONOMICS OF HYDROGEN ENERGY

Although hydrogen can be produced totally pollution free, utilizing water and renewable energy, the majority of today's hydrogen is synthesized from fossil fuels. Methane reformation and coal gasification processes are used to produce the vast majority of hydrogen currently.

Experts expect that around the year 2040, as natural gas and petroleum reserves become depleted, there will be a major shift from our current liquid fossil fuel base to cleaner burning fuels. At this time, the cost of the little remaining liquid fossil fuels shall notably increase, and the only appreciable fossil fuel remaining shall be coal. Fuel will be acquired from renewable energy sources or from coal gasification (14).

Figure 2 compares the end-use and effective costs of various energy alternatives projected for the year 2000. End-use costs, or internal costs, include the production costs as well as the costs of transportation and distribution. The effective costs, on the other hand, take into account the end-use costs, estimated external costs, and fuel utilization efficiencies. The external costs include costs associated with damage to the environment caused by utilization of various fossil fuels, and the energy generation utilization efficiency takes into account the efficiency of converting fuels into various energy forms, such as mechanical, electrical, and thermal. The effective costs are expressed as \$/GJ of fossil fuel equivalent, or, in other words, \$/GJ of fossil fuel required to provide the same service (16). The overall equation for the effective cost is as follows:

$$C_{eff} = (C_{in} + C_{ex})(h_F/h_S), \quad (5)$$

where C_{eff} = effective cost
C_{in} = internal cost
C_{ex} = external cost
h_F/h_S = utilization efficiency

With the advent of the fuel cell, hydrogen is more attractive as a fuel than ever before. For years scientists have been searching for ways to produce hydrogen more economically. In the fuel cell, however, hydrogen is used two to three times as efficiently as in an internal combustion engine. Hence, when utilized in a fuel cell, hydrogen can cost two to three times that of more conventional fossil fuels and still be competitively priced. Today, the market price for hydrogen when used in a fuel cell and produced by electrolysis is, for the first time, competitively priced with gasoline. Perhaps this will be the impetus needed to trigger the movement toward the hydrogen economy.

HYDROGEN STORAGE

To date, numerous methods of storing hydrogen have been proposed. These include solid hydrogen, microspheres, compressed gaseous hydrogen, liquid hydrogen, and metal hydrides. Most commercially viable hydrogen energy applications will employ one or more of these storage technologies.

Solid Hydrogen

Interesting research is currently underway to investigate the potential of forming solid hydrogen (slush). Although opinions vary regarding the potential of this form of hydrogen as a means of storage, the phenomenon is interesting and, if commercially developed, could be the lightest and most compact form of hydrogen storage available. At present the concept is a theoretical/laboratory curiosity and cannot be considered as a viable hydrogen storage alternative.

Microspheres

Researchers have investigated the possibility of storing hydrogen gas trapped inside tiny glass microspheres. This storage method takes advantage of hydrogen's ability to diffuse through hot glass. High pressure hydrogen diffuses through the thin glass walls of the microspheres when the spheres are heated and the hydrogen is supplied at high temperature. The spheres are then cooled under pressure and the hydrogen gas is trapped inside, completing the charging procedure. When a vehicle operates utilizing hydrogen stored in microspheres, the spheres are heated by the exhaust gases. As the glass walls of the spheres become warm, hydrogen gas diffuses through and is released to fuel the vehicle.

Though somewhat functional, microsphere storage of hydrogen has serious drawbacks. The amount of hydrogen that can be stored in a vessel of a specified volume is not great, which causes vessels containing microspheres for hydrogen storage to be unreasonably voluminous. More important, studies have concluded that approximately 12 percent of the microspheres rupture during each charge/discharge cycle. This means that after recharging the vessel just four times the storage capacity is cut in half. Though laboratory prototypes using this technology have been built and tested, the technology has not yet proven feasible.

Compressed Gas

Gaseous hydrogen is the most voluminous and the heaviest of the practical hydrogen storage technologies. It is also the most economical. For these reasons, compressed gas storage is a good method for stationary hydrogen storage installations, but it is not a practical choice for vehicle storage systems. The very high pressures of compressed hydrogen could be dangerous in the event of a malfunction or a collision. Even at high pressures the tanks are heavy and voluminous, making this method of storage impractical for most vehicular applications. A comparison of the relative volume and weight of various storage technologies is presented in *Figures 3* and *4*.

Liquid

Liquid hydrogen is the preferred method of storing hydrogen for space and aircraft applications where light weight is the most important consideration. This method of storing hydrogen involves transforming the gas into its liquid form. This is accomplished by cooling hydrogen to 20.3 Kelvin, a process which is energy intensive and requires specialized equipment. Consequently, liquid hydrogen is considerably more expensive than gaseous hydrogen.

Liquid hydrogen is stored in specially designed insulated flasks called dewars. Any heat which leaks into the liquid hydrogen will cause a proportionate amount of the liquid to "flash" or vaporize into the gaseous form. For this reason, engineers have developed super insulation, which is employed in cryogenic storage containers. The common method of making super-insulated containers involves a lightweight, thin-walled inner vessel, which actually holds the liquid, surrounded by a jacket consisting of alternating layers of reflective mylar and an insulating layer of foam. The mylar and foam layers are encapsulated in a vacuum to further minimize the conduction of heat. The entire vessel is then enclosed in an outer vessel which makes possible the maintenance of the vacuum.

One of the difficult problems associated with the design of a cryogenic hydrogen vessel is the method of supporting the inner vessel. If it were to be supported inside the outer vessel by some type of a metallic bracket, the bracket would conduct a significant quantity of heat into the inner vessel, causing the hydrogen to vaporize (boil off). However, various technologies have been developed and are commercially in use to overcome this problem. In a well-designed cryogenic storage container, a boil-off rate of less than one-half-of-one percent is achieved. This means that the inner vessel is so well insulated that the amount of heat penetrating the insulation is only sufficient to vaporize one-half-of-one percent of the hydrogen every 24 hours. Such a tank would be able to store liquid hydrogen for a maximum of 200 days. After that time, all of the liquid would have been transformed back into the gaseous state. Meanwhile, as the hydrogen changes back into the gaseous form it becomes voluminous and must be vented, or the pressure inside the liquid hydrogen tank begins to increase. In a liquid hydrogen vehicle, some hydrogen should be used almost every day. Alternatively, the vaporized hydrogen could be run through a re-liquefaction machine and put back into the tank -- this approach, however, is more practical for larger-scale applications.

When a liquid hydrogen tank is filled for the first time or is refilled after sitting for some period of time, the inner vessel must be cooled down to liquid hydrogen temperature. This is done by pumping liquid hydrogen into the inner vessel and allowing it to flash-off. The hydrogen which is vaporized into gaseous form by the latent heat of the inner containment vessel is referred to as flash-off and must be dealt with when refueling liquid hydrogen containers. The amount of flash-off is usually equal to only one or two percent of the tank capacity. Fortunately, flash-off only occurs when a liquid hydrogen tank is completely empty and has been empty long enough for the inner vessel to warm up to ambient temperature. Nevertheless, a provision for handling flash-off during recharge of liquid hydrogen containers must be taken into consideration in such systems.

To prevent hydrogen from vaporizing during the refueling process, it is transported in special vacuum-jacketed transfer lines. If liquid hydrogen is allowed to be transported in fuel lines which are not vacuum-jacketed, oxygen and nitrogen from the air, which come in contact with the lines are cooled to below their liquefaction temperatures. As a result, liquid nitrogen and liquid oxygen drip off the line. Liquefied oxygen can be dangerous, causing severe cold burns when brought into skin contact. Furthermore, the density of oxygen in liquid air can cause combustible materials to oxidize at more rapid rates than normal. For these reasons, liquid hydrogen is always transferred in vacuum-jacketed lines.

Metal Hydride

Certain metal alloys have been found to have the property of forming reversible covalent bonds when reacting with hydrogen (17). These alloys, when combined with hydrogen, are referred to as *metal hydrides*. It has been demonstrated that metal hydrides are capable of safely storing hydrogen in a powdered form with an improved weight density as compared to compressed hydrogen and with a volume density similar to liquid hydrogen. Metal hydride technology, which has now been tested extensively, is considered the safest method for storing hydrogen on board a vehicle.

The chemical properties of reaction with hydrogen vary greatly from metal to metal. This reaction becomes especially complex when two or more different metals, each with different hydriding properties, are mixed together into an alloy of varying percentages. Over 200 different alloys have been characterized and studied to determine which would be most applicable to an automotive storage system. Some alloys were found to be extremely heavy, storing only small quantities of hydrogen, which made them too heavy for use in an automobile. Others were extremely expensive and therefore impractical. A third category of metal hydride alloys was classified as dangerous. For example, mischmetal hydride is pyrophoric, spontaneously igniting if brought into contact with air.

The first successful vehicle hydride storage vessel was installed in a Pontiac Grand Ville in 1974, by Billings Research Corporation (18). This vessel used the iron-titanium family of metal hydrides. The powdered alloy was contained in stainless steel tubes which were manifolded together inside a jacket. During vehicle operation, exhaust gases from the engine flowed over the outside of the stainless steel tubes, providing the heat necessary to dissociate the hydride. Technicians at Brookhaven National Laboratory who did some initial research on hydrides suggested that a small gas reservoir would be needed to provide the hydrogen necessary to start the engine and to warm up the metal hydride storage container. In practice, however, such a reservoir was not required. Enough hydrogen was available from even a very cold hydride vessel to get the system started.

In the Pontiac prototype installation, the system functioned properly, but the vehicle had a limited range. To fill the vessel, a hydrogen gas line was connected to the metal hydride storage tank, allowing gaseous hydrogen, which was at 1.8 MPa pressure (18 atm), to flow into the tank where it reacted with the metal hydride alloy. When the hydrogen and the alloy combined, heat was liberated. The hydride storage tank could be charged as quickly as this heat could be dissipated. In the Pontiac Grand Ville the heat was discharged by blowing air through the tank. Utilizing this method, it took about four hours to recharge the vehicle with hydrogen. Refinements to the system, employed in later prototypes, reduced the recharge time to less than 15 minutes.

There were many problems with early hydride systems. One such problem was decrepitation of the alloy. Each time the metal hydride storage tank was recharged, the particles would break down. This effect did not impact the hydride's ability to store hydrogen, but eventually the particles became so small that they began to pass through the five-micrometer sintered metal filter which kept the hydride inside the tank. This problem was addressed by adding a tiny amount of manganese, which caused the decrepitation process to discontinue when the particles reached a size of about ten micrometers.

One of the biggest advantages of hydrides is their safety (19). As part of a study to determine the safety of the iron-titanium-manganese metal hydride storage system, tests were conducted in conjunction with the U.S. Army. The tests consisted of firing armor-piercing incendiary bullets at fully-charged hydride containers, simulating the worst possible conditions resulting from a serious collision. The tests demonstrated that the metal hydride vessels do not explode. As the vessels were ruptured by the impact, a small, pilot-light-sized flame burned for several hours at the hole where the bullets had entered the vessels, as gaseous hydrogen was slowly released from the hydride material.

Another important attribute of metal hydride storage systems is longevity. Metal hydride storage containers have been tested by charging and discharging them over and over in thousands of cycles, equivalent to millions of miles of driving. Tests show that during the first 100 cycles, the hydride actually improves with each cycle. After the first 100 cycles, the improvement phenomenon is no longer observed, but there is no degradation in the performance of the tank even after thousands of cycles. For this reason, used metal hydride storage vessels can be salvaged from old vehicles, reactivated, and used again. Although metal hydride storage systems don't degrade, hydrides can be poisoned by impurities in the hydrogen, including such things as carbon monoxide, oxygen, or even water vapor. However, because of the unique chemical properties of hydrogen, it is very easy to remove these impurities from a hydrogen gas stream even on a small scale, such as in electrolysis. Electrolysis units, which have been built for the past ten years, produce hydrogen directly out of the cell at pressures adequate for recharging hydride vessels. [Although the electrolyzers will work at pressures of up to 4.1 MPa (40 atm), only 1.7 MPa (17 atm) is necessary to recharge a metal hydride storage vessel.] Simple, yet reliable methods of purifying the hydrogen to the hydride grade have been successfully employed in the electrolysis process. These methods include de-oxo catalysts and pressure swing adsorption technologies employing molecular sieves.

Chemical properties of three common metal hydrides are presented in *Figures 5, 6*, and *7* (20-22) by depicting the pressure of hydrogen in a metal hydride storage container, at a specified temperature, as a function of the weight percent hydrogen. In a hydride storage container, the higher the temperature, the higher the pressure. Typical hydriding alloys have a pressure plateau, or, in some cases, several pressure plateaus. As the tank is charged, the pressure rises rapidly until it reaches the plateau pressure for that particular temperature. Then, as additional hydrogen is added to the container, the hydrogen begins to be absorbed by the alloy in the vessel, and the pressure does not increase. This phenomenon continues until the reactive sites within the metal alloy are completely occupied. At this point addition of more hydrogen causes the pressure to increase again, indicating that the tank's storage capacity is saturated.

The iron-titanium family of hydrides store hydrogen with weight percents ranging between 1.6 and 2.3. Other alloys have much higher weight percents. For example, magnesium hydrides have hydrogen weight percents as high as 7.6. Unfortunately, these alloys require very high temperatures to release the hydrogen, making them impractical for most automotive storage applications. The energy required to dissociate various metal hydride alloys is shown in *Figure 8.*

Examples of Metal Hydride Storage in Mass Transit Vehicles
The first metal hydride transit vehicle to go into service was a Winnebago 21-passenger bus, put into public transit service in 1975 on the route connecting the cities of Provo and Orem, Utah (23). The bus was powered by a Dodge engine. To provide the bus with sufficient range, two Fe-Ti hydride vessels were fabricated, each having a total mass of 714.4 kilograms (1575 pounds) for a total hydride system mass of 1429 kilograms (3150 pounds). The bus metal hydride system is depicted in *Figure 9.* The actual amount of hydride alloy contained

inside the stainless steel vessels was 1016 kilograms (2240 pounds), which could store 12.7 kilograms (28 pounds) of hydrogen, representing a 1.25 weight percent alloy and a 0.9 weight percent storage system. The metal hydride performance data is presented in *Figure 10.*

In a later test, an Argosy bus was retrofitted for hydrogen operation in Riverside, California. In the Riverside bus, radiator water was used to supply the heat of dissociation for the metal hydride storage system. The overall weight of the hydride storage system was substantially reduced by substituting aluminum vessels for the stainless steel used in the earlier prototype. The design of the Riverside bus hydrogen system is depicted in *Figure 11.* The aluminum pressure vessels were enclosed in an exterior water jacket, thereby allowing radiator water to flow over the outer aluminum vessel, providing heat to the metal hydride bed. This design is shown in *Figure 12.* The cooling water from the engine provided the heat for metal hydride dissociation, while the metal hydride dissociation provided the cooling for the engine.

To recharge a metal hydride vessel, it is necessary to supply hydrogen to the vessel while dissipating the heat given off by the exothermic reaction. In the Riverside Bus Project, the recharge waste heat was dissipated by connecting an external heat exchanger with two "quick" connects. A diagram of the system is presented as *Figure 13.* After the lines were connected to the bus, a coolant circulation pump caused the fluid to circulate in a loop through the metal hydride vessel and then through an external finned-tubing-to-air heat exchanger.

HYDROGEN ENERGY APPLICATIONS

The Hydrogen Internal Combustion Engine

Standard gasoline and diesel-powered internal combustion engine vehicles can be converted to run on hydrogen. These vehicles have often been found to be considerably less polluting, safer, and more efficient than their fossil fuel-burning counterpart.

Backfire

Early hydrogen engine conversions were typically affected by an unexpected occurrence known as backflash. *Backfire*, or *backflash,* as it is also known, is a serious problem associated with hydrogen engines that must be taken into consideration when designing an engine conversion system.

Backfire is the occurrence of one or the other (or sometimes both) of two distinguishable phenomena. A true backfire occurs when an explosive mixture of fuel and air builds up in the exhaust system of an engine and is then ignited, with the explosion coming out the tailpipe of the vehicle. This type of backfire is common in poorly-tuned engines running on gasoline and is fairly rare in hydrogen engines. The second type of backfire, which is more properly referred to as "backflash", is that phenomenon which occurs when a combustible mixture of hydrogen and air in the intake system of the engine is ignited. In this case the explosion comes out of the carburetor of the vehicle, very often damaging the equipment. Backflash is a very common problem in hydrogen engines.

In a current-model, high-performance, gasoline-burning engine, fuel is ignited at the appropriate time in four, six, or eight alternative cylinders, usually by means of a high voltage ignition system which is directed to the cylinder ready to be fired by a distributor cap and a rotor. Careful experimentation on a conventional ignition system reveals that when the ignition system fires a spark in the proper cylinder, simultaneously a much smaller, induced spark is created in other cylinders. These induced sparks are usually caused by electromagnetic interference between the cables running from the distributor cap to each individual spark plug. Laboratory bench tests show that an induced spark can be reliably produced at the end of an ignition cable located as far away as 48 centimeters (19 inches) from the cable where the primary spark has been fired. (See *Figure 14.*) These induced sparks are sometimes present in a cylinder during the intake stroke and come in direct contact with a flammable mixture of fuel and air. In a conventional gasoline engine, induced sparks are too weak to ignite the gasoline/air mixture. In the hydrogen engine, however, since less than one-tenth as much energy is required for ignition, the hydrogen/air mixture ignites readily. If the cylinder is in the intake cycle, the flame burns up through the intake valve and into the intake manifold, causing backflash.

Another significant cause of backflash in the hydrogen engine is from pre-ignition sources. This phenomenon can be produced in the test laboratory where the hydrogen engine is connected to a dynamometer to simulate normal automobile operating loads. When the intake filter is removed and a small quantity of very fine dust particles introduced into the engine intake, backflash occurs. This procedure will create no extraordinary results if the engine is operating on gasoline, but when the engine is operating on hydrogen, the result is an almost immediate backflash. The reason, illustrated in *Figure 15*, is that the particles are drawn into the combustion chamber during the intake stroke (A). The fuel is then compressed (B), and the charge is ignited, resulting in the power stroke (C). During the exhaust stroke, the exhaust valve opens and the burnt charge, along with most of the particles, is pushed out into the exhaust system (D). The cycle repeats itself with the intake valve opening and a new intake stroke beginning (E). At this point, however, some of the dust particles, which are still hot from the last cycle, remain in the combustion chamber. As the hydrogen fuel comes into contact with these hot particles, the incoming fuel charge is ignited and burns quickly back into the intake manifold, resulting in the detonation. In the case of the hydrocarbon engine, these small particles do not have sufficient thermal mass to ignite the hydrocarbon fuel.

Other sources of pre-ignition backflash are carbon buildup within the combustion chamber (caused by previous operation of the engine on hydrocarbon fuels) and rusting spark plugs. If carbon buildup comes to a sufficiently high temperature during engine operation, it can ignite the incoming charge of hydrogen and air during the intake stroke, causing backflash. In the same way, iron oxide on the spark plugs is an excellent catalyst for the ignition of hydrogen at high temperatures. In this case, the incoming hydrogen/air charge comes into contact with the hot spark plug, which provides a source of catalytic ignition and the resulting backflash. To prevent these problems, several things can be done. First of all, in the case of carbon buildup, operating an engine on hydrogen will quickly burn out any carbon deposits left over from gasoline operation, and the problem, essentially, takes care of itself. For the iron oxide source of backflash, the best solution is to exchange conventional spark plugs for plugs made of non-corrosive material such as stainless steel.

Water Induction

One interesting method of controlling backflash, and a host of other problems associated with hydrogen engine conversions, is water induction (24). The technology employed in water induction consists of inducting tiny droplets of liquid water into the combustion chamber with the air/fuel charge. When the charge is compressed and ignited, the droplets of water become involved in the ensuing chemical reaction. Consider the combustion chamber depicted in *Figure 16*. Here, the charge inside the combustion chamber has been ignited by the spark plug and is beginning to propagate through the combustion chamber. Outside the arc of flame, the air/fuel mixture is filled by a mist of water. As the flame front encounters these droplets, it instantly transforms them from liquid water into water vapor. In the process, a significant amount of heat energy is consumed. Since the peak combustion temperature is substantially reduced, auto-ignition sources are not heated to a high enough temperature to cause the auto-ignition phenomenon to take place. For this reason, water induction virtually eliminates backflash from this source.

Water induction is also very effective in reducing the formation of nitric oxide pollution (25). Nitric oxide is formed when nitrogen and oxygen in the air are heated to high temperatures. The formation of nitric oxide in an uncontrolled hydrogen engine is presented in *Figure 17* as a function of hydrogen equivalence ratio. If a water induction system is installed on the engine and adjusted to provide sufficient droplets of water so as to maintain the peak combustion temperature below the nitric oxide formation threshold of 700 degrees Centigrade, this undesirable chemical reaction is virtually eliminated, as shown in *Figure 18*. As a result, the nitric oxide pollution coming from a hydrogen engine with water induction properly installed is one part per million or .01 grams per kilometer.

The water induction technique of hydrogen engine conversion has other benefits. For example, there is an increase in brake thermal efficiency of the engine as water induction is applied. There is also a corresponding increase in horsepower output of the engine. (See *Figure 18*.) This occurs as the propagating flame encounters the water droplets, causing them to vaporize into steam and expand dramatically. (Refer to *Figure 16*.) Although the vaporizing of the water droplets reduces the peak combustion temperature, the amount of energy consumed as the water undergoes a phase change generates a greater increase in pressure than would result if that same energy were utilized to heat and expand the air and combustion products normally encountered in the combustion chamber. It is the pressure of the gases which force the piston down and not the temperature. Therefore, if a high pressure can be achieved inside the combustion chamber without high temperature, the hydrogen engine will have increased horsepower and no backflash.

There are some problems with water induction. One difficulty is how to accomplish the water induction itself. In order to be effective, the water droplets must be uniformly dispersed throughout the combustion chamber. Any pockets of combustible product within the chamber devoid of the water droplets will generate high combustion temperatures, nitric oxide, and possibly even cause backflash. *Figure 19* shows a typical intake manifold system with a water-atomizing spray nozzle on the right side of the common air inlet. The stream of incoming air sweeps the water droplets into the intake manifold, with the majority of them ending up in cylinder number four. A similar nozzle installed over the inlet to each cylinder caused the droplets to be evenly distributed between each of the cylinders, but test results indicated that there were pockets within the cylinders devoid of the water droplets. Nitric oxide concentrations averaged 100 to 500 times higher than in better-designed systems where water droplets were evenly distributed within each of the cylinders.

The converted Provo-Orem bus previously mentioned employed a specially designed water induction system as a means of increasing the water induction mass ratio when operating at or near full-power setting. Engine performance data as a function of the equivalence ratio and water injection mass ratio are presented in *Figures 20* and *21*, respectively.

Direct Cylinder Induction

One of the major disadvantages of converting an engine to hydrogen is the resulting loss of 25 to 35 percent of the power from the engine. Interestingly, this loss of power is not an intrinsic characteristic of hydrogen engines. To the contrary, hydrogen, with its higher flame speed, actually has the potential of producing more power than the conventional gasoline fuels. This loss of power phenomenon may be understood by examining the operation of the internal combustion engine. During the intake stroke, a mixture of fuel and air is drawn into the combustion chamber. The carburetion system on the engine adjusts the ratio of fuel to air for proper and complete combustion. In the case of gasoline, the droplets of fuel occupy such a small area that during the intake stroke they constitute less than one percent of the total volume of fuel and air drawn into the combustion chamber.

The case with hydrogen is very different. If the carburetor is set to mix hydrogen and air at 0.8 stoichiometry (which is a reasonable setting), then 25 percent of the volume of gas in the intake manifold is hydrogen and only 75 percent is air. In a cylinder having 1000 mL of displacement, 990 mL of air would be drawn into the chamber during the intake stroke if the engine was operated on gasoline. If the same engine was converted to hydrogen, the amount of air drawn into the cylinder during intake would be reduced to just 750 mL—a reduction of one-fourth. The rest of the combustion chamber would already be filled with the voluminous hydrogen. This phenomenon is depicted in *Figure 22*. Since one-fourth of the air is displaced by the hydrogen, there is not as much oxygen available to react with the fuel, and the amount of power is proportionately reduced by a fourth. Thus, the reason that many hydrogen engines are less powerful than their gasoline counterparts is the lack of a full charge of air in the engine during the intake cycle. The remedy to this problem is a technique known as *direct cylinder injection* (DCI).

In a DCI system there is no throttle limiting the amount of gases going into the intake port of the engine and no fuel pre-mixed with air. During the intake stroke, the engine draws in a full charge of 1000 mL of air. After the chamber is filled with air, while the intake valve is closing and at the beginning of and during the compression stroke, hydrogen is injected into the combustion chamber under pressure. By this method the reduced power problem is completely overcome. In fact, the pressurized hydrogen entering the cylinder increases the pressure over that drawn in naturally by the engine, giving a sort of turbo-charged effect to the system, which results in not only equivalent, but even enhanced power output as compared to operation on gasoline.

There are many benefits of a DCI system. Since there is no throttle on a DCI engine, the pumping loss at part throttle settings is eliminated. In fact, this engine has the main efficiency advantage of a diesel engine. Furthermore, since the hydrogen is injected into the combustion chamber during the intake stroke, or, in some cases, at the end of the compression stroke, the gases in the chamber do not have an opportunity to mix completely. As a result, the charge is stratified, resulting in a much slower flame speed and consequently a lower peak combustion temperature. In this way the tendency of the engine to detonate is greatly diminished and nitric oxide formation is reduced. Another major advantage of DCI is its effect on backflash. When the intake valve opens in a DCI system, a cool charge of pure air is drawn into the combustion chamber. This cool air quenches any residual flame, hot catalytic particles, or any other source of ignition within the combustion chamber —- all before the hydrogen fuel is introduced into the system. As a result, DCI systems do not have the backflash problem.

There are, however, some problems associated with the implementation of a direct cylinder injection system. The first is that the injection of the hydrogen gas must be carefully synchronized or controlled to be in perfect phase with the internal combustion engine. This is accomplished by placing a position sensor on the crankshaft which is used to electronically trigger a computer control system, or, alternatively, to mechanically actuate an injector system. The more difficult problem is to inject the large volume of hydrogen fuel required for each combustion cycle. To accomplish this requires an injector having a substantial orifice size and which operate fast enough to keep up with the engine at high speeds. These problems are readily solved with newer engines having two intake valves -- one for hydrogen, the other for air. Hydrogen-fueled engines of the future will be based on DCI technology. Often these conversions will also employ water induction to minimize nitric oxide formation and to increase efficiency.

Improved Vehicle Performance
One of the advantages of operating engines on hydrogen, as compared to hydrocarbon fuels, is the higher utilization efficiency which can be achieved. In a testing program sponsored by the United States Postal Service, two identical vehicles were operated on a rural mail route (26). One vehicle was the conventional gasoline version used extensively by the Postal Service. The other vehicle was retrofitted for hydrogen operation. The engine conversion method employed in this Postal Jeep was water injection. The hydrogen was stored in an iron-titanium- manganese hydride. The resulting pressure vessel was the same dimension as the spare tire, and therefore could be mounted in the cavity in the vehicle frame which had been made for that purpose.

During the testing program careful data was collected to determine the efficiency of a hydrogen engine in a mail delivery application compared to a gasoline vehicle. Each day the miles traveled and the fuel consumed was monitored for each vehicle. The results of these tests are presented as *Figure 23* and show that the hydrogen-fueled vehicle proved more efficient than the gasoline counterpart.

Hydrogen in Aviation
Hydrogen is of considerable interest in aviation because of its extremely high energy content per unit mass -- it is the lightest fuel available and, in use, hydrogen could greatly increase the aircraft payload. Hydrogen-powered supersonic aircraft would have a range of 1-1/2 to 3 times that of the hydrocarbon-fueled plane. The doubtful economic viability of supersonic aircraft would greatly improve if redesigned to run on liquid hydrogen; hypersonic aircraft, on the other hand, could not be operated unless they ran on hydrogen (27).

Another advantage of using liquid hydrogen in aircraft is the airframe cooling which such a system would offer. Presently titanium and other heavy alloys are required for supersonic aircraft frames because they are the only materials having the strength needed at the high temperatures to which the exterior surface of supersonic aircraft are exposed. With such cooling, lightweight alloys such as aluminum and boron could be used, further increasing the range and payload (28).

Presently, the absence of the necessary infrastructure is the major obstacle preventing the widespread conversion to hydrogen-powered aircraft. It is expected by many that there will be a movement toward liquid hydrogen in aviation.

The Hydrogen Homestead

Hydrogen is also an ideal fuel for a wide variety of residential applications. This versatility was demonstrated in the Billings Energy Corporation Hydrogen Homestead Project in the mid-70's (29). The Hydrogen Homestead was a high-tech home in which the appliances were converted to operate on hydrogen. The hydrogen system employed is presented as *Figure 24*. The fuel for the project was produced by electrolysis based on solid polymer electrolyte technology and was stored in an iron-titanium-manganese metal hydride vessel. This 1791 kilogram stainless steel vessel is described in *Figure 25*. The vessel was instrumented with temperature probes, micro-expansion measurement transducers, and an internal vessel viewing window — all for the purpose of evaluating the feasibility of metal hydrides for bulk hydrogen storage. The stainless outer vessel was encapsulated in a ten-centimeter (four-inch) layer of foam to insulate it from the external environment. Heating and cooling of the vessel was accomplished with an internal heat exchanger. The vessel stored 31 kilograms of hydrogen, a quantity adequate to maintain operation of the Hydrogen Homestead for about ten days.

Hydrogen Range

The natural gas range was modified for operation on hydrogen. Hydrogen contains approximately one third the energy per unit volume of natural gas, yet its low viscosity allows it to flow through an orifice approximately three times as fast as natural gas. Thus, the natural gas orifice sizes on the range were suitable for hydrogen conversion without any modification. Hydrogen has such a high flame speed that it is better if it is not pre-mixed with air as is done in the case of natural gas. Therefore, in such a conversion, the primary air intake is eliminated. The secondary air of the natural gas appliance, which mixes with the fuel and supports combustion after the fuel has passed through the burner head, (see *Figure 26*), is more than adequate for hydrogen combustion. When natural gas is burned without primary air, the results are incomplete combustion, carbon buildup on cooking pots, and dangerous quantities of carbon monoxide in the air. With a hydrogen burner there is no carbon, and no problem with incomplete combustion. In fact, the problem with hydrogen is the other extreme. Even without primary air, the hydrogen burns too quickly, resulting in regions of high temperature in which nitrogen and oxygen from the air combine to form nitric oxide. In a conventional range burner, a nitric oxide concentration of 35 parts per million was measured when the burner was operated on natural gas. When the burner was converted to operate on hydrogen and the primary air was eliminated, the nitric oxide concentration jumped to 250 parts per million. Either concentration is too high considering that these stove-top burners are operating in the enclosed environment of a home.

A simple solution is available to eliminate this problem. At high temperature, stainless steel is an excellent catalyst for hydrogen combustion. Utilizing this property, stainless steel catalyst can be carefully fitted around the outside of the burner head. (See *Figure 27*.) The catalyst performs three functions: First, it inhibits the mixing of hydrogen and air, thereby causing a zone immediately around the burner head where the concentration of hydrogen is very high and the concentration of air is very low. Second, it catalytically supports the reaction between hydrogen and air at concentrations which are not flammable. Since there is no high temperature zone, nitric oxide formation is virtually eliminated. Third, it produces a visible flame. Normally when hydrogen burns, the flame is invisible. The stainless steel catalyst glows a bright orange, making it apparent that the flame is burning, while at the same time giving some indication of how high the burner is set. Catalytic burners were utilized in virtually all of the natural gas-to-hydrogen conversions in the Hydrogen Homestead. Careful measurements indicated that only one to five parts per million of nitric oxide were generated by such a burner.

Hydrogen Water Heater and Furnace

One significant advantage of using hydrogen instead of conventional fuels, which was demonstrated by the Hydrogen Homestead experiments, was the increase in utilization efficiency when appliances were converted to hydrogen. One such case was the hydrogen water heater. When this appliance was converted, the stainless steel catalyst was not necessary because it was possible to control the mixing of hydrogen and air in such a way that the actual combustion took place on the surface of the water heater's internal heat exchanger. The burner consisted of a piece of stainless steel tubing which sprayed the hydrogen against the metal wall of the heat exchanger. (See *Figure 28*.) Since the metal was a good conductor of heat and was constantly being cooled by the water on the other side of the wall, the flame was effectively quenched with virtually no nitric oxide formation. (Nitric oxide levels of three parts per million were measured from this appliance.) In the hydrogen water heater, it was possible to cool the steam in the exhaust gases until it condensed on the sides of the internal heat exchanger, draining into a pan underneath. Thus, the hydrogen water heater was able to take advantage of the higher heating value of the hydrogen fuel. This advantage was also realized in the conversion of the furnace. With a conventional furnace, during the cold months of the year the humidity inside a home becomes uncomfortably low. Expensive heating systems employ a power humidifier to put additional moisture into the air, thereby increasing the comfort level. Since the only by-product of burning hydrogen (with the catalyst) is water vapor, it was possible to modify the furnace so that all of the exhaust gases went directly into the house. This eliminated the waste of heat going up the chimney and also took advantage of the higher heating value of the fuel. In addition, all of the steam was captured within the living environment of the home, increasing the humidity and lowering the fuel bill.

The Hydrogen Fuel Cell

As the world begins to move closer to using vehicles which are fueled by hydrogen, some real technological opportunities become available as a result of the unique chemical properties of hydrogen. One of the most interesting of these new technologies is the hydrogen fuel cell. A hydrogen fuel cell is an electrolytic cell in which hydrogen and oxygen combine to form water (30). The resulting energy is liberated as electricity. Unless a blower is used to supply air to the fuel cell; it has no moving parts, it makes no noise, has absolutely no air pollution, and utilizes the hydrogen fuel very efficiently. *Figure 29* is a diagram of a solid polymer electrolyte fuel cell. In the fuel cell, a solid polymer membrane (31) is located between two electrodes. On the side of the negative electrode (the anode), hydrogen gas is supplied to the cell. On the opposing positive side (the cathode), oxygen is supplied. The anode reaction is:

Anode Reaction: H_2 ® $2H^+ + 2e^-$ *(6)*

Hydrogen dissociates in the presence of the catalyst, forming hydrogen ions (protons) and giving up electrons to the anode. The hydrogen ions are transported across the membrane to the cathode. At the cathode, hydrogen ions react with oxygen to form H_2O.

Cathode Reaction: $2H^+ + 2e^- + 1/2\ O_2$ ® H_2O *(7)*

The electrons, which are liberated at the anode, travel by electrical cable through the external load (such as an electric motor) to the cathode where they become part of the water-forming reaction. If the external circuit is open, not allowing the electric current to flow, then the reaction is stopped. No fuel is consumed and no power is generated. The electrolytic reaction, then, is controlled by the load connected to the cell.

The overall fuel cell reaction is:

Overall Reaction: $H_2 + 1/2\ O_2$ ® H_2O *(8)*

By this simple chemistry, each cell generates a maximum potential of just over one volt. The cells are stacked in series to achieve higher voltages as required.

Figure 30 shows a typical fuel cell output versus current density at 25°C. As can be seen from this figure, the output voltage of a fuel cell decreases as the electrical load is increased. The theoretical polarization voltage of 1.23 volts per cell (at no load) is not realized in real life due to various losses. Typically, solid polymer electrolyte fuel cells operate at .75 volts per cell under peak load conditions or at about a 60 percent efficiency. The efficiency of a fuel cell is a function of such variables as catalyst material, operating temperature, reactant pressure, and current density. Typical operating efficiencies of 60 percent are normal, but at low current densities, efficiencies as high as 75 percent are achievable.

Some fuel cells have the capability of being operated in reverse as an electrolyzer. The dual-role capability of these devices makes them a contender for off-peak electric power storage applications. In the case of the fuel cell car, it is possible to operate the fuel cell in reverse at night by connecting a source of electricity and water to the unit. In this electrolysis mode, hydrogen is generated which is used to recharge the hydrogen storage vessel. The next day, the electric power and water are disconnected, and the unit switches over to the fuel cell mode, generating the electricity to power the vehicle.

The efficiency advantage of the hydrogen fuel cell system must be taken into account in considering today's popular movement towards retrofitting internal combustion engines to operate on methane as an answer to the pollution/ energy crisis. Typically, 75 percent of the energy derived from burning a fuel in an internal combustion engine is wasted as hot gases escaping through the exhaust system or as heat lost to the environment through the engine radiator. Only one fourth of the energy is actually utilized to turn the wheels. It becomes apparent, then, that there is a major advantage to converting the natural gas into hydrogen and then using the hydrogen in fuel cells where 60 percent of the energy from the hydrogen is utilized to power the vehicle. If the natural gas necessary to power a fleet of 1000 cars were reformed into hydrogen and then utilized in hydrogen fuel cell cars, the same amount of gas would power a fleet of 2400 vehicles, resulting in a proportionate reduction in CO_2 emissions. *Figure 31* compares the use of one gigajoule of natural gas directly as a fuel and then converted to hydrogen to power a fuel cell vehicle. As a fuel, the natural gas will provide 209 kilometers (130 miles) of driving and will release 282 grams of carbon dioxide per kilometer. By reforming the same amount of natural gas into hydrogen for a fuel cell vehicle, the range is extended to 547 kilometers (340 miles) and the carbon dioxide is reduced to just 108 grams per kilometer.

Hydrogen Fuel Cell Vehicle Demonstrations

The International Academy of Science in Independence, Missouri, is currently testing a prototype automobile powered by an SPE fuel cell (32). Hydrogen is stored on board the prototype in an iron-titanium hydride vessel. The pressure of hydrogen gas, evolved from the hydride, is utilized in a diaphragm compressor to compress the air to the cell. *Figure 32* is a simplified block diagram of the system installed in the prototype vehicle. Thermoelectric heat pumps were included in the heat exchanger design to boost heat transfer in either direction should it become necessary during the operation of the vehicle. With a metal hydride mass of 136 kilograms, the vehicle has a total capacity of 2.3 kilograms of hydrogen resulting in a range of 300 kilometers.

Ballard Power Systems, in conjunction with the province of British Columbia and the government of Canada, have recently converted a diesel bus for the Vancouver BC Transit. (33). This 30-foot vehicle is powered by a 105-kW fuel cell. Gaseous hydrogen, stored on-board the bus in DOT-approved glass-wound composite cylinders operating at 3000 psi provides the necessary fuel required for the 150-km projected vehicle range.

As hydrogen fuel cell technology comes of age, offering a pollution-free, economical, and highly efficient solution to many of our energy problems, the hydrogen economy is beginning to become a viable option for helping to meet the world's growing energy demands.

BIBLIOGRAPHY

1. G.R. Davis, *Scient. Am.,* **263**, 55-62, *(1990).*

2. W. Fulkerson, R.J. Judkins, and M.K. Sanghvi, *Scient. Am.*, **263**, 129-135, (1990).

3. F. Barbir and T.N. Veziroglu, *Int. J. Hydrogen Energy,* **17**, 301, (1992).

4. R.H. Perry and C.H. Chilton, *Chemical Engineers' Handbook,* 5th Ed., McGraw-Hill Book Company, New York, New York, 1973, p. **9**-19.

5. p. **9**-19 of ref 4.

6. E. Obert, *Internal Combustion Engines and Air Pollution,* Intext Educational Publishers, New York, New York, 1973, p. 96.

7. p. **9**-16 of ref. 4.

8. p. **9**-4 of ref. 4.

9. United Catalysts, Inc., Product Bulletin entitled "Feedstock Purification & Manufacture of Synthesis Gases: Hydrogen, Ammonia, Methanol".

10. R.E. Billings, *Hydrogen From Coal: A Cost Estimation Guidebook,* PennWell Publishing Company, Tulsa, Oklahoma, 1983.

11. G.D. Smith, G.D. Ewart, and W. Tucker, *Int. J. Hydrogen Energy,* Vol. **17**, pp. 695-698, (1992).

12. D. Kumar and H.D. Kumar, *Int. J. Hydrogen Energy,* Vol. **17**, pp. 847-852, (1992).

13. D.A. Mathis, *Hydrogen Technology for Energy,* Noyes Data Corporation, Park Ridge, New Jersey, 1976, pp. 24-25.

14. pp. 299-300 of ref. 3.

15. pp. 299-308 of ref. 3.

16. p. 303 of ref. 3.

17. K. Sapru, "An Elementary Approach to Designing Metal Hydrides for Practical Applications", *Project Hydrogen '91 [Conf.],* American Academy of Science, Independence, Missouri, (1991).

18. D.L. Hendriksen, et. al., "Prototype Hydrogen Automobile Using Metal Hydride", *World Hydrogen Energy Conference [Conf.],* Miami Beach, Florida, (1976).

19. R.L. Wooley, and G.J. Germaine, "Dynamic Tests of Hydrogen-Powered IC Engines", *World Hydrogen Energy Conference [Conf.],* Miami Beach, Florida, (1976).

20. J.J. Reilly and R.H. Wiswall, Jr., *Inorg. Chem.* **13**, 218, (1974).

21. K.C. Hoffman, J.J. Reilly, F.J. Salzano, C.H. Waide, R.H. Wiswall, and W.E. Winsche, "Metal Hydride Storage for Mobile and Stationary Applications", *SAE Fuels and Lubricants Meeting [Conf.]* St. Louis, Missouri, (1976).

22. J.H.N. van Vucht, F.A. Kuijpers, and H.C.A.M. Bruning, *Philips Res. Rep.* **25**, 133 (1978).

23. R.E. Billings, "A Hydrogen-Powered Mass Transit System", *1st World Hydrogen Energy Conference [Conf.]* March, (1976).

24. R.L. Wooley and D.L. Henriksen, "Water Induction in Hydrogen-Powered IC Engines", *World Hydrogen Energy Conference [Conf.]*, Miami Beach, Florida, 6C, 3-24, (1976).

25. R.L. Wooley, "Hydrogen Engine NO_x Control by Water Induction", *NATO/CCMS Fourth International Symposium on Automotive Propulsion Systems [Conf.]*, Washington D.C., (1977).

26. V.R. Anderson, "Hydrogen Energy in United States Post Office Delivery Systems", *World Hydrogen Energy Conference [Conf.]* **2**, Zurich, Switzerland, (1978).

27. J. O.'M. Bockris, *Energy: The Solar-Hydrogen Alternative,* John Wiley & Sons, New York, New York, 1975, p. 292.

28. J. O.'M. Bockris, Energy : *The Solar-Hydrogen Alternative,* John Wiley & Sons, New York, New York, 1975, p. 293.

29. R.E. Billings, "Hydrogen Homestead", Billings Tech Library, 78005, American Academy of Science, Independence, Missouri, (1978).

30. J. O.'M. Bockris and S. Srinivasan, *Fuel Cells: Their Electrochemistry,* McGraw-Hill, New York, New York, 1970.

31. W.G.F. Grot, et. al., *Perfluorinated Ion Exchange Membranes,* Plastics Department, E.I. DuPont de Nemours and Company, Wilmington, Delaware, 1972.

32. R.E. Billings, M. Sanchez, P.A. Cherry, and D.B. Eyre, "LaserCel1 Prototype Vehicle", *Project Hydrogen '91 [Conf.],* American Academy of Science, Independence, Missouri, (1991).

33. B.N. Gorbell and K.B. Prater, "The Ballard/BC Bus Demonstration Program", *Project Hydrogen '91 [Conf.]*, American Academy of Science, Independence, Missouri, (1991).

FIGURE 1: COMBUSTION PROPERTIES FOR SEVERAL COMMON HYDROCARBON FUELS#

	Quantity of Fuel per GJ (LHV)	Flammability Limit in Air (vol % gas)		Maximum Flame Speed (cm/s)	Spontaneous Ignition Temperature (°C)	Ignition Energy## (mJ)
		Lower	Higher			
Hydrogen	98 m^3*	4	75	265	571	.02
Methane	29 m^3*	4	16	34	632	.47
Propane	11 m^3*	2	11	40	504	.31
Butane	9 m^3*	2	10	37	431	.76
Methanol	47 kg*	6	50	49	470	-
Gasoline	22 kg	.8	6	34	447	1.35
Diesel (heptane)	22 kg	1	8	40	247	.70
Anthracite Coal**	30 kg	-	-	-	-	-
Bituminous Coal**	36 kg	-	-	-	-	-
Lignite Coal**	58 kg	-	-	-	-	-

#Ref. 5 *Ref. 7
##Ref. 6 **Ref. 8

FIGURE 2. PROJECTED SYNTHETIC AND FOSSIL FUEL PRICES -- ESTIMATES FOR 2000 (in 1990 U.S. $/GJ)

Fuel	End-Use Cost	Effective Cost (IC Engine)
Gaseous Fuel		
Coal-Hydrogen	12.08	20.86 10.17*
Hydro-Hydrogen	15.01	12.31 6.00*
Solar-Hydrogen	18.94	15.53 7.58*
Synthetic Natural Gas	11.45	24.81
Natural Gas	7.04	-
Liquid Fuel		
Coal-Hydrogen	14.23	25.36
Hydro-Hydrogen	17.89	14.67
Solar-Hydrogen	22.80	18.70
Synthetic Gasoline	19.85	34.97
Synthetic Jet Fuel	15.65	30.77
Fuel Oil	9.14	-
Gasoline	11.23	21.40
Jet Fuel	7.42	17.59

See reference 15 *Fuel Cells

FIGURE 3: HYDROGEN STORAGE -- VOLUME COMPARISONS

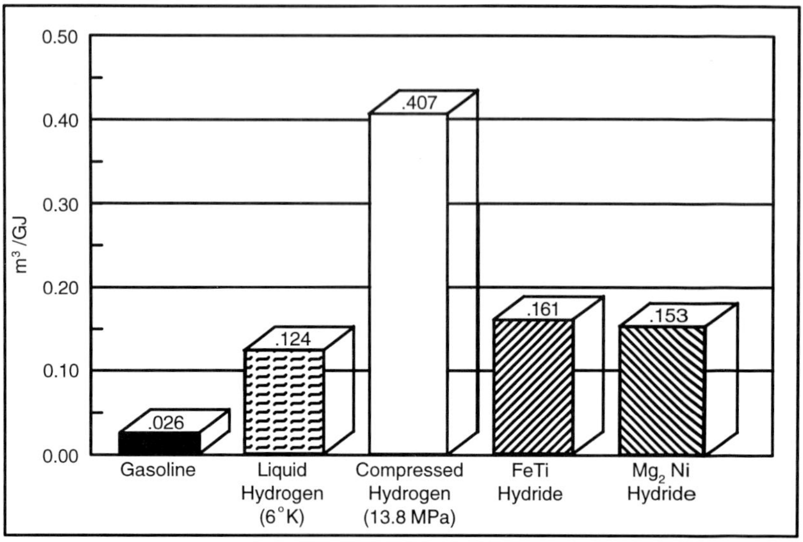

FIGURE 4: HYDROGEN STORAGE -- WEIGHT COMPARISONS

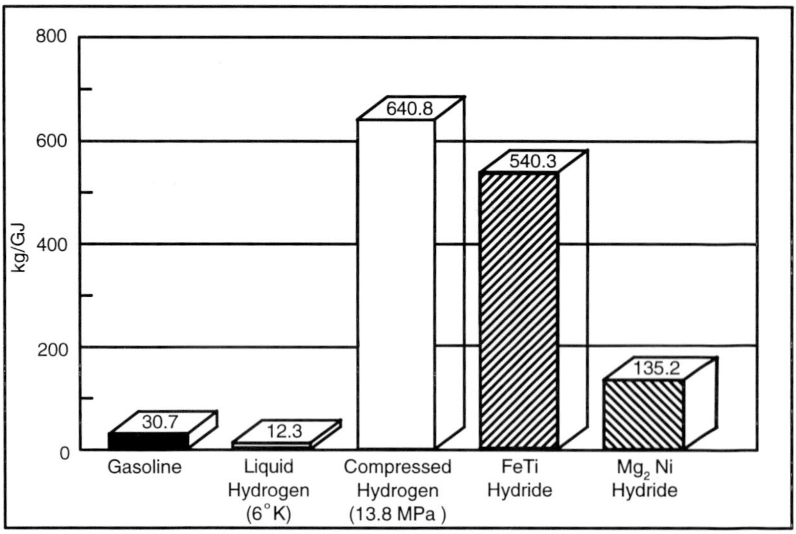

FIGURE 5: IRON-TITANIUM-MANGANESE HYDRIDE

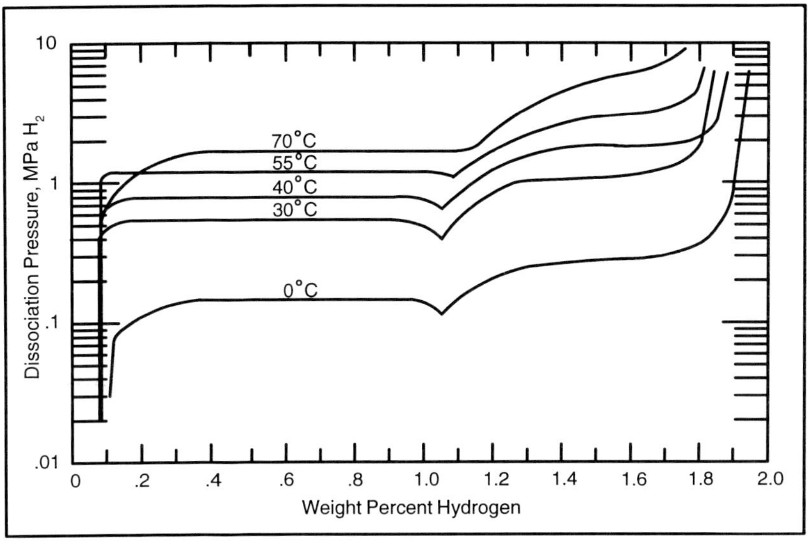

FIGURE 6: MAGNESIUM-NICKEL HYDRIDE

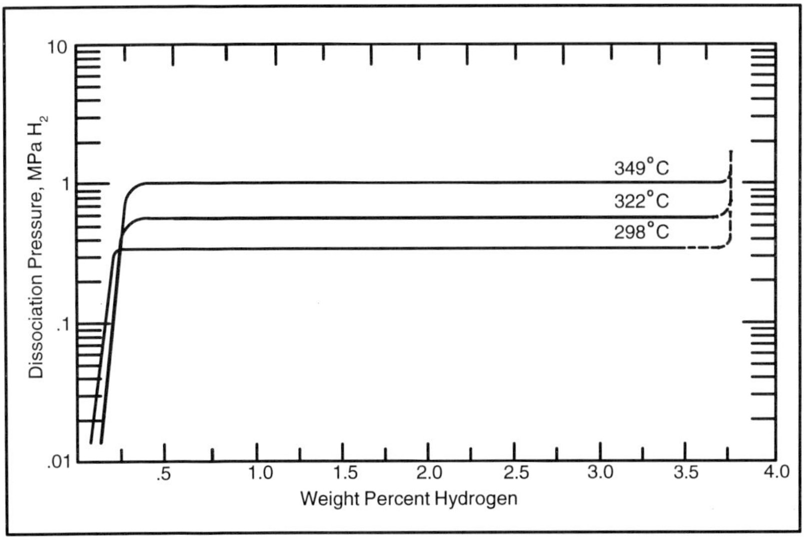

FIGURE 7: LANTHANUM-NICKEL HYDRIDE

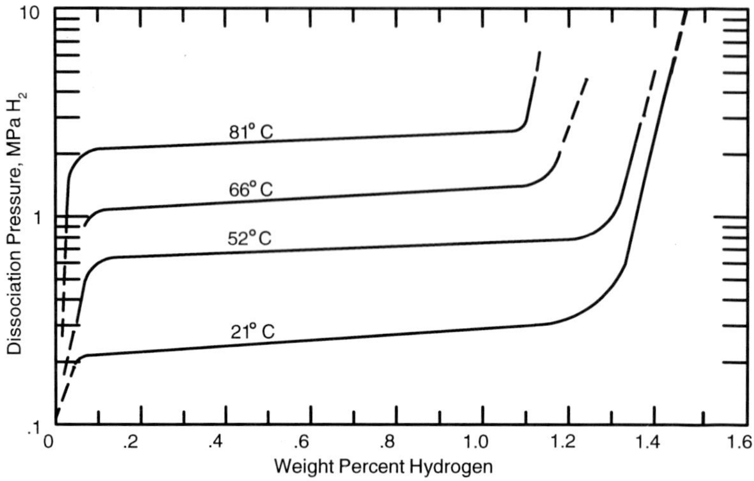

FIGURE 8: HEAT OF FORMATION OF METAL HYDRIDES

Material	Hydrogen Weight Percent	Heat of Formation
FeTi (trace Mn)	1.9	-14.2 kJ/g H_2
Mg_2Ni	3.6	-32.2 kJ/g H_2
Mg	7.6	-38.9 kJ/g H_2
$LaNi_5$	1.5	-15.1 kJ/g H_2

Hoffman and Reilly, BNL, 1976.

FIGURE 9: METAL HYDRIDE SYSTEM -- PROVO-OREM BUS

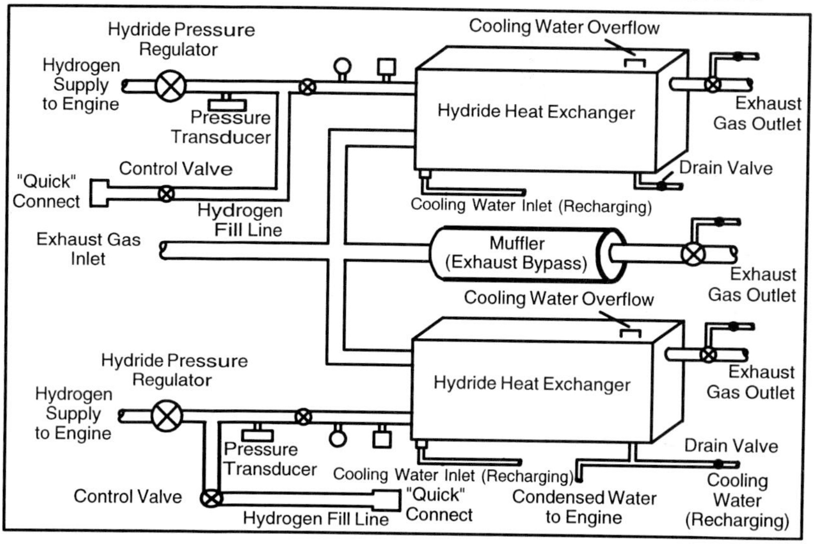

FIGURE 10: METAL HYDRIDE VESSEL PERFORMANCE PROVO-OREM BUS

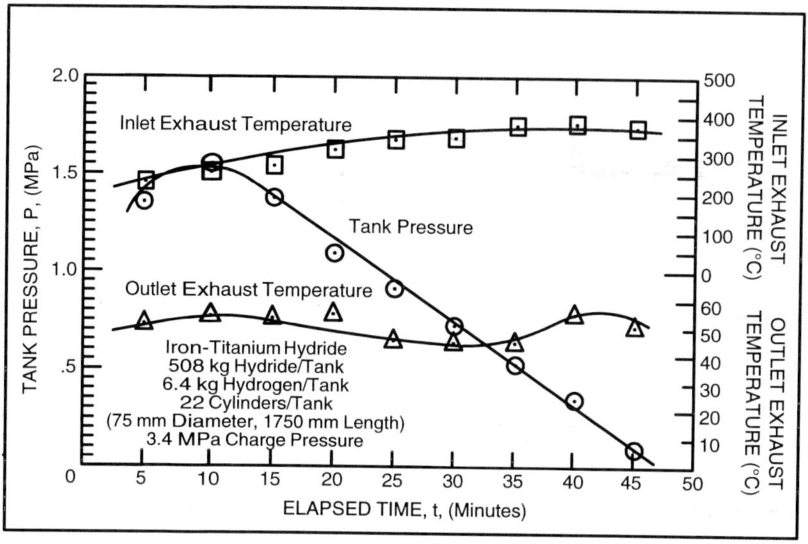

FIGURE 11: RIVERSIDE BUS HYDRIDE SYSTEM

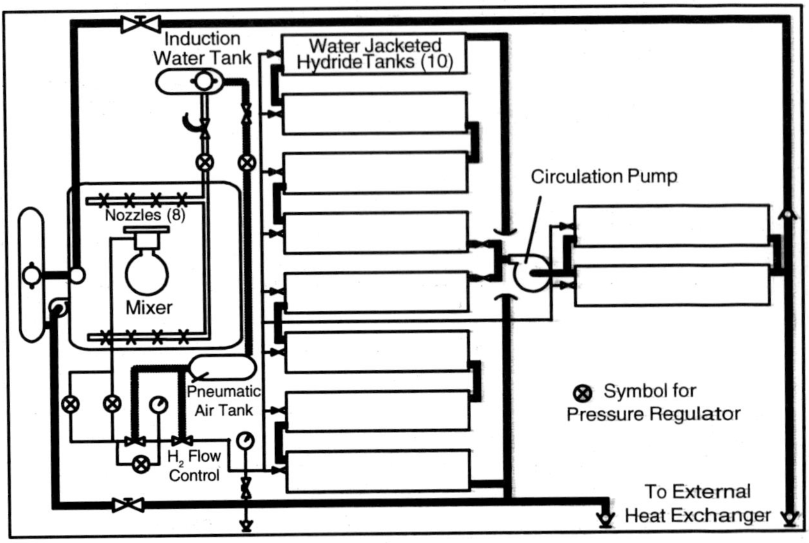

FIGURE 12: HYDRIDE HEAT EXCHANGER DESIGN

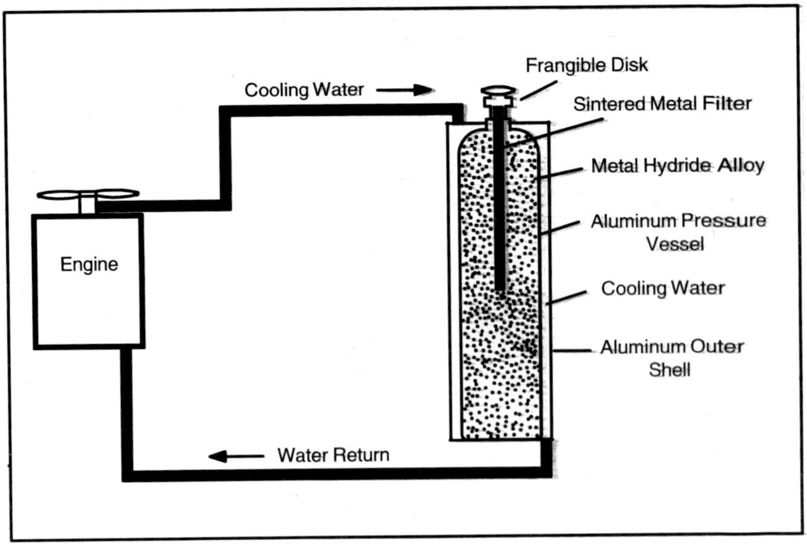

FIGURE 13: COOLING SYSTEM FOR FAST RECHARGE

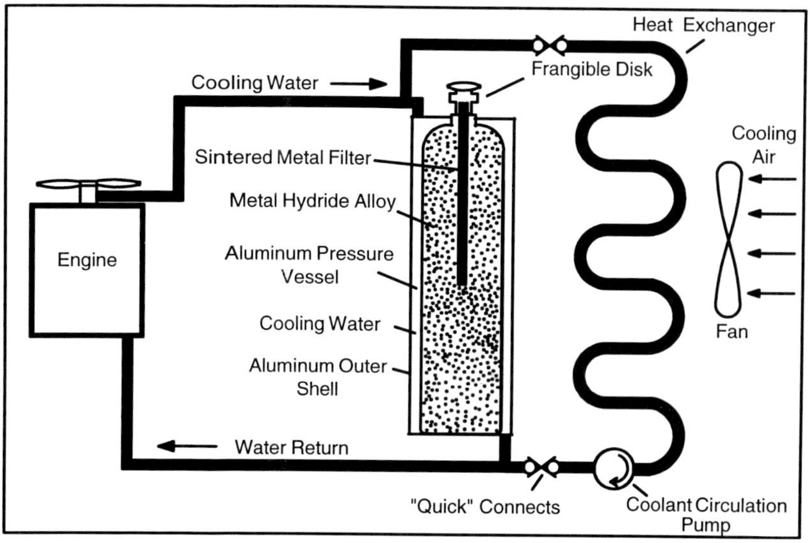

FIGURE 14: INDUCED IGNITION IN HYDROGEN ENGINES

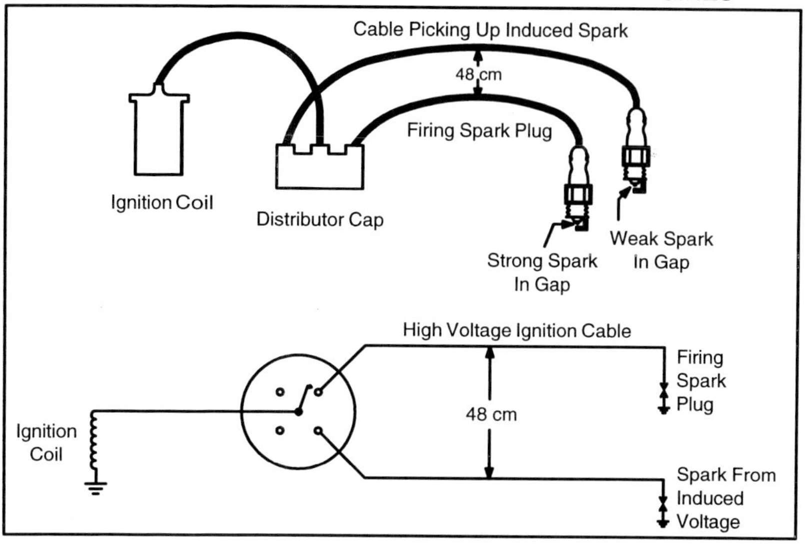

FIGURE 15: DUST PARTICLE PRE-IGNITION MECHANISM

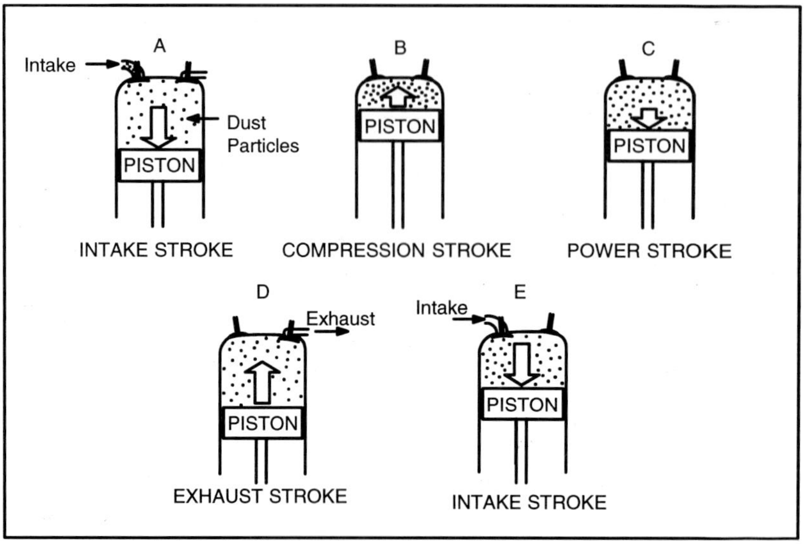

FIGURE 16: WATER INDUCTION MECHANISM

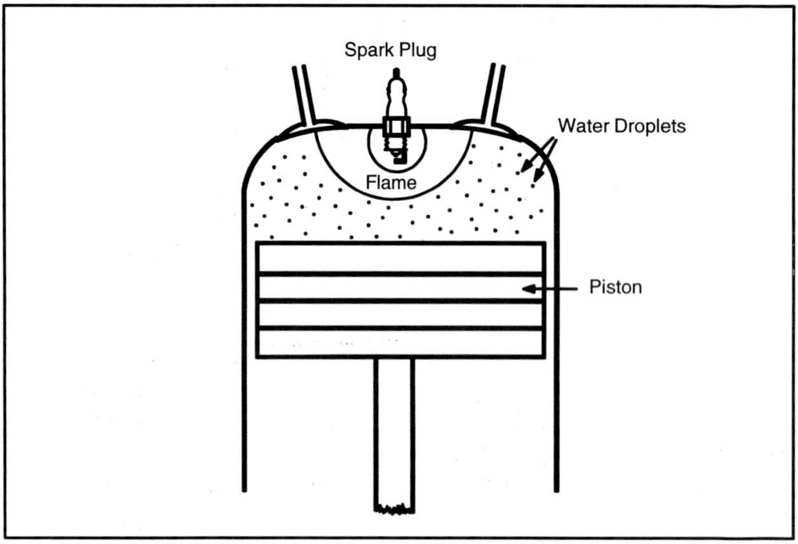

FIGURE 17: NITRIC OXIDE FORMATION IN A HYDROGEN ENGINE

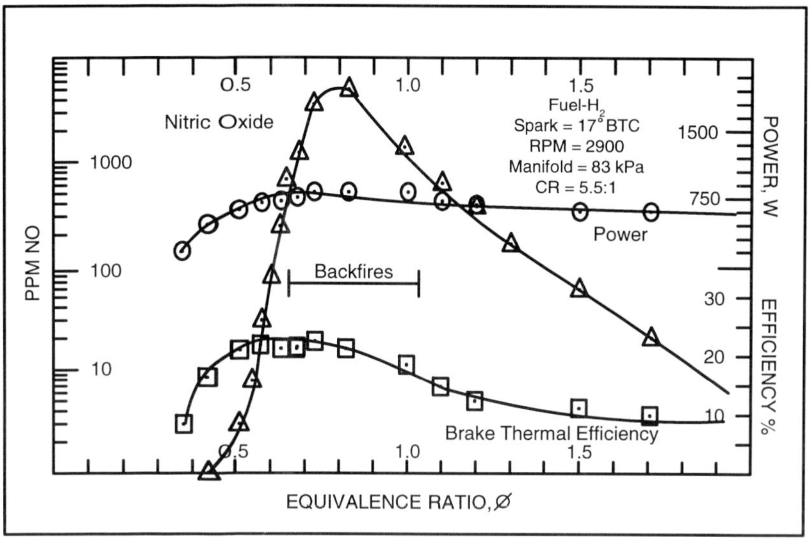

FIGURE 18: EFFECT OF WATER INDUCTION ON THE FORMATION OF NITRIC OXIDE

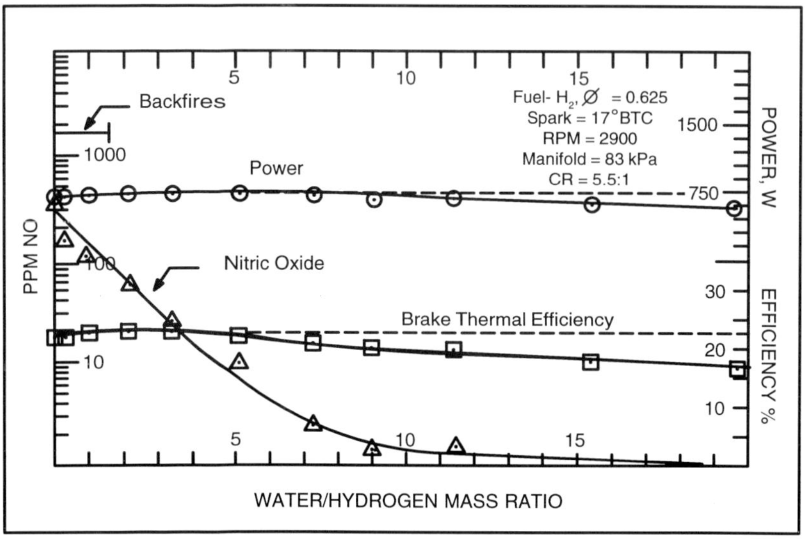

FIGURE 19: IMPROPER WATER DISTRIBUTION

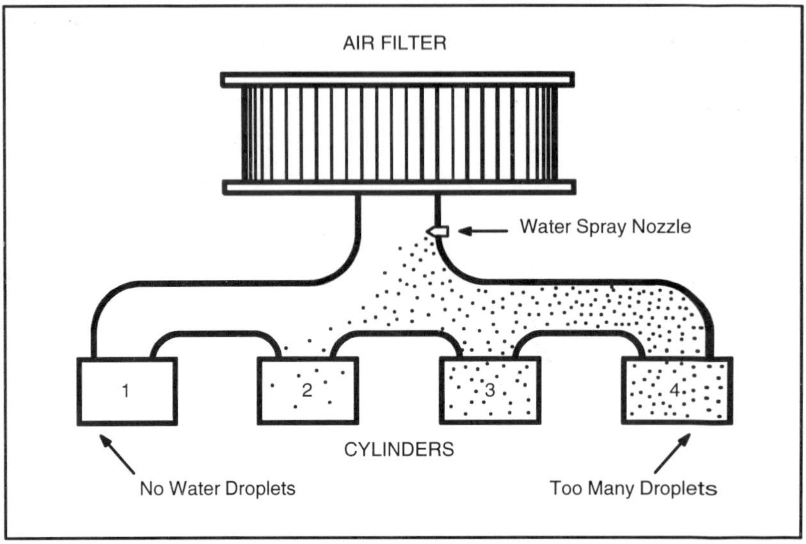

FIGURE 20: EFFECT OF EQUIVALENCE RATIO ON NO_X FORMATION IN PROVO BUS

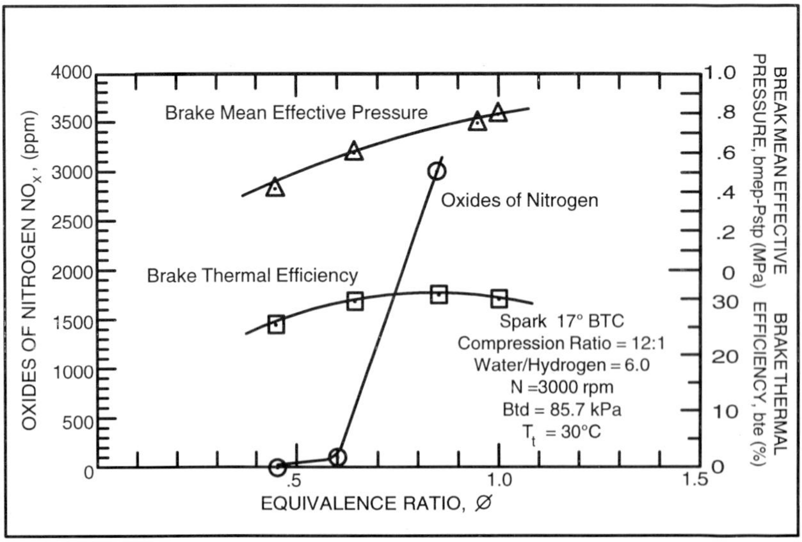

FIGURE 21: EFFECT OF WATER INJECTION ON HYDROGEN BUS ENGINE NO_x

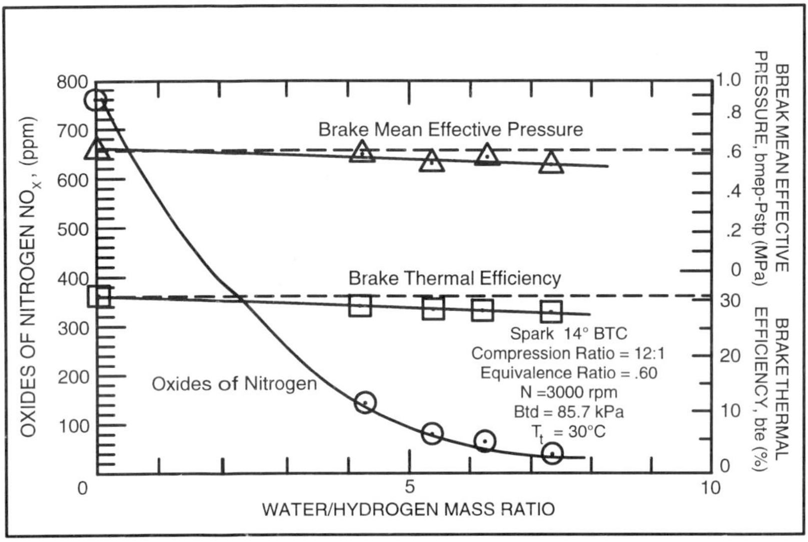

FIGURE 22: LOSS OF POWER IN HYDROGEN ENGINES

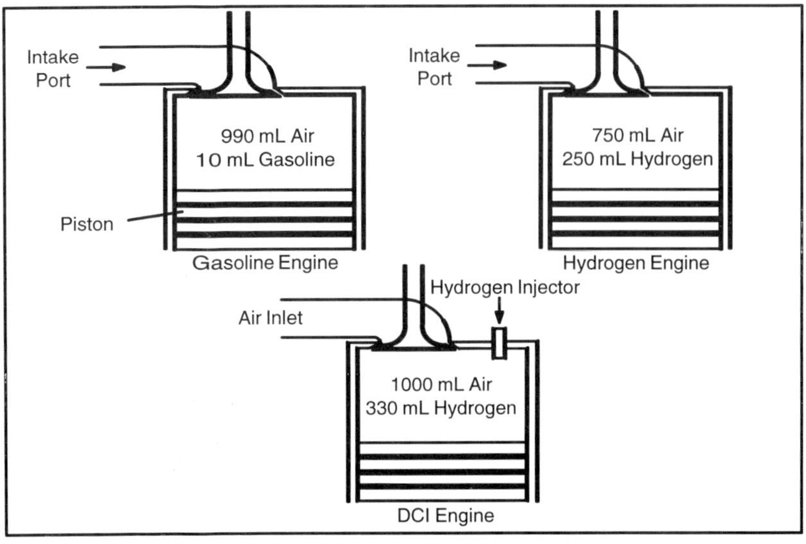

FIGURE 23: POSTAL VEHICLE FUEL EFFICIENCY COMPARISON

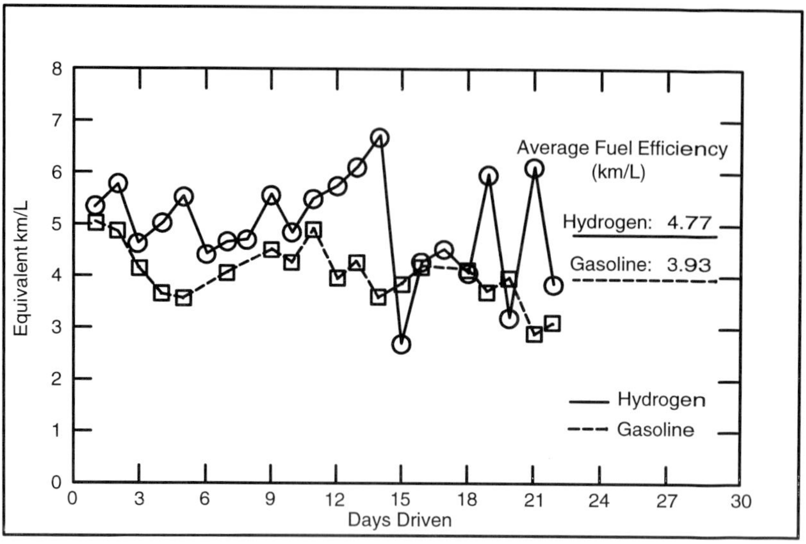

FIGURE 24: HYDROGEN HOMESTEAD ENERGY SYSTEM

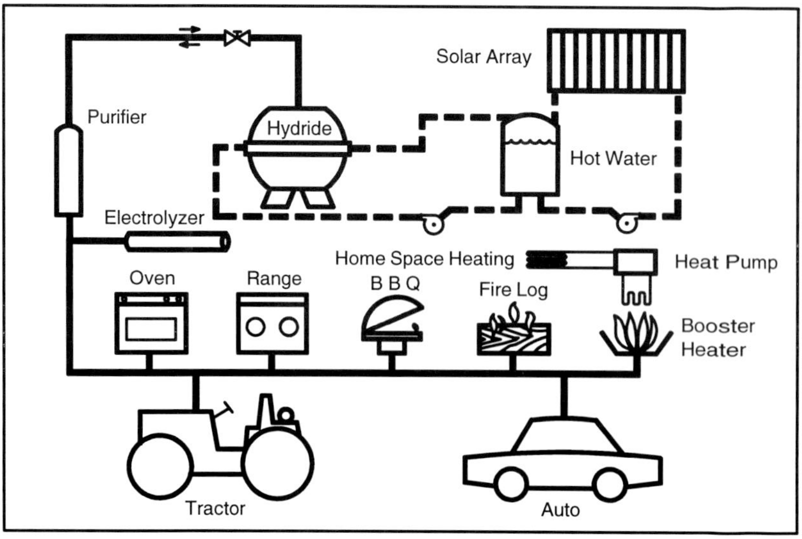

FIGURE 25: HOMESTEAD METAL HYDRIDE VESSEL

Height	123.2 cm
Diameter	97.3 cm
Wall Thickness	2.38 cm
Vessel Material	Mild Steel: 0.15 w/o C, 1.1 w/o Mn 0.09 w/o Cr, 0.005 w/o N
Internal Volume	.5969 m^3
Service Pressure	3447 kPa
Test Pressure	6895 kPa
Hydride Composition	51% Ti, 44% Fe, 5% Mn
Hydride Mass	1791 kg
Service Temperature	55°C
Pressure Excursion	7-3447 kPa
Useable Weight Percent	1.72%
Stored Hydrogen	30.81 kg
Stored Energy (Higher Heating Value)	4.37 GJ

FIGURE 26: NATURAL GAS BURNER DESIGN MODIFICATIONS

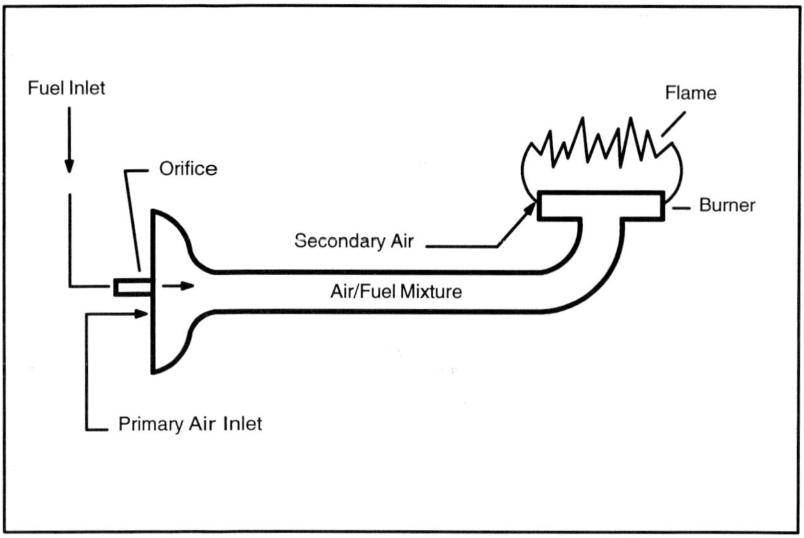

FIGURE 27: STAINLESS STEEL CATALYST BURNER

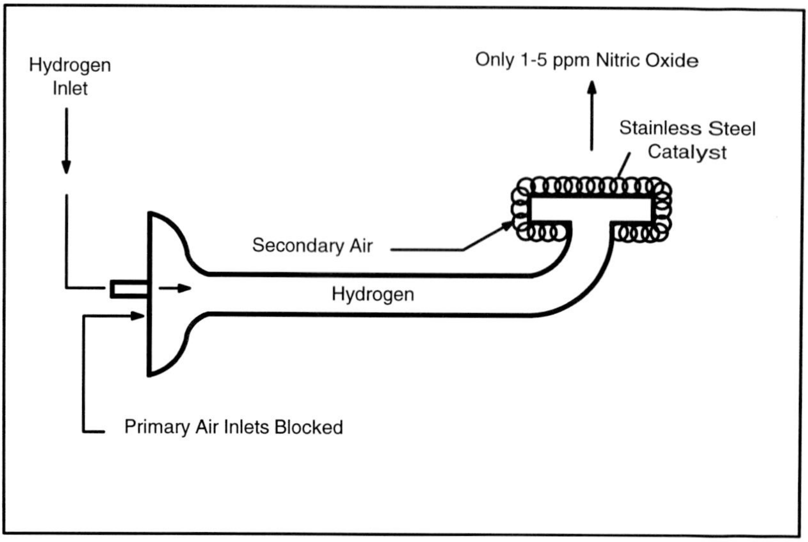

FIGURE 28: HYDROGEN WATER HEATER CONVERSION

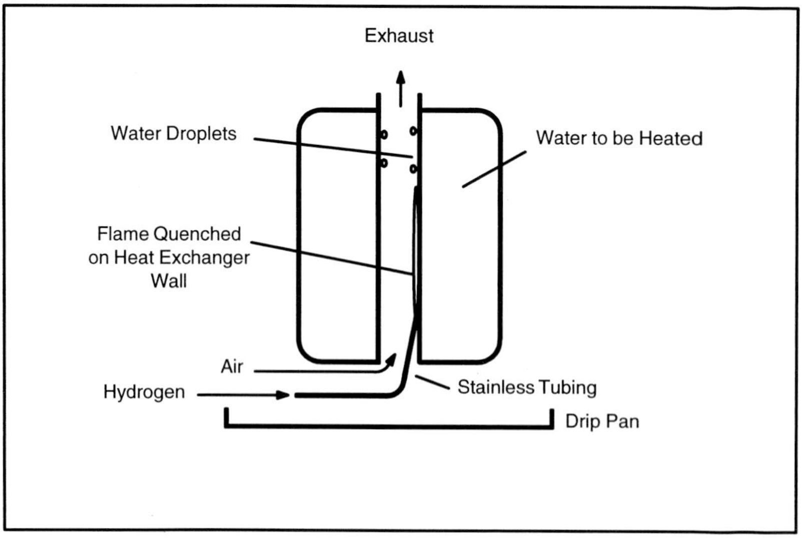

FIGURE 29: SPE FUEL CELL

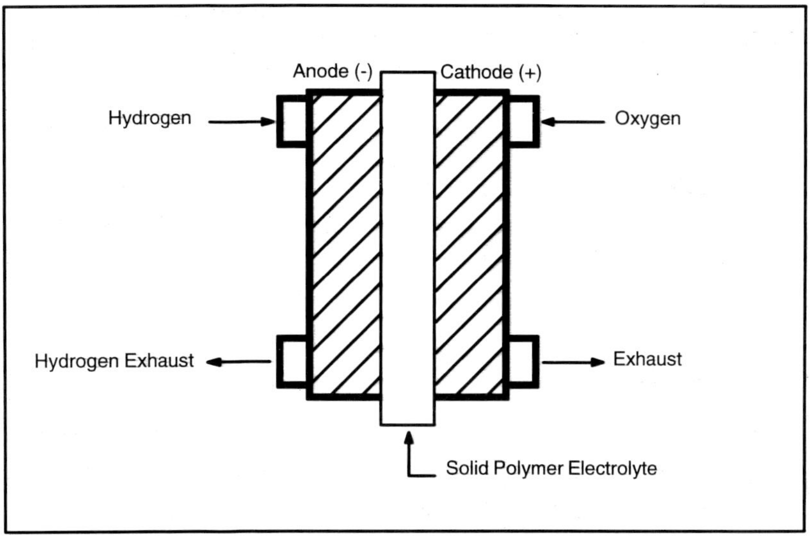

FIGURE 30: SPE FUEL CELL POWER CURVE
HYDROGEN-OXYGEN FUEL CELL

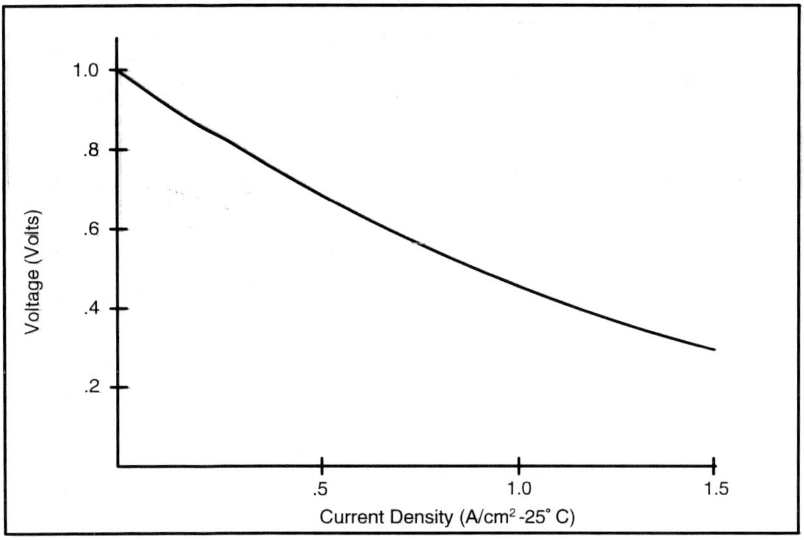

FIGURE 31: COMPARISON OF VEHICLE RANGE AND CO_2 EMISSIONS FOR METHANE- AND HYDROGEN - FUELED VEHICLES

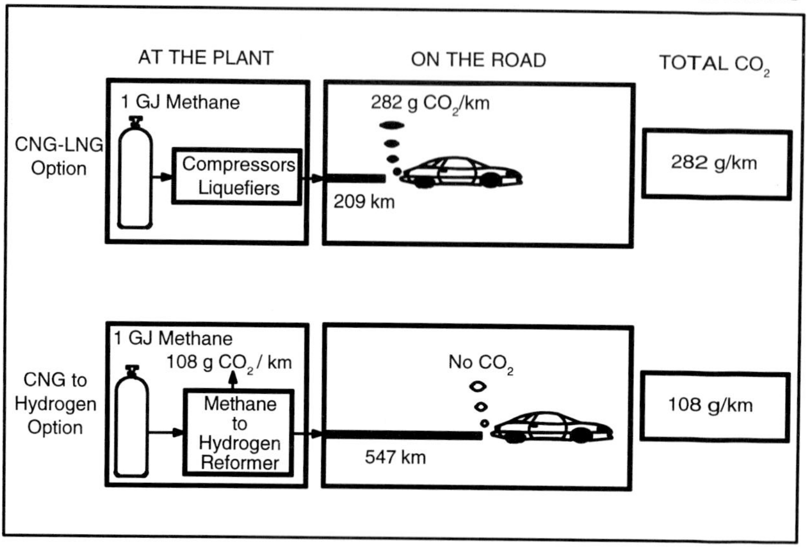

FIGURE 32: FUEL CELL VEHICLE BLOCK DIAGRAM

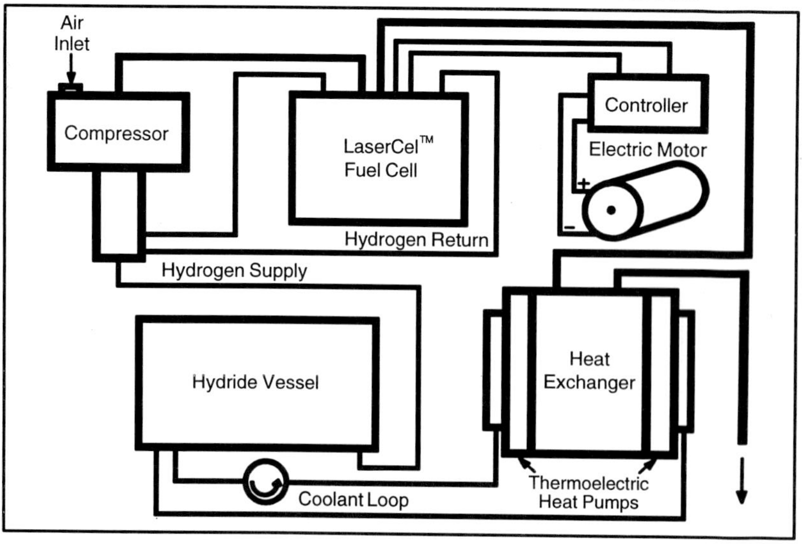

Utilization of Hydrogen or Liquid Natural Gas as an Aviation Fuel

Alexei A. Tupolev
Full Member
Russian Academy of Sciences
Moscow, Russia

Background:

Dr. Tupolev, a full member of the Russian Academy of Sciences, is also noted as the past head of the Tupolev Design Bureau and Chief Designer of the Soviet Supersonic Transport, TU-144. He is a Mechanical Engineer in Aircraft Construction, a Doctor of Technical Sciences, and has taught many years in the Moscow Aviation Institute.

Utilization of Liquid Hydrogen Or Liquid Natural Gas as an Aviation Fuel

Alexei A. Tupolev
Full Member of the Russian Academy of Science
Leninski prospect, 32A
Moscow 117334
Russia

INTRODUCTION

Hydrogen is widely used in many branches of industry -- in metallurgy, in petroleum processing, in ammonia and methanol production, and in space and nuclear applications. Presently hydrogen is considered as a universal energy carrier and energy accumulator.

Demand for energy resources is constantly growing. Meanwhile, most energy is presently obtained by utilizing nonrenewable resources (petroleum, natural gas, coal, shale), and their reserves are not limitless. According to contemporary knowledge, petroleum reserves will be exhausted by the beginning of the next century; the rest of combustible fossils will be depleted within two to three centuries.

The ever-growing energy needs of our society could be most successfully met by the utilization of nuclear energy and transition to thermonuclear power generation. Possibly, such a transition could be realized in the next few decades. Naturally, this would require an effective secondary energy carrier. Hydrogen is the most suitable secondary energy carrier, since it has a high heat of combustion (approximately 3 times higher than those of hydrocarbon fuels) and is practically harmless to the environment. The only pollutants resulting from the utilization of hydrogen as a fuel in heat-producing energy appliances are nitric oxides, which form as the fuel burns in the air when the flame temperature exceeds 1800°C. Otherwise, hydrogen has a distinguishing advantage if compared to organic fuels whose extensive utilization is associated with enormous amounts of carbon oxides, sulfur, and carcinogenic substances escaping into the atmosphere and causing disturbance to the natural ecological balance. For example, accumulation of carbon dioxide in the atmosphere is dangerous because of the emergence of the so-called "greenhouse effect", aggravation of which could lead to catastrophic consequences for our planet. The utilization of hydrogen as a chemical raw material allows transformation of carbon dioxide -- which is, figuratively speaking, the energy's "dead-end" -- again into hydrocarbons suitable for producing the materials and food products necessary for the existence of mankind. Given the nuclear and thermonuclear power industry, hydrogen can be produced from water. Moreover, oxygen produced at the same time can be used to dispose of waste products in an ecologically sound manner.

In the near future, hydrogen can be produced from natural gas practically without any pollution. Hence, hydrogen is not only a prospective universal energy carrier and raw chemical material, but it is also a possible energy alternative for petroleum, coal and other nonrenewable primary energy sources of organic origin.

Hydrogen is particularly attractive as an energy accumulator -- a secondary energy carrier, which could be effectively used at power plants for providing peak energy demands. In addition, the utilization of hydrogen as an energy carrier makes it possible to deliver energy over long distances with a higher efficiency than provided by modern systems, including distribution of electrical energy by power lines.

The utilization of hydrogen as a fuel in transportation is very promising, considering its high heat of combustion. At the same time, it is expedient to use hydrogen in the liquid form at a cryogenic temperature since liquid hydrogen has an enormous cooling capacity which can be used to significantly improve vehicular performance.

Even now liquid hydrogen is utilized in reusable space systems such as the "Space Shuttle", "Buran", the "Adrian" launcher and others. It should be noted that production of aerospace planes is only possible with the use of liquid hydrogen.

Hydrogen utilization has a special significance in motor transport as a means of improving the environment. Today, the amount of harmful substances escaping into the atmosphere with the exhaust gases from automobile engines constitutes approximately 65% of the overall pollution. Consequently, a broad utilization of hydrogen as a fuel in motor transport will cause a significant improvement in our planet's ecology, especially in urban areas and along the congested highways.

The main problems to be solved for widespread utilization of liquid hydrogen in motor transport are: 1) the production of a relatively inexpensive cryogenic fuel tank with an efficient insulation facilitating liquid hydrogen storage on board a vehicle for a long period of time; and 2) modernization of the engine to increase its specific power and to eliminate nitric oxides emissions.

Germany and Japan have recently developed experimental automobiles with cryogenic tanks in which liquid hydrogen may be stored for up to 4 days without being vented. Such tanks, with screen-vacuum super-insulation, are presently very expensive, very fragile and not easily repaired. If the time of liquid hydrogen storage on board a vehicle is reduced, which is probably possible for some types of trucks and buses, then the price of the cryogenic tanks could be considerably lowered through the utilization of less efficient, yet more convenient and cheaper kinds of insulation, such as polyurethane foam.

The utilization of hydrogen in liquid form in the railway system, especially in cargo transportation, also seems to be a reasonable goal. In this case, the liquid hydrogen necessary for the locomotive engine could be transported in a separate cryogenic tanker coupled to the locomotive. Tankers of this kind already exist, for example, ShVTs-100, made in Russia with a capacity of 100 m^3. Naturally, the locomotive engine should be converted for hydrogen service.

Hydrogen usage in river and sea transport will require not only the conversion of the engine, but also the right choice for a safe placement of the cryogenic tanks with proper ventilation of the compartments with cryogenic equipment.

The possibility of liquid hydrogen utilization in aviation is quite convincingly shown by the experimental airplane TU-155 made in Russia, one of whose engines was operated exclusively on hydrogen. The supply of liquid hydrogen stored on board the airplane enabled the engine to run for up to one hour. At that time, the experimental hydrogen engine was functioning as a standard engine, under all operating conditions, during the plane's evolutions, switching on and off in the air. The development and testing of the plane provided the data for further development of the engine and the cryogenic systems to meet the plane's requirements and to develop operation techniques with the same complexity level and flight preparation time as required for analogous traditional kerosene airplanes.

Although the amount of pollution accounted for by aviation transport is not large (only about 2%), this pollution is brought about in the upper levels of the atmosphere and at the airports where there is a large local accumulation of planes. Consequently, we can speak of a strong local pollution, which forms separate ecologically unfavorable regions.

Particular public misgivings are associated with industry's ability to ensure sufficient safety in hydrogen applications. This could be explained by the scholastic ideas about the detonating gas explosion and the memory of the Hindenberg catastrophe. Yet a large number of barrage balloons filled with hydrogen were widely used during World War II without any kind of complications, explosions, or fires. Moreover, at the present time the properties of hydrogen and its mixtures are fairly well explored, and also the means of its safe utilization are developed. If spilled liquid hydrogen is ignited, the resultant human injuries and damage to construction would probably be on a smaller scale in comparison to those from a hydrocarbon fuel fire, since hydrogen evaporates quickly and its flame has a lesser infrared radiation capability.

DISADVANTAGES OF LIQUID HYDROGEN

Hydrogen has a low density, equal to 70.78 kg/m^3, and a low ignition energy, equal to 0.019 mJ, which is a magnitude less than that of kerosene or gas. Before hydrogen can be poured into a tank, the vessel must be treated with neutral gas, the most suitable being nitrogen. Hydrogen has a very low boiling point (-253° C) which requires very expensive and efficient screen-vacuum insulation. Hence the price of hydrogen is more than 10 times higher than that of jet fuel.

Creating an infrastructure for producing, distributing and refuelling transport vehicles with liquid hydrogen or liquid natural gas presents a serious problem. Since this infrastructure should be created to replace the traditional jet fuel infrastructure in the future, it needs to be environmentally clean, efficient and inexpensive in operation. It should be taken into consideration that in the near future hydrogen will be used in the domestic sphere as well as a raw material in chemical engineering applications. Consequently, the steps for establishing hydrogen users and the means of its production, distribution and delivery should be coordinated and develop as a whole.

Liquid natural gas, when compared to hydrogen, has the following advantages:

- Liquid natural gas is six times heavier than liquid hydrogen, with a density equal to 425 kg/m^3.

• Its boiling point is -161.73° C.

• It has an ignition energy of 0.29 mJ.

• The cost is $130-150 for a ton, which is approximately three times less than the price of a ton of liquid hydrogen.

• In general, liquid natural gas is considerably easier to deal with in comparison to hydrogen.

The unique properties of liquid hydrogen (mainly its high specific heat value, which is three times that of hydrocarbon fuels) and the notable advancements in the field of hydrogen energy technology which, in the near future, will allow production of sufficient quantities of liquid hydrogen by economically sound methods, attract the attention of aviation specialists to this alternative fuel.

On the other hand, an increased danger of explosion or fire, cryogenic temperature and a small specific weight create additional problems for the application of hydrogen in aviation. A number of technical research projects on the utilization of liquid hydrogen as a fuel for airplanes were conducted. Still, it is our opinion that such a complex problem should not be approached theoretically.

To solve this problem and to reveal all of its aspects, and to solve some other major scientific problems also, it was necessary to build an experimental prototype -- in this case an experimental airplane.

Developing a plane for liquid hydrogen fuel brought about a series of technical and scientific problems to be solved. Let us consider the main ones.

The first problem was providing explosion and fire safety. In a normal airplane, a hermetic fuel system malfunction leads to a spill of slow-evaporating liquid -- jet fuel -- taking comparatively small volume, with the ignition of the resulting mixture requiring a significant energy source. However, if a hermetic system with liquid hydrogen malfunctions, the liquid hydrogen evaporates almost immediately, fills the overall volume of the airplane compartment -- which, as a rule, is equipped with electronic equipment -- and then is ignited by insignificant energy in a wide flammability range.

The second problem is the utilization of fuel at a cryogenic temperature. A common aviation fuel within the range of operation temperatures and pressures without insulation of tanks and cables does not freeze or boil at the tank pressure, which is approximately equal to atmospheric pressure at the height of the flight. Liquid hydrogen, with a boiling point of -253°C, requires a highly efficient, typically screen-vacuum insulation. Nevertheless, even with this sophisticated insulation, the hydrogen within is constantly boiling and requires a special system to maintain a pressure slightly higher than atmospheric, otherwise most of the fuel would boil off while climbing.

The third problem is to provide large volumes for liquid hydrogen storage. The liquid hydrogen fuel requires approximately 4 times the volume of jet fuel, and the fuel containers should have a good insulation.

On modern airplanes, fuel is placed in the wing. However, the volume of the wing is not sufficient for the liquid hydrogen; in addition, liquid hydrogen also requires extra space for insulation. Consequently, the problem arises of producing a light hydrogen tank to be placed either in the fuselage or on the outer surface.

Let us discuss in more detail how we approached these problems while creating the experimental airplane.

Hydrogen, like no other kind of aviation fuel, is capable of forming explosive mixtures with air in a wide range -- from 4 to 74 volume percent (compared to 0.8% and 7.0% for kerosene). In addition, the energy required for ignition is only 0.02 millijoules, which is approximately an order of magnitude less than that of hydrocarbons. This peculiarity of hydrogen in conjunction with its high diffusion properties and cryogenic condition drastically change the concept of providing explosion and fire safety.

The main danger of explosion on board an airplane is in the areas adjacent to fuel tanks, fuel pipes and fuel apparatus. Hence, there are two major courses of action pursued in order to achieve explosion and fire safety on the plane:

1) Development of constructive measures to prevent or minimize hydrogen leaks, including various cases of fuel system malfunctioning;

2) Minimization of any possible source that could ignite the mixture of hydrogen and air.

1. Configuration

An experimental fuel system, including a fuel tank, fuel supply control units, and pipelines, is situated in a special hermetic compartment.

The engine, NK-88, operating on hydrogen, is situated in a separate nacelle, and the hydrogen propulsion system apparatus and pipelines are placed in a container of hydrogen apparatus situated in the upper part of the nacelle.

The flight and emergency drain pipelines extend from the experimental fuel system compartment to the drainage device, placed at the upper part of the tail assembly.

The emergency drain line is brought outside the fuselage to the nozzle section of the middle standard engine.

The main standard systems of the prototype are separated from the experimental fuel system compartment. Electric cables, hydraulic and fuel lines of the airplane's standard systems are placed outside the experimental fuel system compartment and during the flight are fanned by the incoming stream of atmospheric air.

2. Constructive Features to Prevent Possible Extreme Situations

To prevent leakage from the experimental fuel system compartment into the adjacent fuselage compartments which are not specially protected, the experimental system is separated by special zones with double partitions into which surplus pressure is maintained by pumping in air. The pressure within the special zones exceeds that of the experimental fuel system compartment, thus hindering leakage into the neighboring compartments.

The container of hydrogen units is made out of stainless steel and is hermetically sealed. The joints between the panels of the hydrogen units container and the joints between the container and the casing of the NK-88 engine are sealed. The pipeline inlets and electric cables running into the hydrogen units container are also hermetically sealed, which prevents liquid hydrogen from leaking into the nacelle in the case of a spill.

In the experimental fuel system compartment and in the nacelle, equipment is spark-safe, and the electric circuits are either spark-safe, or have a special protection.

3. Systems of Prevention, Discovery, and Suppression of Hydrogen Leakage Situations

Besides these constructive features, a number of systems were developed in order to discover and suppress the arising danger in case of a hydrogen leak.

The systems of discovery included the following:

- The gas control system, which provided control of the gas environment in the experimental fuel system compartment, the nacelle, the container of hydrogen units, the cockpit, and the passenger and luggage compartments;

- The signalling system to warn of fire in the experimental fuel system compartment, nacelle, and the hydrogen units container;

- The smoke detection system;

- A special vacuum control system in the insulation of the fuel tank and pipelines.

The systems to prevent and suppress extreme situations included the following:

- The ventilation system of the nacelle and the of hydrogen units container, which provided air flow in the under-cowling area of the nacelle and in the space of the hydrogen units container;

- The ventilation system of the experimental fuel system compartment, which provided air flow coming from the air conditioning system on board;

- The nitrogen supercharge system, which replaced the air in the experimental fuel system compartment, the nacelle, and the hydrogen units container with nitrogen;

• The fire-extinguishing system in the experimental fuel system compartment, the nacelle, and the hydrogen units container;

• The system which supercharged the special zones, maintaining a surplus pressure in those areas above the experimental fuel system compartment.

THE PROBLEM OF CRYOGENIC FUEL UTILIZATION

The utilization of cryogenic fuel -- liquid hydrogen -- made it necessary to radically change practically all of the plane's power plant.

To provide the fuel supply with a pressure up to 40 kg/cm^2, the engine was equipped with a two-staged centrifugal pump with a drive from a turbine which was run by the air supplied from the engine's compressor.

Considering that it is very difficult to insure stability of the combustion process as the liquid fuel is fed into the chamber (since it is practically impossible to avoid vaporization in the main pipelines), before being fed into the injector, the liquid hydrogen was gasified and heated to -170°C in the heat exchanger.

To create a cavitation supply for the engine's pump (that is to prevent boiling of liquid hydrogen) a pumping system is used to feed the fuel from the tank into the engine.

To insure cavitation-free operation of the pumping devices without losing pressure, they were placed in a specially designated, hermetic fuel-compartment.

Liquid hydrogen was fed into this compartment from the basic volume of the fuel tank with the help of jet pumps powered by the fuel taken from the main, high-pressure pipeline located behind centrifugal pumps. The capacity of a jet pump was somewhat higher than the decrease of hydrogen from the feeding compartment; consequently the latter was always full and had a higher pressure than the main compartment. In this way, the necessary cavitation supply was created for the pumping devices. At the same time, the constant consumption of hydrogen through the operation of the centrifugal pump to actuate the jet pump did not allow hydrogen in the centrifugal pump to be overheated and gasified even with an idle engine.

The fuel tank and the main pipeline feeding the engine, along with the units coupled to the pipeline, possessed cryogenic vacuum insulation.

To lessen the probability of hydrogen leakage into the airplane's compartments, the main feeding pipeline had two valves, one which was situated inside the tank, and the other in front of the engine. The valves mentioned, and also all of the other cryogenic valves, were actuated by gaseous helium at high pressure, since other gases at this temperature are transformed into a solid state.

The valve inside the tank allowed the cutting of the outer cryogenic pipelines in case of an emergency.

With an idle engine, the main pipeline between the valves mentioned was supplied with helium under pressure several times greater than the pressure in the tank. The existence of such a helium "cork" prevents a thermal contact of liquid hydrogen with air and, hence, its condensation. In addition, improper sealing of the valve will cause helium, and not hydrogen, to escape into the engine feeding system and the airplane's compartment.

The main hydrogen pipeline leading from the turbopump unit into the inner volume of the engine was insulated with polyurethane foam. The turbopump unit and the engine's main feeding pipeline were situated in the upper part of the engine's structure in the hydrogen units container.

Since hydrogen is stored in tanks in a boiling condition, it is expedient throughout the flight that its pressure be maintained at the same pressure as that of saturated vapor. For this reason, the plane was equipped with a pressure-maintaining system, which included the subsystems of supercharging with gaseous hydrogen and controlling hydrogen exhaust, (emergency-preventive device), and the emergency supercharging of the tank with helium.

The pressure in the tank was maintained at 1.3 to 1.5 kg/cm^2 by venting hydrogen from the gaseous environment into the main drainage pipeline. The main drainage pipeline was insulated and had a pressure-maintaining regulator with three valves. Two valves were parallel and one was in tandem with the other two. This placement insured a sufficient reliability of the system's functioning. The valves were controlled by a signal from pressure sensors in the tank by means of gaseous helium at high pressure. Hydrogen was vented through a safe drainage device, whose function was based on the difference between the higher speed of escaping hydrogen and the lower speed of flame propagation. To provide safety during transitional stages of operation, helium was fed into the main drainage pipeline.

The emergency-preventive device consisted of parallel-positioned safety valves tuned to different pressures. Once the fuel tank's thermal insulation was disrupted, the safety valves were actuated and hydrogen was discharged from the tank, through a heat-insulated main pipeline, into the upper part of the tail assembly.

The subsystem for emergency supercharging was designed to supercharge the tank with helium in case the hydrogen supercharge system malfunctioned or for emergency fuel drainage.

Emergency drainage of liquid hydrogen during a flight was accomplished through the main fuel pipeline which, once the tank was filled with hydrogen, was coupled with a pipeline leading to the airplane's tail assembly.

All the main fuel pipelines were joined by means of hermetic welding, except where they connected with the engine. The main pipelines of the engine and of the pressure maintaining system had detachable joints.

THE PROBLEM OF PROVIDING LARGE VOLUMES FOR LIQUID HYDROGEN STORAGE

The problem of providing large volumes for liquid hydrogen storage turned out to be easy for the experimental airplane, because the chosen base airplane, TU-154B, had large passenger compartments. The tail passenger compartment was chosen as the place for the liquid hydrogen tank. In this compartment it was possible to accommodate the amount of fuel necessary for conducting test flights of a one to one-and-a-half hours duration.

For the first liquid hydrogen tank we chose the concept of the most structurally simple, inserted tank suspended at 6 points by means of a statically defined support system.

This decision allowed for the evaluation of the design of the inserted tank and its support system -- the development and testing of a large number of systems to provide the proper function of the inserted tank and to insure explosion and fire safety with the given structural design of the fuel tank.

GROUND-BASED EXPERIMENTAL FILLING STATION

To refuel the airplane with liquid hydrogen, a filling station was built, located according to the requirements of labor safety for an isolated station utilized for parking and technical maintenance of an airplane.

Here is the description of the station and the refueling process (however, it should be noted that with the mass utilization of the hydrogen, the station would be different).

The delivery of liquid hydrogen to the grounds was performed by mobile bowsers. The machines were connected to the plane by means of stationary cryogenic pipelines, with check-valves. The pipelines were mounted on the grounds and provided the necessary fire-safe distances between the plane, bowsers, and the place where the gaseous hydrogen was vented into the atmosphere.

The stationary network of pipelines allowed for up to 4 bowsers to operate simultaneously.

The pipelines had a screen-vacuum thermal insulation and were joined to the airplane and the bowsers with metal sleeves, fixed on a mobile wagon.

Refuelling of the airplane was achieved by transferring liquid hydrogen from the bowser's reservoir where an excessive pressure was created by means of a built-in liquid hydrogen gasifier which utilized the heat supplied from the atmosphere.

The check-valves of the airplane, bowsers, and the ground-based station were remotely controlled from a control station during the refuelling process. The adjustment of the refuelling parameters was semiautomatic.

To insure labor safety and reduce the accumulation in the tank aboard of oxygen and other gases transforming into a solid state at the cryogenic temperature of the fuel, the technical maintenance routine included a thorough cleaning of the reservoir and the pipelines before refuelling, and their regular warming to the temperature of sublimation of impurities to provide control with a gas chromatography method.

Gaseous hydrogen necessary for preliminary cleaning and warming was taken from the above-fuel volume of a bowser and was heated above 0°C in a heater.

At the end of the main filling pipeline there was a filter rated at 5 μm.

At the location of possible hydrogen leaks a provision was made for stationary devices to spray the airplane, bowsers, and the ground-based station with water to protect them from the burning spilled liquid hydrogen.

LIQUID NATURAL GAS

The second alternative cryogenic fuel is liquid natural gas. Its main constituent (more than 85%) is methane -- the hydrocarbon closest to hydrogen according to its physicochemical properties. Its specific weight is 425 to 480 kg/m^3, its boiling point is -160°C, its heat of combustion is approximately 15% more than that of jet fuel, it has an ignition energy of 0.29 millijoules and forms dangerously explosive mixtures with air in the range between 5 and 15 volume percents.

Technical problems arising when creating airplanes to operate on liquid natural gas are analogous to those encountered when utilizing liquid hydrogen, but there is a significant quantitative difference.

Liquid natural gas still requires a larger tank volume, but if tanks for liquid hydrogen are four times the size of traditional ones, liquid natural gas requires only twice the size.

In contrast to hydrogen, the energy required to ignite liquid natural gas is 15 times greater, and the range of dangerous concentrations of liquid natural gas vapors in air is drastically narrowed. All of these factors essentially reduce fire-explosion danger and significantly simplify maintenance of airplanes on liquid natural gas in comparison with liquid hydrogen.

Ground-based infrastructure is also significantly simplified. In developed countries, pipelines with natural gas are connected to practically all airfields, which takes care of the distribution problem.

While hydrogen and jet fuel must be produced at specialized plants, liquid natural gas should only be purified and liquefied, and natural gas from any natural deposit can be used as aviation fuel.

Considering that the potential supplies of natural gas in Russia amount to 270 trillion cubic meters, the annual consumption of only approximately 0.7 trillion, and the world's potential supply amounts to 600 trillion cubic meters, with an annual production of only two trillion, we see that, in contrast with crude oil, the supplies of natural gas will last for more than 150 years.

A Solar-Hydrogen Clean Energy Environment

John O'M. Bockris
Distinguished Professor of Chemistry
Texas A&M University
College Station, U.S.A.

Background:

Dr. Bockris has been at Texas A&M University since 1983. Previously he was Professor of Chemistry at the University of Pennsylvania, and Chair of Physical Chemistry at Flinders University in South Australia. He co-founded the International Association for Hydrogen Energy and made some of the earliest contributions to the idea of a hydrogen economy. He is the author of the book *Energy: The Solar-Hydrogen Alternative.*

HOW DOES THE SOLAR-HYDROGEN ECONOMY RATE IN THE 1990's?

J. O'M. Bockris
Department of Chemistry
Texas A&M University
College Station, TX 77843

EXHAUSTION AND POLLUTION

The idea that was behind many of the energy initiatives of the 60's and 70's was that we were to run out of fossil fuels by the end of the century. This referred to oil and retains some truth. Liquid oil, used up at the present rate, will not last for more than another twenty years or so. However, there is certainly more natural gas than we had thought at the time of the earlier estimates and, therefore, it is worth asking the time at which it will be necessary to convert to an energy base other than fossil fuels.

The answer to this is quickly given by a glance at Diagram 1, which shows the steady gain of volatile organic material in the atmosphere including nitric oxide and CO_2. These are given out at rates which, although they slowed down for a little while, are now again increasing. There is no way to stop this unless we abandon the use of fossil fuels. This, then, is the basis for our advocacy of a Solar-Hydrogen Economy, the word solar being meant to include hydro and wind.

HOW MUCH WOULD CONSERVATION HELP?

The most frequently recommended course for the troubles of the polluted world is to conserve, i.e., to cut back on energy use. Whilst there is no doubt that such a move would help, it would not help much or for very long as can be seen by looking at Diagram 2, which is an interesting plot of income versus energy. It can be seen that as the energy per head passes about 6 kW per head (smoothed out continuous use throughout the year), then a further increase in income does not occur.

To begin with, this seems like good news because in the United States and Canada we are using more like 10-11 kW per person and so a certain downtrend, i.e., conservation, wouldn't hurt. But, on the other hand, the graph teaches another lesson which is very severe. It shows what would happen when one goes over the mark into that part of the graph below 6 kW per person when a rapid discontinuity occurs and the living standard falls off dramatically.

So, it looks as though we can undergo about 40% conservation without changing our life style too much (i.e., by using small cars instead of big cars, and etc.). This is confirmed by the fact that the big countries of Europe, such as France, Germany, Italy, do well on about 5 to 6 kW per person although the living standard there hardly differs from ours here.

But note that were we to go to less than 5 kW per person then the decline in living standard is faster than the decline in energy use. 5 kW/person is hence our benchmark.

HOW MUCH DOES POLLUTION COST?

This is easily seen from Diagram 3 where the cost of pollution -- about \$10 a gigajoule (or \$1.30 per gallon of gasoline) -- is broken down. We are, in fact, implicitly paying more for our fuels than we think because when we use gasoline, we incur more than a dollar a gallon of pollution, so that a fuel which costs \$2.00 for the same energy is now contained in a gallon of gasoline but does not pollute, would be cheaper than the \$1.25 we are now paying for gasoline.

COST OF TRANSMISSION OF ENERGY

In an age in which energy is got increasingly from renewable resources such as wind, hydro, and finally solar itself, then the transfer of the energy will be an important component of the final cost. Indeed, for the United States the heavy use areas are in the east coast around ±400 miles from New York; then the midwest and ±300 miles from Chicago, and, finally, in the area of Los Angeles and San Francisco. It is to these great centers that we have to bring the energy, and, therefore, it is interesting to note (Diagram 4), that the cost of transmitting energy becomes less in hydrogen (compared with electricity), when the distance is about 400 miles or 600 kilometers. It transpires, then, that if we are going to collect energy in the southwest and bring it to the high energy use areas, then clearly we have to do it not in electricity but in hydrogen.

Here is where a cheap superconducting cable working at room temperatures would transform the situation. One is not in sight. High temperature superconductivity (even up to 100°K, say) would certainly not be cheap because one would have to surround the cable with a cooling gas.

THE HYDROGEN ENERGY SYSTEM

This is well known now among people who study energy but just to make sure that we know what we are talking about we show Diagram 5.

Briefly, we would go on using electricity as we use it now, but an increasing amount of the electricity would be produced on the spot via fuel cells, and for space heating and cooling and for many of those functions now run by oil and gasoline, we would use hydrogen itself produced from far off solar, hydro, or wind sources.

RIVALRY BETWEEN NUCLEAR ELECTRIC AND SOLAR-HYDROGEN POWER

There is one alternative to the Solar-Hydrogen Alternative. This is the Nuclear-Electric Economy. This has been on the back burner since Chernobyl and, indeed, most people in the world probably would vote against it anyway. But we must always allow for the fact that there will be fail-safe reactors produced, perhaps twenty years from now, and that an attempt will be made once more to push this source of energy forward.

It is, therefore, interesting to note that according to calculations made and published by Bockris, Dandapani, and Wass, in 1987 (Ref), the cost of the nuclear electricity would become more expensive than solar about 2020 to 2030. The turning point will be the massive conversion to breeder reactors which will be far more expensive in terms of the nuclear option than the present fission reactors. Breeder reactors are bound to be substituted for fission reactors at some time in the 25 year future because of the exhaustion of uranium resources. Breeder reactors need more complex machinery, but the major point is that they need much more safeguards because they produce plutonium.

On the other hand, the cost of renewables seems headed downwards because the efficiency can be upped by two times (as seems probable with amorphous silicon); and further, the mass productive component should bring it down to another fifty percent. Ogden and Williams have calculated that by 2010 the cost of electricity from solar photovoltaics should be about 3 cents per kWh and the corresponding cost of hydrogen then $15.00 per gigajoule (1990 $).

The rivalry between solar and nuclear sources is seldom seen in the light of economics because there is a tendency to see everything in terms of the "cost now" (the businessman's attitude) whereas, of course, the important thing we are talking about in our ten to twenty year future is to try to estimate relative cost among fuels in that time scale. We need to calculate in this way because of the large lead times which are necessary in contemplating the introduction of a new energy system.

COST OF SOLAR-HYDROGEN FROM VARIOUS SOURCES

In the diagrams which follow, an attempt has been made to bring forth the cost of hydrogen from various sources, both of hydro power and wind power.

The costs of hydro power are variable, depending on the size of the plant, height of the source, etc. They go from 1 to 10 cents per kWh. The amount of hydro, counting the Canadian and American resources together, is large, and would supply about one-third of North America's needs if fully utilized.

Now, the cost of wind power is dependent upon the average rate of passage of air at a given geographic location and it can be shown that this is constant on a yearly basis. But the cost of the electricity which one can get from it varies with the cube of the velocity so that for low velocity winds, such as 10 kilometers per hour, the electricity would be expensive, but for the higher winds, such as 20 kilometers per hour, upwards, then the costs fall within an acceptable norm.

It is interesting to contemplate the production of hydrogen from Antarctic sources where the available winds are very high and quasi continuous. A 40 kph average wind would be ~ 8 times cheaper than that obtainable from a normal 20 kph.

METHODS OF PRODUCING HYDROGEN FROM SOLAR

There are several ways of doing this and one way is shown in Diagram 9 which shows the $(PIN)^3$ concept. Solar light impinges upon a sandwich structure of photovoltaic cells and these, then, are composed of three layers, -

one with an energy gap larger than the average (that in the middle), and finally, the last sandwich where the smallest energy gap is, the idea being to capture a large swath of all the solar photons.

Diagram 10 comes from several sources. One can see that the extrapolation to an asymptote which would be relevant to, say, 2010, gets right down to a cost of electricity in the region of 4 cents per kWh which would really transform the production of electricity and make us able to be independent of the present polluting sources. But, note that one still has to transfer the electricity from the sunny places in the form of hydrogen to somewhere where the electricity is going to be used (as natural gas is now transferred from natural gas fields in Texas to New York City).

WHAT ABOUT MEANS OF ELECTROLYZING WATER?

There is confusion about the source of hydrogen to be produced from this photoelectrochemical electricity. The National Committee overseeing the tiny U.S. Hydrogen Progam even suggested that one might make the hydrogen from natural gas, but, of course, this is a businessman's suggestion -- selling the natural gas commodity -- and has no application to an environmentally clean future because it would obviously create CO_2 in the process.

Therefore, the only raw material from which we can get the hydrogen is water, itself. Diagram 11 shows the dependence of the reversible thermodynamic EMF for hydrogen evolution upon temperature and one can see that it goes on going down. When one gets to relatively high temperatures like 1500°C it has sunk down to about half of what it was at 25°. The overvoltage would be correspondingly reduced.

Some people see a paradox here: By heating it one seems to get much more than one did -- something for nothing -- but the point is that one is assuming that there is an outside power source heating the vessel to a high degree and, therefore, needing a small amount of electrical energy from the decomposition of water.

Although there is no breakdown of any laws of thermodynamics by the lowered potential here, there is a cost advantage because, in a sense, one is getting the heat to go to hydrogen, and, of course, heat is 2-3 times cheaper than the corresponding electricity. One doesn't need so much electricity.

One must be cautious before implying that one is advocating 1500°K production of hydrogen because it may well be that at that rather high temperature, the zirconia solid electrolyte which is used in such cells may be relatively unstable. More research is required here.

Corresponding to these high temperatures there is already an electrolyzer. The Germans call it "Hot Elly" and it is shown in Diagram 13. Here the zirconia is used with electrodes on each side which could be perhaps uranium oxide but could also be a noble metal.

Then, Diagram 14 illustrates a method by which one roughly halves the potential, and thus the cost needed for splitting water by eliminating the oxygen half of the reaction and utilizing the anodic reaction to create something valuable. The simplest thing to create earlier was oxygen, but the price at which one could sell this was small and the energy needed to produce it large, so now the concept is to produce sulfuric acid, or nitric acid, by using SO_2 or NO, respectively, as depolarizers.

Then Diagram 15 shows a novel method of hydrogen production. It goes back to an old principle, namely the Faraday homopolar motor, whereby one uses mechanical energy to rotate a wheel, and this wheel is kept in a strong magnetic field. So far as the wheel cuts the lines of force in the magnetic field it induces a potential difference across it, and if this is more than the needed electrolysis voltage for a certain current density, then one gets hydrogen at the center and oxygen at the edge of the wheel. Separating these is fairly easy and one can, after that, obtain hydrogen.

What is the advantage of this approach? It is that one avoids stepping down the voltages in which power arrives in the grid in an alternating current form, to direct current and at low volts.

It is also amenable to direct use of the flow of water as mechanical energy, - hydro power, - and to the flow of wind.

Homopolar generators are one of the missed technologies of electrochemistry. They should be used in all electrochemical procedures in industry because they are more flexible and cheaper for the low volts consideration than are ordinary devices which involve rectification and transformer step down.

Diagram 16 shows the status of water electrolyzers. Right now one envisages the network at the top on the left, expensive, too), but the major thing is that if one goes to the Hot Elly, as shown at the bottom left, the potential for hydrogen evolution, at, say, 100 milliamps per square centimeter is reduced right down to about 1.3 volts, hardly more than the decomposition potential at room temperatures.

What then, of the efficiency of conversion of these various machines? When the hydrogen comes to the towns, it will be necessary. Several possibilities are seen in Diagram 17 and it will be admitted that the best cell is the Hot Elly cell. But the other manufacturers don't have to give up. It may well be that they can beat Hot Elly in overall costs, e.g., stability, and cheapness, versus short life and expense.

THE MONOLITHIC OXIDE FUEL CELL

The best way to convert back to electricity would be via the monolithic fuel cell (Diagram 18) because this fuel cell, by its nature, can be built on and on to any size.

Thus, the fuel travels down tubes through which it permeates to the outer layer which is in contact with the electrolyte. If one utilizes other types of arrangements, there are many in which there is a large build up of potential in the thin tubes so that only part of the tube is utilized. Here, the whole of the tube can be utilized for the production of electricity. Furthermore, this tube can be made big and long (monolithic!) and hence, the utility of the concept. Monolithic fuel cells are now being built by Ballard and should be soon available.

SAFETY ASPECTS OF HYDROGEN

These are presented in Diagram 20. Hydrogen is said to be an unsafe fuel but this is hardly a fair designation. It has to be compared with something. If it is compared with methane the answer is that it is more dangerous and less dangerous. The greater danger arises because the ignition range in which hydrogen can explode with oxygen is larger than it is for, say, methane (or natural gas), the obvious thing with which to compare.

But hydrogen has some counter advantages and one of them is that it has a lesser explosion energy -- about a third of that with methane. This would result in a less damaging explosion for a given volume. Again, hydrogen escapes upwards in an accident and would not form noxious fumes.

THE STORAGE OF HYDROGEN

There is no one method to store hydrogen -- most methods have some part to play. If one wants to run an airplane on hydrogen, then it certainly has to be liquified, an expensive process, probably doubling the cost of the gaseous hydrogen. There again, if one wants to build a city on hydrogen, then one would have to take to natural gas fields, or something analogous, of very large volume, to hold the hydrogen. In transportation on land, light weight cylinders would do.

TIME AT WHICH TO AIM COST PROJECTIONS

This is an important and neglected aspect. The way people who control the funding of hydrogen seem to think is as businessmen would think, namely, they are only interested in hydrogen coming down in price enough to fit a market. Then, they will start to build. However, I think one has to think far ahead of this, for the lead times are certainly in decades. Let's imagine that we could make a changeover in fifty years. Then twenty-five years from now, or about 2018 is the midpoint, and it is the time at which we should aim the cost estimates.

How long would it take to carry out the unprecedented task of changing the energy system? It is a matter of choosing some arbitrary fraction of the Gross National Product -- some people take 5% -- for the money set aside to build the system. Taking into account a 4 kW/person need and then, say $3,000 per kW, the number of years needed to make a changeover is 20-50 years.

SUMMARY AND GENERAL STATEMENT

The future of our energy supply in some way is rather simple. To maintain our living standards we cannot let the energy per head slip by more than around one third of what it is now in the United States. If we are going to face the increase in population which will come from China, India and South

America before they realize that they have to control their birth rate, we have got to face a world population of at least 5 billion people.

Now, these certainly cannot be supplied with fossil fuels for more than 10 years. But long before exhaustion comes the CO_2 danger is upon us.

The time to change will be several decades, perhaps as much as four or five, depending on how much we can canalize our resources in one direction, and for this reason the time to start the conversion is right now, not at some future time.

Thinking in the energy business is distorted by the fact that those who control the purse strings for research and development are themselves businessmen, or at least have a businessman's attitude towards the future, investment, etc., so that they are really only interested in when the price is right for them, they are not interested in any future costs.

This gives rise to a distortion because, of course, at any given moment the cheapest fuels in a First Law sense and not allowing for pollution costs, are fossil fuels. If one doesn't think about the future. about exhaustion, 50 year lead times, or gross pollution, then one gets the wrong impression.

Another distortion arises with the businessman because he refuses to look at the <u>cost component</u> of the pollution which is taking place all the time now. Thus, he comes out with costs of solar hydrogen which might be twice as high as gasoline and thinks that that is a true comparison. The appropriate comparison, of course, is the cost of the gasoline at some future time, 20-30-40 years in the future (if there is any gasoline then) plus the cost of the pollution it will cause.

There does not seem to be any alternative but the Solar Hydrogen Economy, so long as one realizes that one takes the word solar in a broad context. The only possible thing would be a clean fusion reactor but we haven't even seen sustained fusion at this time. Unless that comes about we are in for a Solar-Hydrogen Economy, whatever the short term thinkers say.

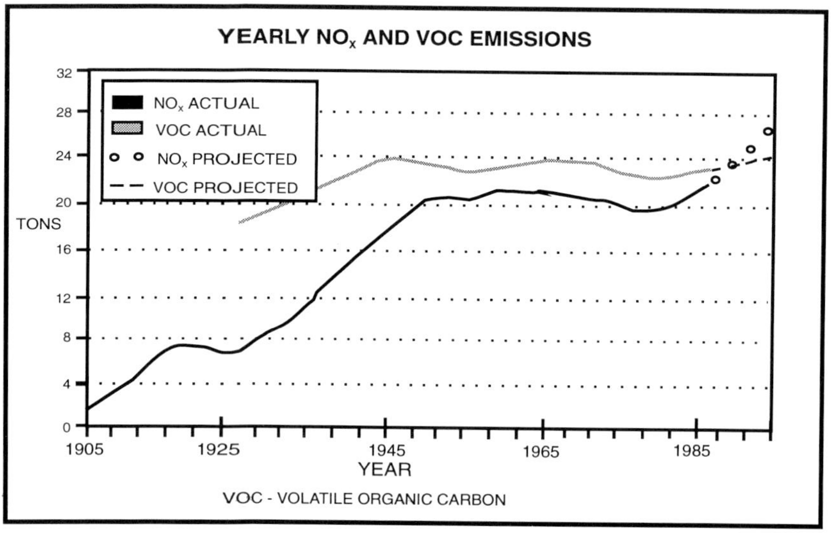

Diagram 1: The Growth of Oxygen Containing Pollutants in Recent Years

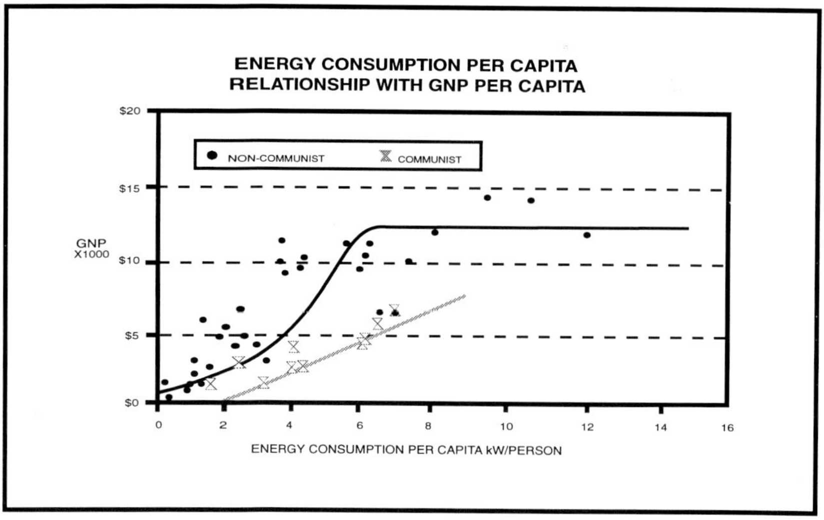

Diagram 2: The Relation of Income to the Consumption of Energy per Head of the Population for Various Countries

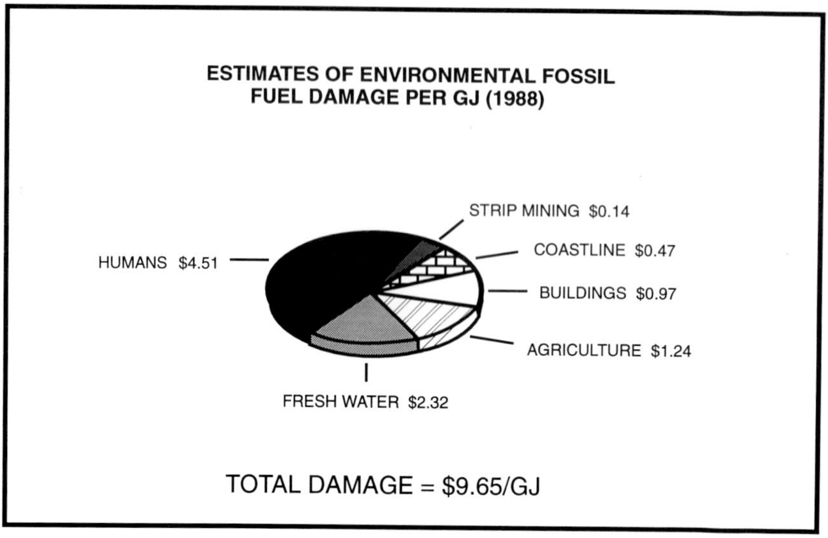

Diagram 3: The Cost of Pollution Broken Down

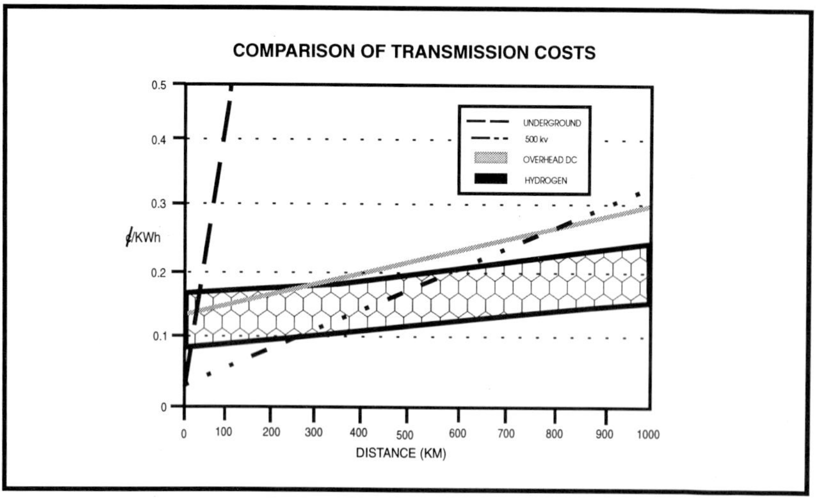

Diagram 4: The Cost of Transferring Energy in Hydrogen or in the Form of Electricity

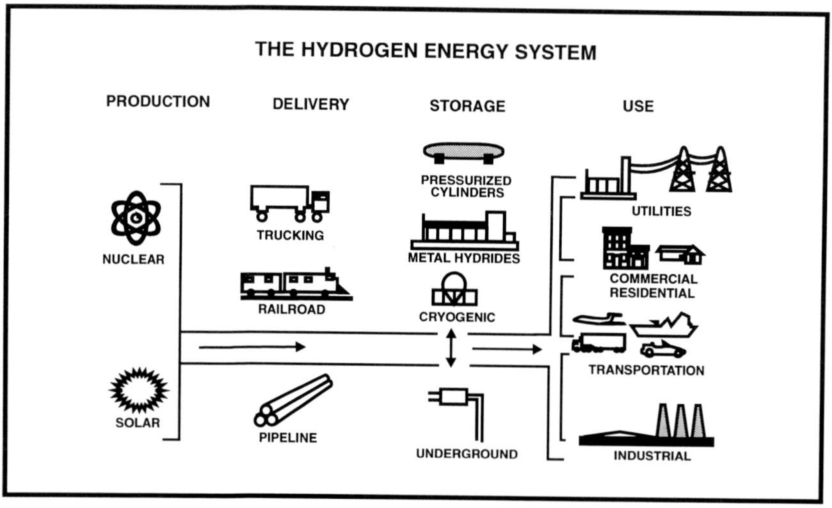

Diagram 5: The Solar-Hydrogen Economy

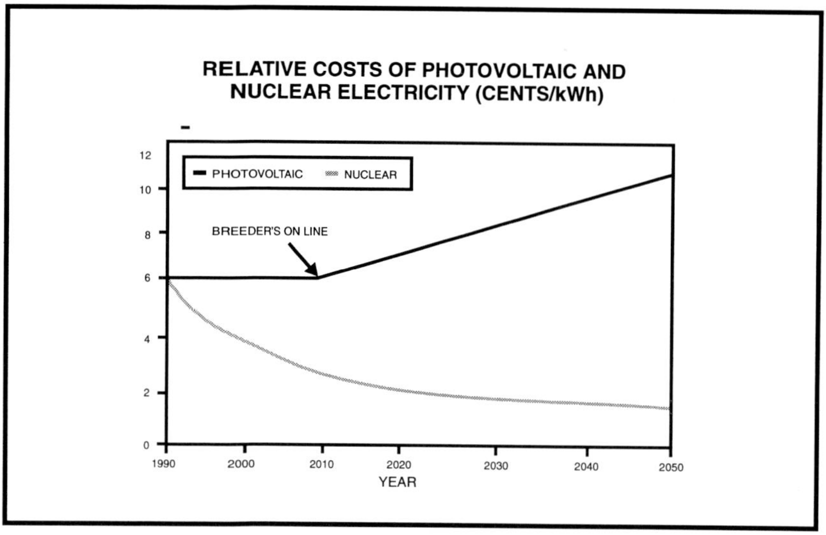

Diagram 6: A Schematic of the Change of the Cost, in Constant Dollars, of Nuclear and Solar Energy

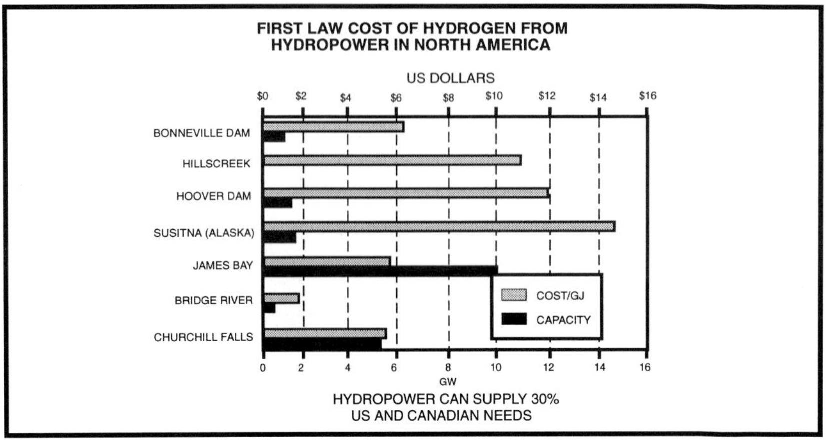

Diagram 7: The Cost of Hydrogen from Hydro Power in North America Together with the Amount Available from Various Sources

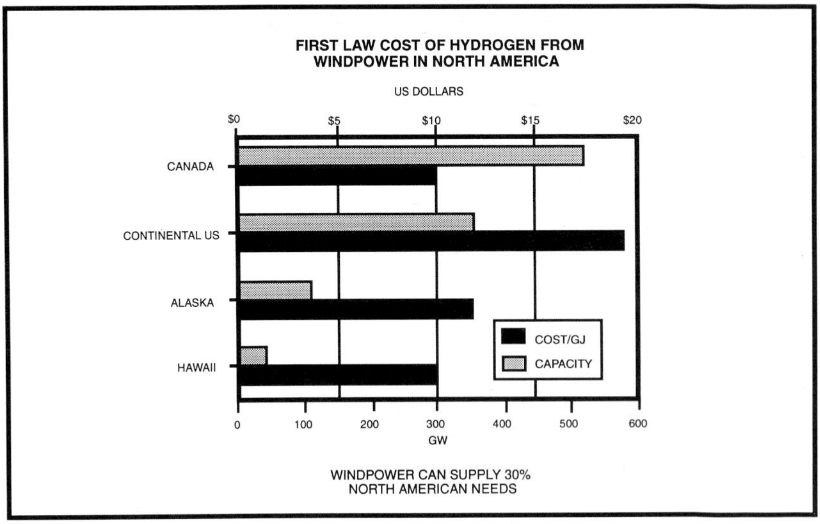

Diagram 8: The Cost of Hydrogen from Wind Power in North America

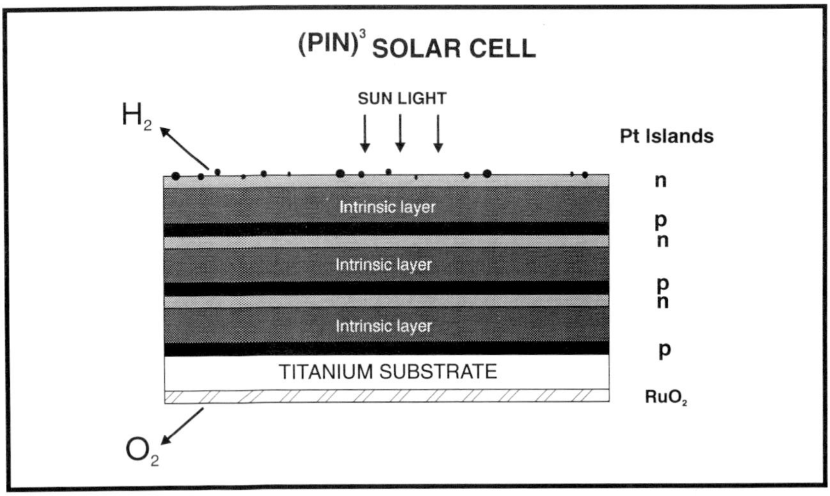

Diagram 9: The (PIN)^3-Solar Cell, The Various Layers Have Different Energy Gaps in Order to Optimize the Adsorption of Solar Light.

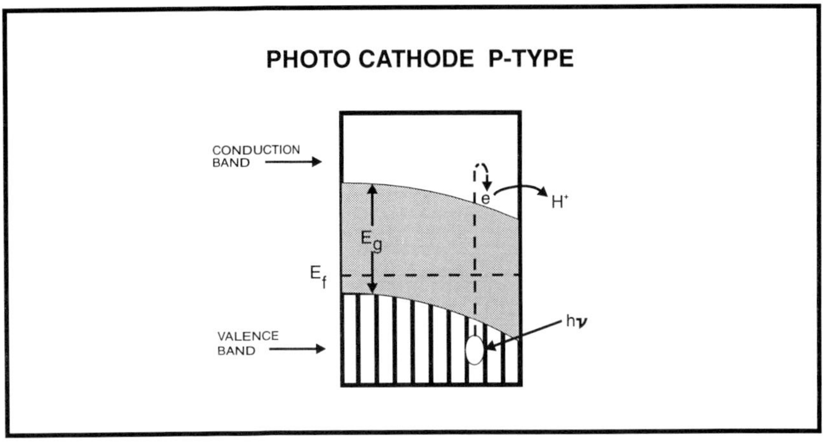

Diagram 10: Photoelectrochemical Evolution of Hydrogen. Here the advantage is that there is no p-n junction, but rather p-solution junction. The hydrogen is evolved in the solution and this may lead to lower costs.

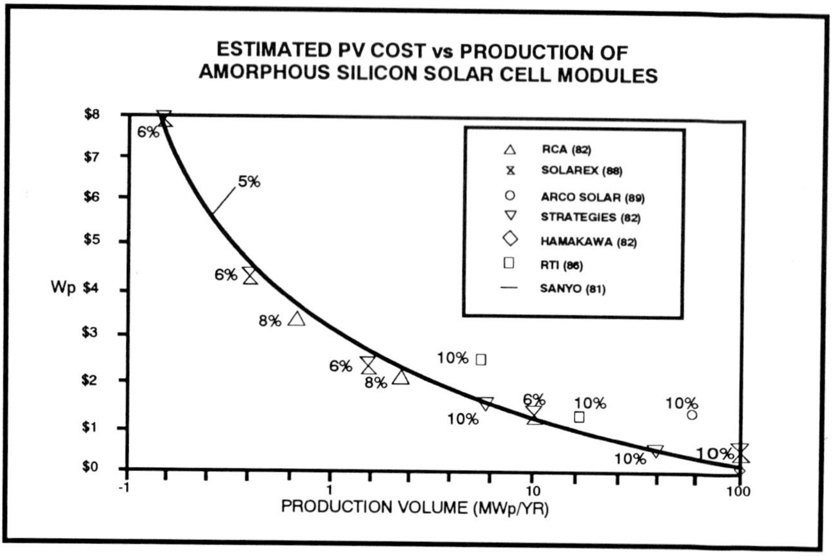

Diagram 11: The Project Cost of Hydrogen in 1989 Dollars from Solar Sources. It will be seen that the costs are extrapolated on a rational basis forward to where they are competitive even with present polluting fossil fuel costs.

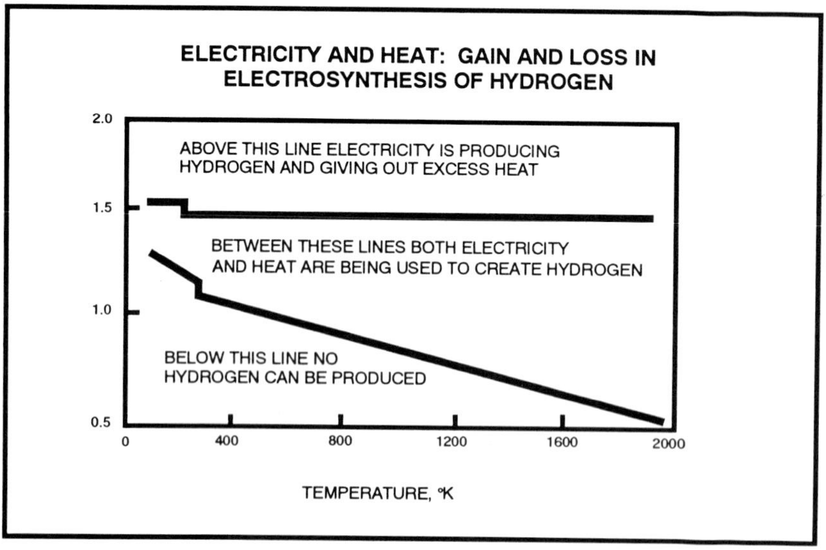

Diagram 12: The Dependence Upon Temperature of the Basic Thermodynamic Energy to Produce Hydrogen.

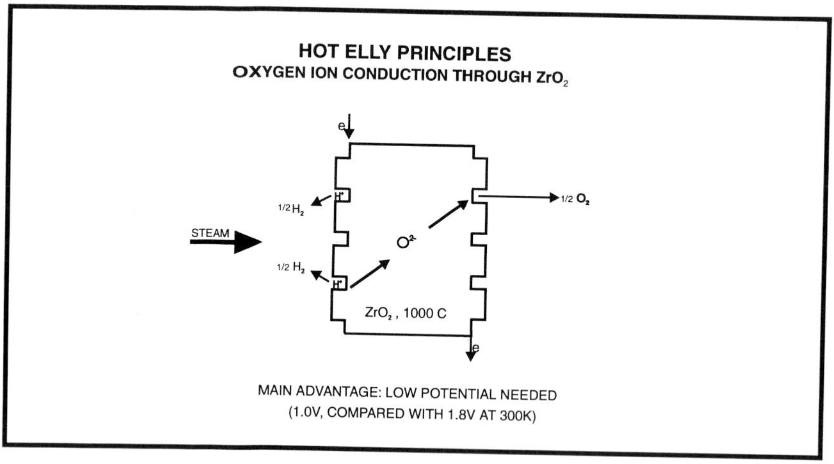

Diagram13: The So-called Hot Elly Cell, in which steam is projected onto one side of the zirconium membrane and which is a conductor. The zirconia acts as an oxygen ion electrolyte and on the other side the oxide ions form molecular oxygen, giving their charges to a conductor.

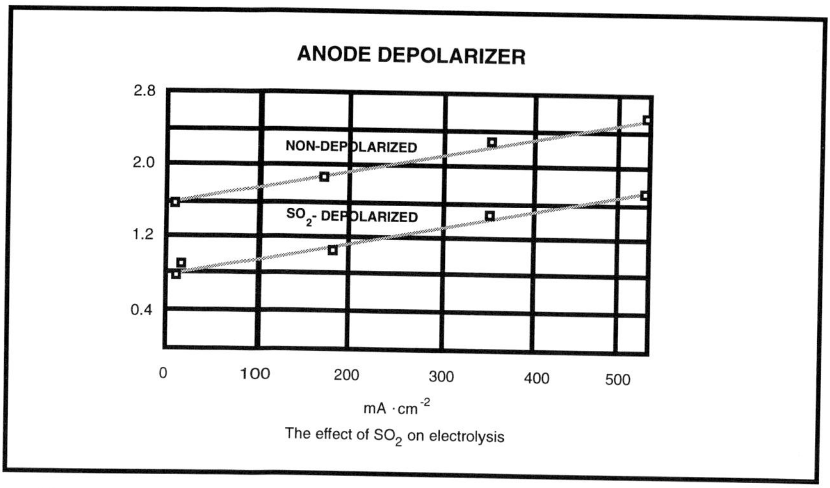

Diagram 14: Anode Depolarization As A Method of Reducing Costs of Hydrogen The anode can be depolarized by using various waste materials such as sewage or sulfur dioxide or nitric oxide or etc.

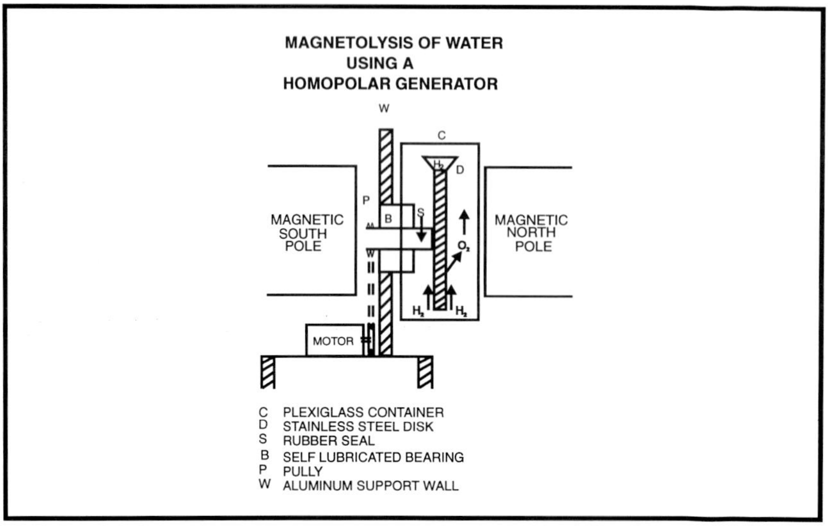

Diagram 15: The use of the homopolar motor to reduce the volts needed to produce hydrogen. In this way one can utilize mechanical energy without actually producing overtly electricity. Rotation of the wheel causes the different potentials to be produced in different parts of it and these then give rise to the production of hydrogen and oxygen.

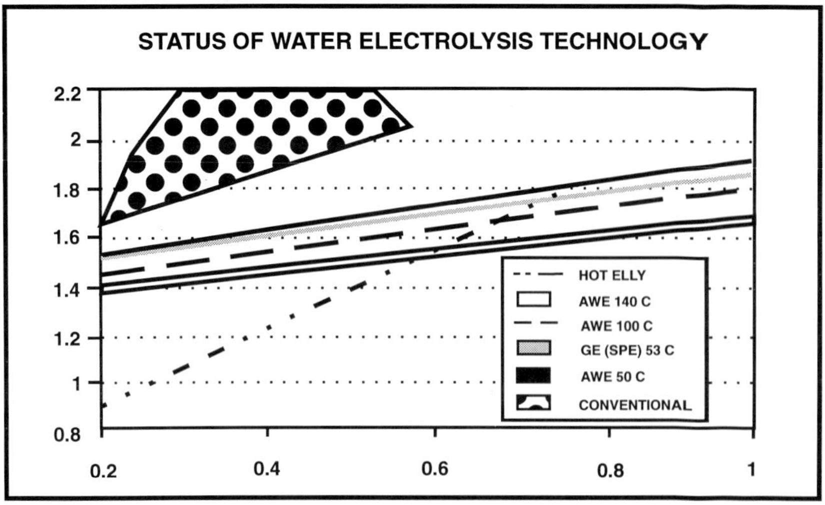

Diagram 16: The Cost of Hydrogen by Electrolysis. It is seen that the costs vary very greatly but the cheapest are with the high temperature processes.

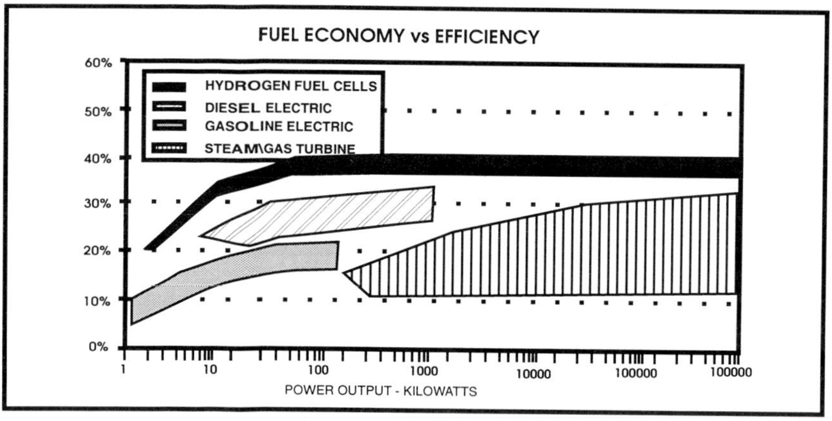

Diagram 17: The Efficiency of Various Ways of Converting Hydrogen Back to Electricity. The fuelcell way is by far the most efficient.

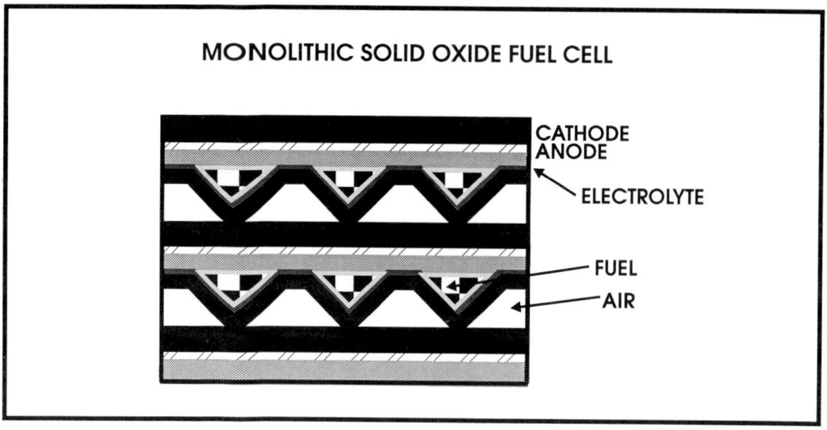

Diagram 18: The MonolithicFuelCell. Here is a fuelcell not yet completely made but which looks very good indeed from the point of view of the future because of its very high power per unit weight.

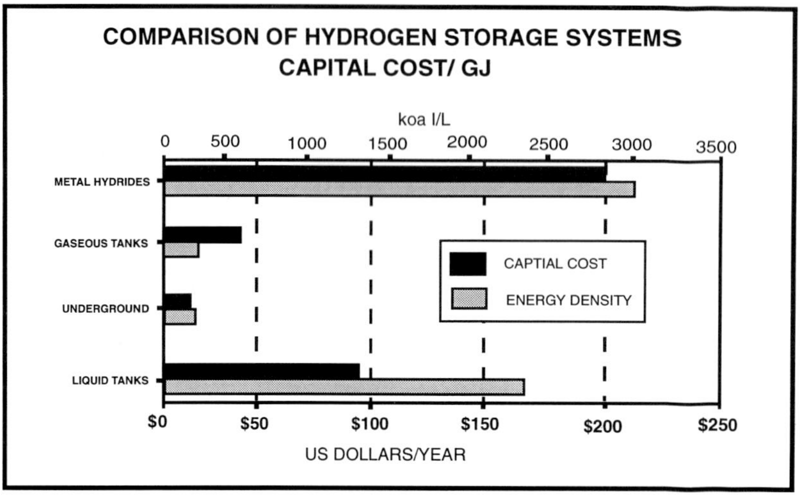

Diagram 19: The Cost of Storage of Hydrogen

SAFETY ASPECTS OF H_2 FUEL
(COMPARISON WITH CH_4)

WORSE	BETTER
LOWER FLASH POINT	EXPLOSIONS 1/4 AS VIOLENT
EXPLODES OVER GREATER RANGE OF COMPOSITION WITH OXYGEN	LIQUID BURNS TO STEAM, NO NOXIOUS GASES
LEAKES 2.8 TIMES FASTER	LIQUIDS BURN UPWARDS, NO LOW LEVEL VAPORS TO IGNITE

H_2 IS A DANGEROUS FUEL

AS DANGEROUS AS OTHER CHEMICAL FUELS

Diagram 20: The Safety Features for Hydrogen

TIME AT WHICH TO AIM COST PROJECTIONS

NEW SYSTEM WILL TAKE ~ 50 YEARS

LIKELY STARTING DATE ~ 2000

HENCE: AIM PROJECTIONS TO 2025

Diagram 21: The Time At Which to Aim Cost Projections. Costs should not be calculated as they are now, but they should be calculated at some point half way between the present time and the time at which the building of the Solar-Hydrogen Economy would have to be finished.

CHAPTER 3 -- SOLAR ENERGY

Hydroelectric Hydrogen -- A Promising Alternative for Brasil

Marieta Mattos
Brasilian Ministry of Mines and Energy
University of Sao Paulo
Sao Paulo, Brasil

Background:

Dr. Mattos has served as an advisor to the Secretary of Technology in the Brasilian Ministry of Mines and Energy, and as Coordinator of Research Activities at the Laboratory of Electrochemistry, Institute of Physics and Chemistry of Sao Carlos University of Sao Paulo.

BRAZIL'S ENERGY PANORAMA: EXCESS ENERGY UTILIZATION FOR PRODUCING ELECTROLYTIC HYDROGEN AS A PROMISING ALTERNATIVE IN BRAZIL

Marieta C. Mattos* and Demostenes B. da Silva**

* Instituto de Fisica e Quimica de Sao Carlos/USP
Laboratorio de Eletroquimica
Caixa Postal 369
13560 Sao Carlos - SP, BRAZIL

** Companhia Energetica de Sao Paulo/CESP
Al. Ministro Rocha Azevedo, 25 - 17a.
01410 Sao Paulo - SP, BRAZIL

ABSTRACT

Brazil is a world leader in the development of renewable resources for energy consumption. A overview of Brazil's energy panorama is presented, emphasis being given to hydroelectrical energy . A hig percentage of the installed generating capacity - 91% - is composed of hydroelectric powerplants, and this pu Brazil in a very favourable energy position. Hydropower plants are planned for operation under differe hydraulic circumstances. A situation might arise necessitating the diversion of excess water, more tha necessary for maintenance and optimization of reservoir levels. An evaluation of this excess energy or unuse off-peak electrical energy is presented, a situation most favourable for hydrogen production by electrolysi Utility companies operating in the North, Northeast and Southeast envision electrolytic hydrogen productic as a market during hydrologically favourable periods. This hydrogen could have several utilizations, enabli thus the amortization of investments in electricity generating capacity.

1. INTRODUCTION

Considering in Brazil the primitive Amazon region and the thriving Sao Paulo, the strong contrast evident. In the Southeast and South regions is 75% of the national industrial capacity, and 59% of the Brazili population. Moreover, these two regions account for 80% of the gross national product (GNP) meaning, thu that the largest national economic concentration and dynamism are restricted to about 18% of the nation territory.

From the 70's important changes occurred in the treatment of energy questions, at world level as w as in Brazil, motivated by concerns associated not only to availability of energy sources and to their geograph distribution, but also to growing concerns related to the environment. The planning of the energy sector resu in coordinated actions signalizing the perspectives in the sector, on the demand side as well as on the supp side, globalizing the different forms of final use and several alternatives for supply. This planification stuc as well as the general characterization of the electric sector that has been adopted in Brazil, follows t systematic evaluation done by the National Energy Balance (Balanço Energetico Nacional) edited by t Ministry of Mines and Energy (MME). Scenarios presented in Table 1 were developed by ELETROBRA (Centrais Eletricas Brasileiras S.A.), that starting from premises and established parametres delineated t consumption evolution of the different forms of energy from 1990 up to 2015 in Brazil.

Brazil is a privileged country in having the third largest hydropower potential in the world - 261.4 G (129.5 GWyear of firm energy). One third of this potential is situated in the far Amazon region. It is in this regic in the North of Brazil, that are 100 GW of hydroelectric power available and already evaluated, but just 5%

this potential installed whereas in the Southeast and South regions water resources have already been exploited and utilized to their maximum. And it should be noted that it is in the Southeast and South regions that electricity consumption centres are concentrated. Figure 1 presents the hydropower potential areas in Brazil. Just to figure out the hydropower potential of Amazonia (Amazon region), the potential of the Amazonas and Tocantins rivers basins, the latter just partially Amazonic, is evaluated as 68,623 MWyear of firm energy, which means 53% of the total Brazilian hydropower potential, and about 6% of the world hydropower potential estimated as 1,100,000 MWyear. The magnitude of those figures indicates the importance of water resources in Amazon region.

The availability of a hydropower potential advantageously competitive to other alternatives maintains the expansion of the electricity supply mostly based on exploitation of water resources. This is reinforced by considering that only 25% of this potential is utilized now or is in implantation. In 1991 the installed generating capacity, in Brazil, was of 57,146 MW, waterfalls corresponding to 52,376 MW (91.7%) of this total. It is concluded, then, that the 100 GW yet to be installed in the North represent almost twice the installed generating capacity in Brazil, at present.

This paper presents an updated overview of Brazil's energy panorama, which has already been presented [1], and an evaluation of the excess energy or unused off-peak electrical energy, a situation most favourable for hydrogen production by electrolysis.

2. ENERGY PANORAMA IN BRAZIL

In 1987 ELETROBRAS, the holding company responsible for the coordination of the electrical sector all over Brazil, and several regional utility companies, made a planification study entitled Plan 2010 [2]. This plan, which studies the technically and economically most interesting alternatives for electrical generation systems in the long term, has chosen the expansion for Brazilian electrical generation systems as shown in Table 2. This table presents the minimum generating capacity for Brazilian evaluated needs. It can be noted in this table that up to the year 2010 waterfalls should go on contributing with a high percentage of 91 to 89% of the installed generating capacity, in Brazil.

In 1992, ELETROBRAS updated Plan 2010 [2], and so Plan 2015 [3] was prepared, which has shown in Table 3 the evolution of the nominal installed capacity in Brazil since 1962; in the period considered in this table, it is noticed that this capacity presented an average annual growth rate of 8.2%, and in the last decade a growth rate of 6.6%. It was pointed out in this report [3] that the Brazilian electric system is characterized by the following aspects:

- great distances between generating and consumption centres;
- hydrographic basins with hydrological diversity, or that present variations in discharge distributions throughout the year;
- high degree of electrical interconnections between hydrographic basins systems;
- reservoirs with pluriannual regulation, or with capacity for storing water for utilization during several years, in hydrologically unfavourable periods;
- large fraction of hydropower potential still available.

An important point that should be stressed is that Plan 2015 was very much concerned about environmental impacts involved in the future construction of hydropower plants. Environmental pressures favour a larger participation of electricity replacing other forms of energy, because it is a clean solution even when it utilizes as primary source fuels having poluent characteristics, once it is easier to control emissions from thermal powerplants than in direct burn by the final consumer.

3. ELECTRICAL ENERGY GENERATION AND TRANSMISSION IN BRAZIL

A high percentage of the installed generating capacity - 91.7% - is composed of hydroelectric powerplants, and this is a very favourable energy situation. Thermoelectric powerplants were installed for supplying electricity to small isolated communities such as the capitals of the states of Amazonas, Rondonia and Acre, or for utilizing mineral resources found in extremely favourable conditions such as Candiota thermoelectric powerplant, in the State of Santa Catarina, in the South of Brazil. A small use of thermoelectric powerplants is made for complementing hydropower generations in hydrologically unfavourable periods, or for complementing peak demands.

Brazilian transmission system consists of two large network systems, not yet interconnected: 1) Southeast - Middle West-South and 2) North-Northeast. Besides these main systems, there are a few isolated systems in the area of ELETRONORTE (Centrais Eletricas do Norte S.A.): 1) Manaus and 2) Acre-Rondonia, in which electricity is generated by thermal plants, but that is receiving now some complementation from hydraulic sources.

Electrical energy transmission from the Amazon region will have an important role in integrating or interconnecting these two large network systems in the near future. In a second stage, the interconnection is planned for isolated systems as the right and left margins of the Amazon river known as the Manaus and Acre-Rondonia systems.

Electrical generation areas in the North Region and the interconnecting prevision for electrical systems as shown in Figure 2 will meet national energy demands; in this figure, the utilization of the Amazon region as exporter of energy is noted. Although distances involved are so great, these two regions have the same rain seasons.

Transmission lines of great extension and great capacity like the ones previewed to flow the Amazonia energy are new in Brazil. In the world, there is parallel only in the Community of Independent States (ex-Soviet Union) that has long lines of great capacity, even though presenting intermediary derivations for supplying charges along the course.

In the case of transmission from Amazonia, distances involved are about thrice the ones of the Itaipu project (Binational hydropower plant in Foz de Iguacu, in the South of Brazil and frontier with Paraguay), having in this case a substantial difference in terms of the scale of the problem, that presents new aspects to be taken into account, mostly related to losses of energy.

Plan 2015 [3] stressed the necessity for establishing regional interconnections as it is presented in Figure 2 for transporting power as shown in Table 4. The chief possibilities for the transmission trunks from Amazonia are presented in Figure 3.

4. EXCESS ENERGY

Hydropower plants are planned for operation under different hydraulic circumstances. Situations might arise [4] as follows, always having in mind the optimization of electrical energy generation and utilization:

1) possibility of diversion of excess water in the spillway, water that could be turbinated but that is not because the electricity thus generated is not required for consumption;

2) necessity of diversion of excess water for maintenance or optimization of reservoirs levels;

3) reservoirs with pluriannual or annual discharges, that can attenuate the variability effect of natural affluences, storing water in periods of great hydrological affluences and diverting in dry periods;

4) reservoirs that are able to regulate temporal (i.e. monthly or annual) discharges taking into consideration the control of floods in the hydrographic basin and of flows;

5) periods of unused off-peak electrical energy consumption.

An evaluation of this excess energy or diverted energy that could be turbinated, or unused electrical energy, is presented in Table 5 by ELETRONORTE regarding the North region. This table also presents electricity generation and guaranteed energy data. Similar data are presented in Table 6 by CESP (Companhia Energetica de Sao Paulo) for the State of Sao Paulo. Other regions in Brazil present excess energy, but in these two tables are shown the most impressive data, concerning a period of 6 years. It is noticed how great is Tucurui turbinable diverted energy, presented in Table 5.

5. ELECTRICITY COSTS AND TARIFFS

From the 30's a growing centralization of the policy for electrical energy prices or tariffs occurred in Brazil. In 1974 the equalisation of tariffs was determined, fixing thus the same prices for electricity all over Brazil. A crossed subsidy was then established among electric utility companies, causing inefficiency: as the ones that had to transfer resources to other utilities were not encouraged to reduce costs, and those that received resources could count on additional revenues to cover their costs.

The inflationary process has assumed great dimensions in Brazilian economy since the 80's, and the government adopted the strategy of containing tariffs for the control of the inflation rate. This resulted in a growing deterioration of the utilities' economic-financial equilibrium and hardly contained inflation. Actually there was a great subsidy of the electric sector for other sectors of the economy.

To change this situation, a new policy was established last March, that transfers the responsibility for fixing electricity tariffs to the states. These tariffs determined by regional utilities should cover costs but imply in costs reduction through improvement of efficiency and rationalization of consumption. This policy of desequalisation of tariffs meant a deep modification, with more liberality in fixing tariffs but involving more responsibility in doing so, once regional utilities are supervised by DNAEE (Departamento Nacional de Aguas e Energia Eletrica) and ELETROBRAS.

Tariffs are being readjusted gradually so that the economy could absorb this new policy of arriving at realistic tariffs, but even then average tariffs now in vigour are about 55% lower than effective costs and this results in strongly weakening the electric sector, in terms of its inner generation resources, necessary for expansion of the system. Although considering that the situation is better for the electric sector now, these electricity tariffs still result in adding to the inflationary process. In this regard, it should be remembered that the capital invested in the electrical system originates in its greatest part from foreign loans, thus meaning an impact on the external debt and the public deficit. Electrical energy tariffs are in the average value of US$ 46.00 per MWh, with the recent readjustment. Table 7 presents marginal costs of electricity supply at long term, taking into account the expansion of the installed capacity of the system, for different regions, and on the other hand, presents costs of production at short term in periods of high hydraulicity that are very low. These low costs presented in the lower part of Table 7 are the average for any time of the day, peak as well as off-peak hours. However, tariffs based on these costs are established just for interchange among electric utilities, and not for supply to end-use consumers. For these, the lowest tariffs in vigour for off-peak hours, in tensions of 230 kV, are US$ 20.13 per MWh, but there is the possibility of negotiating with DNAEE so that industries could have the low tariffs based on the costs presented in Table 7. A comparison with tariffs charged abroad shows that Brazilian tariffs are lower than in different countries [5].

6. CONCLUSION: HYDROGEN AS A PROMISING ALTERNATIVE

The above discussed excess energy and unused off-peak electrical energy define a situation most favourable for hydrogen production by electrolysis, in Brazil. Hydroelectric utility companies, operating in the North, Northeast and Southeast, interested in optimizing electricity uses envision ammonia production from electrolytic hydrogen as a market during hydrologically favourable periods, and this would enable them to amortize investments in electricity generating capacity. Hydrogen has multiple utilizations: as a feedstock, hydrogen is a major industrial chemical commodity.

REFERENCES

1. A. D'Ajuz, E. H. Ellery Filho, A. M. Conti and M. C. Mattos, Electrical energy transmission from the Amazon region: hydrogen as a promising alternative in Brazil. *Int. J. Hydrogen Energy* **14**, pp. 515-523 (1989).

2. Centrais Eletricas Brasileiras S. A., *Plano Nacional de Energia Eletrica (Plano 2010)*, p. 264. Eletrobras, Rio de Janeiro (1987).

3. Centrais Eletricas Brasileiras S. A., *Plano Nacional de Energia Eletrica (Plano 2015)*, Preliminary Report. Eletrobras, Rio de Janeiro (1993).

4. L. A. M. Fortunato, T. A. Araripe Neto, J. C. R. Albuquerque and M. V. F. Pereira, *Introduçao ao Planejamento da Expansao e Operaçao de Sistemas de Produçao de Energia Eletrica*, pp. 35-36. EDUFF, Niteroi (1990).

5. Centrais Eletricas Brasileiras S.A., *Comparaçao Internacional de Tarifas de Energia Eletrica 1973/1989*. Eletrobras, Rio de Janeiro (1989).

TABLE 1: STRUCTURE OF FINAL ENERGY CONSUMPTION BY TYPE
BRAZIL
1990 - 2015

		1990	2015			
			I	II	III	IV
Electricity	(%)	36.5	46.2	46.0	44.8	45.5
Natural Gas	(%)	2.0	4.1	5.3	6.9	8.0
Oil and Alcohol	(%)	36.8	36.4	36.0	35.6	34.5
Coal	(%)	5.2	4.2	4.4	5.3	5.2
Wood and Charcoal	(%)	13.0	5.4	4.9	4.4	4.2
Bagasse and Others	(%)	6.4	3.7	3.3	2.9	2.6
TOTAL (million tonnes oil equivalent)		166.9	353.1	398.1	473.7	527.1

Source: Plan 2015, by ELETROBRAS [3]

TABLE 2: EXPANSION FOR BRAZILIAN ELECTRICAL GENERATION SYSTEMS INSTALLED POWER (GW)
HYDRO DOMINATES ELECTRICITY GENERATION

Type	1986	1990	1995	2000	2005	2010
Hydro	38.5	53.4	73.9	93.3	116.9	141.8
Thermal	4.2	5.1	7.4	9.7	13.8	18.2
Coal	0.7	1.3	1.9	3.0	4.6	6.5
Nuclear	0.6	0.6	1.9	3.1	5.6	8.1
Other	2.9	3.2	3.6	3.6	3.6	3.6
Total	42.7	58.5	81.3	103.0	130.7	160.0

PARTICIPATION (%)

Type	1986	1990	1995	2000	2005	2010
Hydro	90.2	91.3	91.0	90.6	89.4	88.6
Thermal	9.8	8.7	9.0	9.4	10.6	11.4
Coal	1.6	2.2	2.3	2.9	3.5	4.1
Nuclear	1.4	1.0	2.3	3.0	4.3	5.1
Other	6.8	5.5	4.4	3.5	2.8	2.2
Total	100.0	100.0	100.0	100.0	100.0	100.0

Source: Plan 2010, by ELETROBRAS [2]

TABLE 3: INSTALLED NOMINAL CAPACITY (MW)

	Hydraulic	Thermal	Total
1962	4,126	1,603	5,729
1966	5,524	2,042	7,566
1971	10,244	2,426	12,670
1976	17,675	3,385	21,060
1981	30,600	3,656	34,256
1986	38,682	4,483	43,165
1991	52,376	4,770	57,146

Source: Plan 2015, by ELETROBRAS [3]

TABLE 4: REGIONAL ELECTRICAL INTERCONNECTIONS FOR TRANSPORTING POWER

North-Northeast	from	6,000	to 9,000 MW
North-Southeast			
(from Maraba)	from	8,000	to 15,000 MW
(from Cuiaba)	from	5,000	to 10,000 MW
South-Southeast	about	4,000 MW	

Source: ELETROBRAS, Transmission Department (1992)

TABLE 5: EXCESS ENERGY OR DIVERTED ENERGY IN THE NORTH REGION

Year \ GWh	Electricity generation	Guaranteed energy	Excess energy or Diverted energy
1987	13,298,812.500	15,181,080.000	3,707,510.000
1988	15,835,425.000	20,673,600.000	2,455,536.000
1989	17,068,080.000	19,315,800.000	3,199,560.000
1990	15,468,731.862	23,687,040.000	4,561,136.000
1991	18,033,918.000	23,932,320.000	3,545,232.000
1992	8,099,257.000	10,155,600.000	5,021,424.000

Notes: a) These data refer to Tucurui Hydropower Plant, and represent more than 95% of the ELETRONORTE electric system.

b) The Electricity Generation and Guaranteed Energy data for 1992 are up to June.

c) 1 GW = 1000 MW.

Source: ELETRONORTE

TABLE 6: EXCESS ENERGY OR DIVERTED ENERGY IN THE STATE OF SAO PAULO

GWh / Year	Electricity generation	Guaranteed energy	Excess energy or Diverted energy
1987	43,874.087	38,999.520	1,249.429
1988	46,431.988	39,844.224	1,791.862
1989	45,239.727	39,446.280	2,378.807
1990	41,630.087	38,000.880	1,220.667
1991	48,549.461	40,996.800	2,874.041
1992	49,547.728	41,172.000	3,351.430

Source: CESP

TABLE 7: MARGINAL COSTS FOR SUPPLY

At long term

Region	US$/MWh
Southeast/Middle West	89.56
North/Northeast	69.02
South	85.49
Brazil	85.25

At short term

Year	US$/MWh
1990	15.09
1991	4.80
1992	8.31
1993	9.41

Source: DNAEE (1993)

FIGURE 1: BRAZIL - HYDROPOWER POTENTIAL AREAS

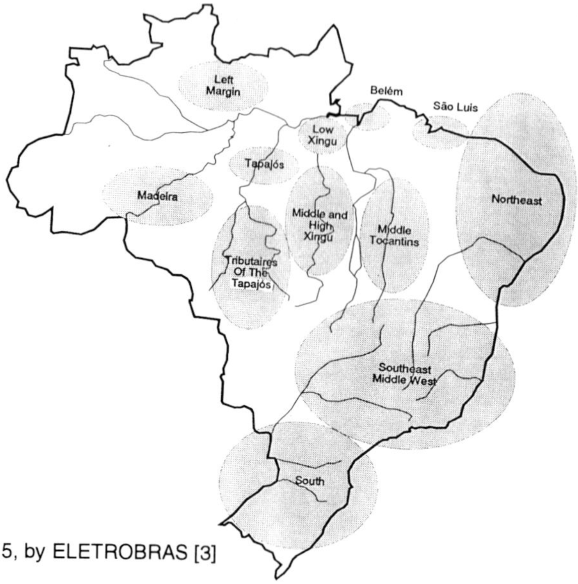

Source: Plan 2015, by ELETROBRAS [3]

FIGURE 2: ELECTRICAL GENERATION AREAS IN THE NORTH REGION AND DISTANCES TO LOAD CENTRES

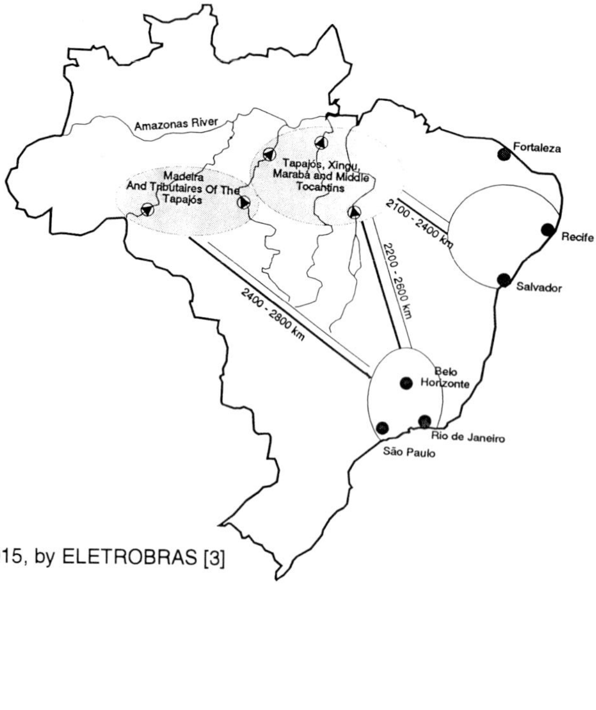

Source: Plan 2015, by ELETROBRAS [3]

FIGURE 3: CHIEF POSSIBILITIES FOR THE TRANSMISSION TRUNKS FROM AMAZONIA

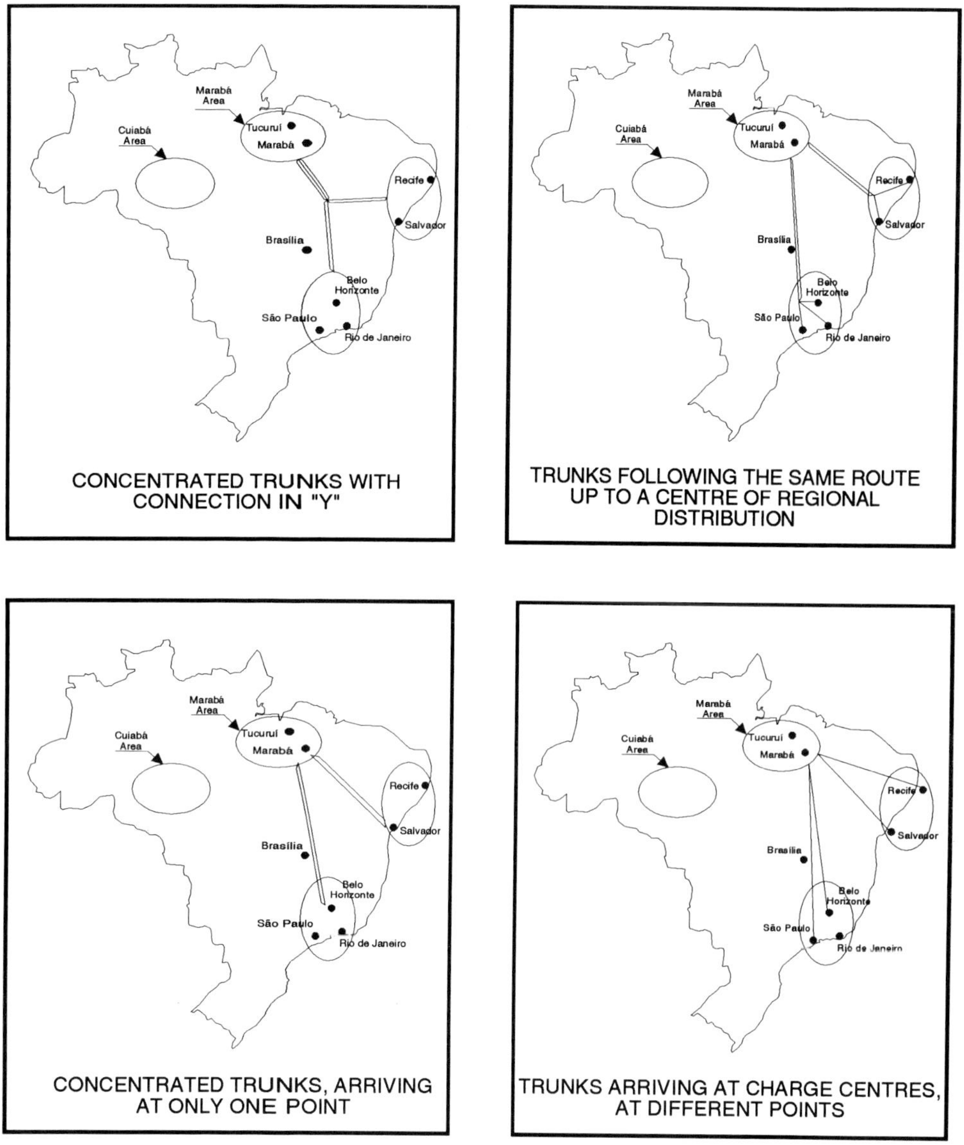

Source: Plan 2015, by ELETROBRAS [3]

The Potential of Renewable Energy

Robert Stokes
Deputy Director
National Renewable Energy Laboratory
Golden, Colorado, U.S.A.

Background:

In addition to his responsibilities managing NREL research and technical programs and commercialization efforts with U.S. private industry, Dr.Stokes is also a VP of Midwest Research Institute of Kansas City, the operating contractor of NREL. Prior to working at NREL, Dr. Stokes held a number of management positions at Battell, Pacific Northwest Laboratories, and was an Associate Professor of Physics at the University of Kentucky.

Abstract:

The implications of the Energy Policy Act of 1992 and other factors accelerating renewable energy use in the U.S. which could result in renewable energy providing one third of U.S. energy supply requirements by the year 2030.

No paper. Oral presentation only.

Solar Electric Systems and the Environment

Paul D. Moskowitz
Scientist
Brookhaven National Laboratory
Upton, New York, U.S.A.

Background:

Dr. Moskowitz has been an Environmental Health Scientist at Brookhaven National Lab for 16 years. He is also Head of National Photovoltaic Environmental, and Health and Safety Information Center.

Abstract:

The myths and realities associated with environmental, health, and safety aspects of solar electric systems.

No paper. Oral Presentation only.

Solar Energy and the Reindustrialization of the Nations

Scott Sklar
Executive Director
SEIA, US/ECRE
Washington, D.C., U.S.A.

Background:

Mr. Sklar serves as Executive Director of Solar Energy Industries Association (SEIA), and the U.S. Export Council for Renewable Energy (US/ECRE). Mr. Sklar has also served for two years as Political Director of the Solar Lobby, as both Washington Director and Acting Research Director at the National Center for Appropriate Technology (NCAT). He is currently serving on the U.S. Department of Energy Advisory Committee on Renewable Energy and Energy Efficiency Joint Ventures.

Abstract:

The Solar Energy Industries Association, the national trade association of the photovoltaic and solar thermal manufacturers and component suppliers, welcomes the opportunity to convey the extraordinary opportunities which these industries offer to our country by way of clean, indigenous energy production, domestic job creation, and U.S. technology exports.

After two decades and millions of dollars of R&D investments, U.S. solar energy technologies are either commercially competitive today or on the cusp of becoming so. The obstacles in the way of the solar industries to the widespread commercialization of their technologies are no longer purely technical in nature; rather, they are related to institutional, financial, and policy biases which must be corrected.

While R&D-generated technological advances will continue to lower costs and improve reliability, the U.S. solar industries have reached the point where their greatest potential for cost-reduction will come from the production side -- through more efficient, advanced manufacturing techniques and economies-of-scale both in production and deployment. However, to justify scaling up their manufacturing operations, these firms must have a demonstrable market for their technologies.

These markets are disparate and often ill-defined. Consequently, U.S. solar technology manufacturers face a classic "catch-22" problem: to penetrate these new markets, they must lower their costs, but to lower costs they need to scale-up manufacturing (to attain economies-of-scale); however, they cannot attract private capital to scale-up manufacturing without a demonstrable market for their technologies.

By streamlining and coordinating its existing programs, the federal government can implement an effective renewable energy commercialization strategy that will bring on line more than ten new manufacturing facilities in the near future and get on with the business of restoring the U.S.'s industrial base and global competitiveness.

Six existing federal programs can be used to create an inexpensive, yet effective commercialization program:

1. Federal Energy Management Program
2. Department of Energy Grant Program
3. Department of Energy Joint Venture Program
4. Advanced Research Projects Agency
5. Defense Conversion Programs
6. Federal Loan Access (FmHA, REA, FHA/VA, SBA)

CONCLUSION

Solar energy technologies represent an industry in which the U.S. holds a technological lead. However, these technologies' deployment will be dispersed throughout the United States, creating significant barriers to their commercialization. To ensure the necessary market aggregation measures that will allow U.S. solar industries to ram-up manufacturing, lower costs, and rapidly penetrate markets, the federal government must choreograph its existing programs so as to focus *related* economic activities with respect to these technologies.

No paper. Oral presentation only.

CHAPTER 4 -- WIND ENERGY

Wind Energy: A Resource for the 1990's -- and Beyond

Randall Swisher
Executive Director
American Wind Energy Association
Washington, D.C., U.S.A.

Background:

Dr. Swisher has been Executive Director of the American Wind Energy Association since 1989. Previously he was legislative representative to the American Public Power Association. He has been Director of Energy and Environment Programs of the National Association of Counties, and Adjunct Professor at Georgetown University.

WIND TECHNOLOGY AFTER A DECADE OF DEVELOPMENT

Randall Swisher, Executive Director
American Wind Energy Association
777 N. Capitol Street #805
Washington, D.C. 20002 USA
June, 1993

INTRODUCTION

After a decade of development in California, wind energy technology has matured to the point where it is ready for widespread application to meet utility needs in many areas of the United States and the world. The inclusion of a 1.5 cent per kWh production incentive for wind-generated electricity in 1992's National Energy Policy Act could be the stimulus to extensive wind development.

Over the last decade, California windfarms have played a key role in the evolution of wind technology, with dramatic gains in both cost-effectiveness and technical performance since the first windfarms were established in 1981.

This progress has begun to be recognized by the U.S. utility industry and by public policy makers, and after a number of years in which major market activity for the wind industry was focused in Europe one can begin to see the opening of opportunities for wind technology in the U.S. in regions outside of California. Specifically, wind energy development is beginning to take place in the Pacific Northwest, the upper Midwest and the Northeast. The speed and scale of that development is still unclear, but this expansion of the market for windpower will be significantly influenced by federal and state policy decisions that will be made over the next few years.

CALIFORNIA WINDFARMS: A DECADE OF PROGRESS

Harnessing the wind for utility bulk power production proved to be much more challenging than renewable energy advocates expected, but windfarm performance has improved steadily since the first wind power plants began operation in California in 1981. In fact, the evolution of wind technology has been so rapid that there is now a significant gap between the perception of wind technology by many utility personnel and its current technical performance. That gap has been significantly narrowed over the last few years due to the active efforts of knowledgeable utilities such as Pacific Gas & Electric and the Electric Power Research Institute.

Production of wind-generated electricity in California has risen from 6000 kWh in 1981 to almost 2.8 billion kWh in 1992. Capacity factors have increased from an average 3 percent in 1982 to 20 percent in 1991, with turbines installed since 1985 averaging 26 percent. Some prime sites have exceeded 40 percent.

Availability factors have also steadily increased from a dismal 50-60 percent at the beginning of the decade to over 95 percent since 1985. Availabilities exceeding 98 percent are not uncommon at a number of wind power plants containing hundreds, even thousands, of wind turbines.

Despite lack of federal leadership throughout most of the decade, installed wind capacity in California reached 1,619 MW by the end of 1991. Installations have been uneven, peaking to meet the expiration of the federal investment tax credit in 1985, then gradually trending downward as financeable Standard Offer 4 (SO4) contracts were completed in the following six years.

California's base of approximately 16,000 wind turbines, concentrated primarily in mountain passes at Altamont, Tehachapi and San Gorgonio, has provided a laboratory for technology development. The costs of wind-generated electricity have declined approximately 75 percent since the early 80s, primarily because of operational learning by windfarm developers and their O&M personnel. Costs of O&M have declined from approximately four cents per kWh to around a penny per kWh, although costs vary considerably depending primarily upon the age of the project. Older projects involving smaller turbines are generally more expensive to maintain than post-1985 projects. Capital costs have likewise declined from $2,500-3,000 per kW of capacity to around $1,000 (with considerable variation depending upon project specifics). Costs of existing projects are typically six to nine cents per kWh depending upon wind regime, size of the project, project financing and transmission investment required.
Projects for construction in the mid-90s should be available for approximately $800 per kW of installed capacity with resulting costs of power in the 5-6 cents per kWh range for good wind regimes. Excellent wind sites in the Pacific Northwest have actually been bid at less than 4 cents per kWh, as evidenced by Bonneville Power Administration's 1992 wind solicitation.

Wind development flourished in California primarily because of progressive state policy epitomized by the SO4 contracts. These contracts were designed by California regulators to be compatible with the characteristics and unique financing needs of new renewable technologies. Unfortunately, no other state stepped forward with such innovative policies, and wind development has been virtually nonexistent in any other state except Hawaii, which has only about 35 MW of installed capacity. In 1993, however, the wind industry has seen evidence that development will soon be occurring in a number of regions beyond California.

EMERGING MARKETS FOR WIND TECHNOLOGY IN THE U.S.

As the cost of wind-generated electricity continues to decline, and utilities learn more about wind's improved technical performance, substantial progress toward wind energy development has been shown in three regions of the U.S.: the Pacific Northwest, the Upper Midwest and

the Northeast. All three regions share at least a modest need for new sources of electric power, but the more important factors contributing to the movement toward wind are:
1) utilities and state regulatory commissions that are "on the cutting edge" in their openness to new technology and 2) increased concern about the environment within each of the three regions, which has lead to more progressive energy policy decisionmaking.

The Pacific Northwest

In the Pacific Northwest, the Northwest Power Planning Council (NPPC) is made up of two representatives appointed by each of the governors of the four states in the region (Washington, Oregon, Idaho and Montana), and is authorized by federal law to oversee power planning for the region. One provision of the law gives a higher priority to renewable energy development than to conventional power resources. The NPPC laid the foundation for wind development with a "confirmation agenda" designed to identify the steps necessary to foster cost-effective development of wind and other renewable resources in the region. Through a collaborative process involving utilities, regulators and renewable industry experts, the confirmation agenda resulted in recommendations for specific steps to be taken in the areas of resource confirmation, permitting and siting, systems integration and research focussed on issues of concern to regional utilities such as cold weather operations and icing.

Although the NPPC has limited direct authority over the region's utilities, the regional planning process helped develop a consensus on the Northwest's energy future, including a major emphasis on renewable energy development. Many utility personnel retained a negative perception of wind technology based upon project failures such as the federal MOD-2 turbines that were erected in the Northwest, but those perceptions began to fade as utility managers learned more about the performance of newer generations of commercial wind turbines.

Last year witnessed dramatic progress toward wind development in the region as four of the largest utilities in the Northwest (Puget Sound Power & Light, PacifiCorp, Portland General Electric and Idaho Power) announced a commitment to a 50 MW project at Rattlesnake Mountain in eastern Washington in cooperation with U.S. Windpower. Project construction is expected to begin in 1994 and be completed by 1996. In addition, in June, 1992, the Bonneville Power Administration (BPA), a federal power marketing administration that serves about half the load in the region, announced a solicitation for up to 50 MW of windpower, with an expected on-line date of mid-1995. Bonneville received ten proposals totalling over 270 MW of wind capacity, selected two of them for the "short list" for negotiations for power purchase, and is negotiating with a number of others to provide transmission and other services to the utility partners involved.

With the announced 1996 closure of the Trojan nuclear plant and diminished hydro-based generation because of operational changes required to protect the region's remaining salmon, the Northwest is facing a rapidly growing need for power. As a result, the commitment to wind is expected to grow over the next few years. By way of example, PacifiCorp has announced plans to purchase up to 500 MW of wind capacity by 2001 (depending upon cost-effectiveness), and Portland General Electric, having seen the effectiveness of the BPA Wind Solicitation, has completed a renewable energy solicitation that will be used to purchase 200 MW of primarily wind or geothermal capacity. Publicly owned utilities in Washington state have also established the Conservation and Renewable Energy System (CARES) as a joint operating agency which will finance, design and implement conservation and renewable energy projects for its members. Wind projects are high on its list of renewable options, and CARES was one of the utilities to submit a winning proposal to BPA's wind solicitation.

AWEA held Windpower '92 in Seattle, Washington, for the purpose of showcasing wind technology for decisionmakers in the region, adding to the momentum that is already evident.

The Bonneville Power Administration's Wind Solicitation has brought about a fundamental change in the way wind is perceived by utilities in the Pacific Northwest. Bonneville's offer to purchase up to 50 MW of wind capacity from teams of utilities and developers forced utilities in the region to take a close look at the current status of wind technology. Prior to the solicitation, most utilities in the region considered wind an R&D technology, most appropriately treated through very small pilot projects. Virtually every major utility in the region is now involved in actively considering wind projects to provide commercial power.

The Upper Midwest

In the last year, commercial progress by wind technology in the Upper Midwest has been almost as dramatic as in the Pacific Northwest.

One of the most significant steps in the Midwest was taken when U.S. Windpower formed a joint venture with a subsidiary of Iowa-Illinois Gas & Electric to develop wind projects within a multi-state region. The joint venture, WindRiver Power Company, is now actively pursuing projects throughout the Midwest.

Minnesota, which has considerably more wind potential than California, has led the way in the region through years of resource assessment and other policies intended to make the state the center of Midwest wind energy development. Leaders in the state legislature have sponsored a series of pro-wind measures, including exempting wind projects from the state property tax and incorporating the environmental benefits of wind projects into the price paid to project developers. Although the environmental benefits of wind development have been an important justification for the legislation, the key appears to be economic development: making Minnesota the heart of the emerging wind industry in the region, and helping to create new and enduring jobs for the 21st Century.

Minnesota wind development received a boost on October 1, 1991, when Northern States Power (NSP), the state's largest utility, announced plans to construct a utility-owned 10 MW windfarm by 1996. The state government and the environmental community responded by pressuring the utility for a larger commitment of at least 50 MW. In August, 1992, NSP responded by announcing a commitment to at least 100 MW of wind capacity, with 25 MW in 1994, and at least 75 MW more by 1998.

A much smaller project, but significant nevertheless, is the small five-turbine windfarm that began operation in April, 1992 serving the Marshall, Minnesota, municipal utility. The contract with developer Minnesota Windpower provides the utility with only one million kWh annually, but small municipal or cooperative utilities could represent a growing market for wind. In total, municipal and cooperative utilities serve about 25 percent of the U.S. demand.

Minnesota is not the only state in the region that has shown interest in wind. The Wisconsin Public Service Commission modified the long range plans of the state's utilities to require 811 MW of renewable energy by 2010, with wind being one of the most likely renewable options. In May, 1993, the Commission issued an order that would provide wind and solar projects with an additional incentive of .75 cents per kWh for utility shareholders.

In addition to its mini-PURPA law (which requires the state's utilities to collectively purchase at least 105 MW of renewable energy capacity), Iowa passed a law in May, 1993 that exempted wind energy conversion property from property (at local option), sales, service and use taxes. According to state Rep. David Osterberg "the new law made sense because it will make wind cheaper vis-a-vis other energy sources. Hopefully we will help change utility investment behavior. Secondly, the new law is significant because the Iowa Legislature has, in effect, endorsed wind power."

The Northeast

Firm commitments to wind projects in the Northeast lag behind the other two regions, but there has been clear progress in the last year nonetheless. New England Power (NEP) has been in the forefront of many changes sweeping the utility industry, including interest in renewable technologies. In December, 1991, NEP issued a "Green Solicitation" for 45 MW of renewable capacity, specifically identifying wind as one of the priorities for the solicitation.

In response to the solicitation, Green Mountain Power (GMP) proposed three specific projects in Southern Vermont and Western Massachusetts totalling 40 MW. GMP has been the most active utility in the Northeast in terms of involvement with windpower, having monitored wind resources for a decade, and invested in California windfarms through a subsidiary.

According to reports published by the Battelle Pacific Northwest Laboratory for the U.S. Department of Energy, Maine has the largest wind resource in New England. Currently, Endless Energy, a small developer, has a research contract with Central Maine Power (CMP) for 15 MW that must be developed by 1994. U.S. Windpower has also been active in the state, proposing a 50 MW project in western Maine to CMP. A 20 MW portion of the project has been tentatively selected by New England Power as one of the winners for their Green Solicitation.

The state of New York has also begun to take steps to harvest more of its large renewable energy potential. In 1991-92, the state updated its energy plan, incorporating a proposal that requires the state's utilities to purchase a total of 300 MW of renewable energy capacity by

1994, with projects on line by 1998. The proposal has been referred to state regulators for implementation, but has been strongly resisted by the state's utility industry.

AWEA's POLICY AGENDA TO CONTINUE WIND'S U.S. MARKET EXPANSION

AWEA advocated a four part program in 1992 designed to open new markets for wind energy technology:

I. Production Incentives

AWEA's priority has been tax equity--specifically, providing incentives for wind energy development that create more of a "level playing field" with conventional technologies, which receive billions of dollars in tax subsidies annually. The vehicle chosen was a production incentive for each kWh of wind-generated electricity from new projects.

Although few observers gave the production incentives campaign much chance of being successful when it was initiated, a 1.5 cent per kWh production incentive for wind became law through the National Energy Policy Act of 1992. The production incentive will go into effect at the beginning of 1994 and should provide a significant impetus to the market.

II. Transmission Access

The bulk of the U.S. wind resource is in the Great Plains, far removed from load centers on either coast. Access to transmission lines to bring wind-generated electricity to load centers is essential if we hope to see wind energy reach anything close to its potential.

The National Energy Policy Act included transmission access provisions that were actively supported by AWEA that would substantially increase access to transmission and would represent a boost to the wind energy industry.

III. Enhanced Research and Development

The federal wind program has never fully recovered from the 90 percent reduction in funding that it suffered during the 80s. Despite a doubling of the wind budget in the last two years, the federal commitment to wind is only $23.4 million for FY '93, a fraction of the level of the wind technology development efforts that have been funded in western Europe. Sophisticated R&D will be essential for wind technology to achieve the cost reduction targets that are feasible. The Clinton Administration has proposed somewhat greater support, increasing the wind budget to slightly over $30 million for FY '94. In contrast, AWEA recommends funding of $80 million, with approximately half of that devoted to commercialization.

IV. Regulatory Reform

Wind and other intermittent renewable technologies are fundamentally different than conventional technologies. The decisionmaking criteria used by most utilities and utility regulators reflect the characteristics of conventional technologies and don't take into consideration either the full costs of conventional technologies (externalities) or all the benefits that renewable technologies bring to utility systems. Unless there is a systematic redesign of the utility resource planning and selection process, wind and other renewables will continue to be disadvantaged in the energy market.

Regulatory reform has focused primarily on energy efficiency and demand-side management in the last five years, but with so much of that agenda having been accomplished, more attention is being turned to the impact the regulatory process and utility resource planning have on renewables.

With progress on tax equity and transmission access in 1992, regulatory reform has become the heart of AWEA's political agenda in 1993.

V. Commercialization Initiative

In the 1980s, the U.S. made a multi-billion dollar commitment to develop less polluting methods of using coal to generate electricity through a cost-shared "Clean" Coal Technology Program. With only a fraction of the financial commitment, the U.S. can take advantage of a near-term cost effective technology which emits no pollutants--wind energy technology.

Despite its relatively new status (the first windfarms became operational in California in 1981), wind energy has become the most cost-effective, non-hydro renewable energy resource. As part of its greenhouse gas mitigation strategy, the European Community has set a goal of 8,000 MW of wind capacity by 2005. If the U.S. expects to rely on renewables to any appreciable extent in achieving its year 2000 CO_2 emissions reductions, wind energy is the technology of choice and should be the focus of an aggressive collaborative commercialization effort between the public and private sectors.

AWEA proposes that the Clinton Administration set a target of 10,000 MW of wind capacity in place in the U.S. by the end of the decade, and that DOE work collaboratively with the wind industry, utilities and the states to accomplish that objective. A modest 10 percent federal cost share should be sufficient incentive for utilities with good wind resources to get involved in the collaborative.

CONCLUSION

Since the first windfarms were developed in 1981, wind technology has made tremendous progress in terms of both technical performance and cost-effectiveness. Despite this progress the market for wind in the U.S. has been confined almost entirely to California. However, 199[illegible] has provided clear evidence that the market for wind technology is emerging in at least three other regions of the U.S. This trend could be markedly accelerated by more effective policy leadership at the federal and state levels, including an active partnership with the wind industry to win widespread commercial development of the most cost-effective of the solar electric technologies.

Wind Energy Potential in the United States

Dennis Elliott
Staff Scientist
Pacific Northwest Laboratory
Richland, Washington, U.S.A.

Background:

Mr. Elliott has been actively involved in wind energy for 15 years and is recognized internationally for his contributions in assessing the wind energy resource on regional, national, and global scales.

WIND ENERGY POTENTIAL IN THE UNITED STATES

D. L. Elliott and M. N. Schwartz
Pacific Northwest Laboratory
P.O. Box 999
Richland, WA 99352, USA

ABSTRACT

Estimates of the electricity that could potentially be generated by wind power and of the land area available for wind energy development have been calculated for the contiguous United States. The estimates are based on published wind resource data and exclude windy lands that are not suitable for development as a result of environmental and land-use considerations. Despite these exclusions, the potential electric power from wind energy is surprisingly large. Good wind areas, which cover 6% of the contiguous U.S. land area, have the potential to supply more than one and a half times the current electricity consumption of the United States. Technology under development today will be capable of producing electricity economically from good wind sites in many regions of the country.

INTRODUCTION

A wind energy resource atlas of the United States [1] shows that areas potentially suitable for wind energy applications are dispersed throughout much of the United States. Estimates of the wind resource in this atlas are expressed in wind power classes ranging from class 1 to class 7, with each class representing a range of mean wind power density or equivalent mean speed at specified heights above the ground (Table 1). Areas designated class 4 or greater are suitable with advanced wind turbine technology under development today. Power class 3 areas may be suitable for future generation technology (year 2000 and beyond). Class 2 areas are marginal and class 1 areas unsuitable for wind energy development. Maps of the areal (percentage of land area) distribution of the wind resource digitized in grid cells of 1/4° latitude by 1/3° longitude (Figures 1 and 2) show that exposed areas with moderate to high wind resource are dispersed throughout much of the contiguous United States.

METHOD OF CALCULATING THE WIND ELECTRIC POTENTIAL

Several factors determine the amount of land area suitable for wind energy development within a particular grid cell in a region of high wind energy potential. The important factors include the percentage of land exposed to the wind resource and land-use and environmental restrictions. The land area exposed to the wind for each grid cell was estimated based on a landform classification and ranged from 90% for relatively flat terrain down to 5% for mountainous terrain. Estimates of land area excluded from wind energy development, in percent per grid cell, were made for various types of land-use (e.g., forest, agricultural, range, and urban lands). Environmental exclusion areas were defined as federal and state lands (including parks, monuments, wilderness areas, wildlife refuges, and other protected areas) where wind energy development would be prohibited or severely restricted. Finally, additional land exclusions were estimated to account for transportation right-of-ways, locally administered park land, privately administered areas, and proposed environmental lands.

The wind electric potential per grid cell was calculated from the available windy land area and the wind power classification assigned to each cell. The amount of potential electricity that can be generated is dependent on several factors, including the spacing between wind turbines, the assumed efficiency of the machines, the turbine hub height, and the estimated energy losses (caused by wind turbine wakes, blade soiling, etc.). The assumptions used for calculating the wind energy potential per unit of windy land area is given in Table 2. (Estimates of wind turbine efficiency and power losses are based on data from existing turbines. For advanced turbines, efficiency is projected to be 30-35% and power losses 10-15%).

RESULTS

A study in 1991 [2] used a variety of land exclusion scenarios to estimate the available windy land and resultant wind electric potential for several levels of wind resource, both for the contiguous United States as a whole and for each of the 48 contiguous states. Figure 3 shows the overall wind energy potential based on terrain exposure and several land exclusion scenarios. Even under the most restrictive (severe) land exclusion scenario, the wind energy potential for areas with power class 3 and above is greater than total U.S. electricity consumption was in 1990. Figure 4 shows the contribution that the wind energy of each state could make to meet the total electrical needs of the nation, assuming a moderate land exclusion scenario. North Dakota alone has enough potential energy from windy areas of class 4 and higher to supply 36% of the total 1990 electricity consumption of the 48 contiguous states.

A study in 1992 [3] recalculated estimates of windy land area and wind electric potential based on a more accurate mapping of environmental exclusion areas and a moderate land exclusion scenario. These new estimates were about 1 to 2% higher than the preliminary estimates made in 1991 based on the same land exclusion scenario. Overall, even with land being excluded from wind energy development for environmental and land-use considerations, the amount of windy land available and potential electric power from wind energy is surprisingly large. The amount of windy land available for power class 4 and above is approximately 460,000 square kilometers, or about 6.0% of the total land area in the contiguous United States. The potential average power from areas with class 4 and higher, which are suitable for development with advanced wind turbine technology, is estimated at 500,000 MW. If future generation technology is utilized to take advantage of areas with wind resource class 3 and higher, than the amount of windy land available is over 1,000,000 square kilometers, or almost 14% of the land area in the contiguous United States. Moreover, the estimates show that a group of 12 states in the midsection of the country have enough wind energy potential to produce nearly four times the amount of electricity consumed by the nation in 1990.

Although the nation's wind potential is very large, only part of it can be exploited economically. The economic viability of wind power will vary from utility to utility. Important factors not addressed in this study that influence land availability and wind electric potential include production/demand match (seasonal and daily), transmission and access constraints, public acceptance, and other technological and institutional constraints.

To provide 20% of the nation's electricity, only about 0.6% of the land of the lower 48 states would have to be developed with wind turbines. Furthermore, less than 5% of this land would be occupied by wind turbines, electrical equipment, and access roads. Most existing land use, such as farming and ranching, could remain as it is now.

CONCLUSION

The considerable wind electric potential has not been tapped before because the wind turbine technology was not able to utilize this resource. However, during the past decade, increased knowledge of wind turbine behavior has led to more cost-effective wind turbines that are more efficient in producing electricity. The price of the electricity produced from wind by these advanced turbines is estimated to be competitive with conventional sources of power, including fossil fuels. Because of the increasing competitiveness of wind energy, wind resource assessment will become essential in incorporating wind energy into the nation's energy mix. For example, the 1992 study [3] also produced gridded maps of wind electric potential and windy land area for the 48 contiguous states that show the distribution of these features within individual states, thus enabling utilities and wind energy developers to identify promising areas for wind energy. The importance of accurate wind resource assessment is also recognized in other parts of the world.

Detailed wind resource assessments have been proposed or are being considered as part of a plan to increase the use of wind energy in Europe, Asia, Latin America, and other regions. The decreasing cost of wind power and the growing interest in renewable energy sources should ensure that wind power will become a viable energy source in the United States and worldwide.

ACKNOWLEDGMENTS

We wish to thank Gene Gower for his efforts on the development of the computer software for the project. This work was supported by the U.S. Department of Energy (DOE) under Contract DE-AC06-76RLO 1830 at the Pacific Northwest Laboratory, which is operated for DOE by Battelle Memorial Institute.

REFERENCES

1. Elliott, D. L., C. G. Holladay, W. R. Barchet, H. P. Foote, and W. F. Sandusky. 1987. *Wind Energy Resource Atlas of the United States.* DOE/CH 10093-4, Solar Energy Research Institute, Golden, Colorado.

2. Elliott, D. L., L. L. Wendell, and G. L. Gower. 1991. *An Assessment of the Available Windy Land Area and Wind Energy Potential in the Contiguous United States.* PNL-7789, Pacific Northwest Laboratory, Richland, Washington.

3. Schwartz, M. N., D. L. Elliott, and G. L. Gower. 1992. "Gridded State Maps of Wind Electric Potential." In Proceedings Windpower'92, pp. 50-58. American Wind Energy Association, Washington, D.C.

TABLE 1: CLASSES OF WIND POWER DENSITY

Wind Power Class	10 m (33 ft)[a]		30 m (98 ft)[a]		50 m (164 ft)[a]	
	Wind Power Density, W/m^2	Speed[b], m/s (mph)	Wind Power Density, W/m^2	Speed[b], m/s (mph)	Wind Power Density, W/m^2	Speed[b], m/s (mph)
	0	0	0	0	0	0
1						
	100	4.4 (9.8)	160	5.1 (11.4)	200	5.6 (12.5)
2						
	150	5.1 (11.5)	240	5.9 (13.2)	300	6.4 (14.3)
3						
	200	5.6 (12.5)	320	6.5 (14.6)	400	7.0 (15.7)
4						
	250	6.0 (13.4)	400	7.0 (15.7)	500	7.5 (16.8)
5						
	300	6.4 (14.3)	480	7.4 (16.6)	600	8.0 (17.9)
6						
	400	7.0 (15.7)	640	8.2 (18.3)	800	8.8 (19.7)
7						
	1000	9.4 (21.1)	1600	11.0 (24.7)	2000	11.9 (26.6)

[a]Vertical extrapolation of wind power density and wind speed are based on the 1/7 power law.
[b]Mean wind speed is estimated assuming a Rayleigh distribution of wind speeds and standard sea-level air density. The actual mean wind speed may differ from these estimated values by as much as 20%, depending on the actual wind speed distribution and elevation above sea level.

TABLE 2: WIND ENERGY POTENTIAL PER SQUARE KILOMETER OF LAND AREA

Assumptions: 50-m hub height, 10 D x 5 D spacing, 25% efficiency, and 25% power losses

Power Class	Wind Power Density, W/m^2	Average Power Intercepted, MW/km^2	Average Power Output, MW/km^2	Annual Energy Production million kWh/km^2
3	350	5.50	1.03	9.02
4	450	7.07	1.33	11.65
5	550	8.64	1.62	14.19
6	700	11.00	2.06	18.04
7	900	14.14	2.65	23.21

FIGURE 1: PERCENTAGE OF LAND AREA WITH CLASS 4 OR HIGHER WIND PESOURSE

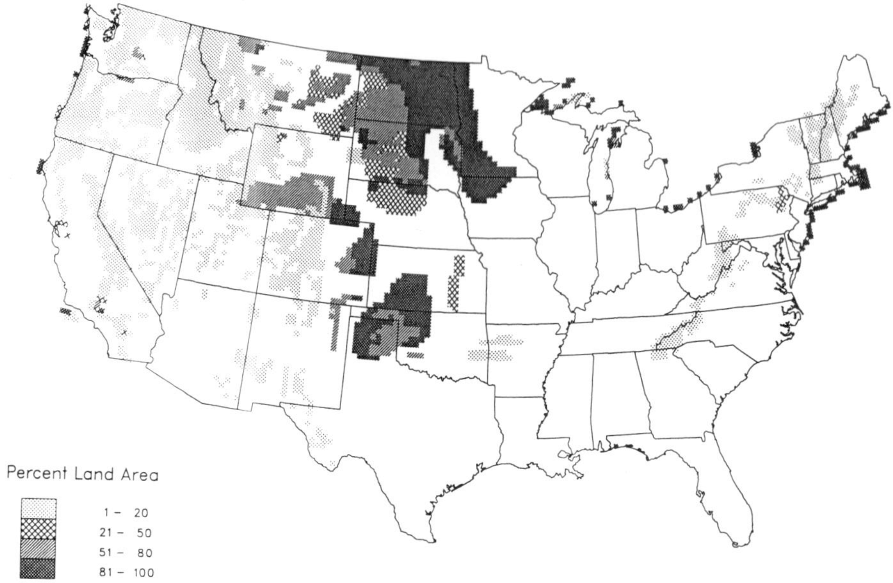

FIGURE 2: PERCENTAGE OF LAND AREA WITH CLASS 3 OR HIGHER WIND PESOURSE

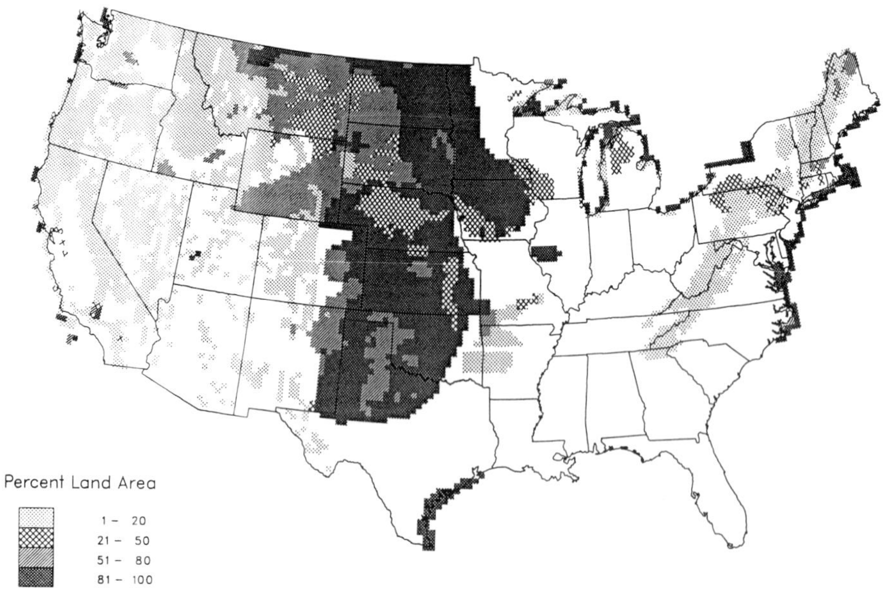

FIGURE 3: WIND ENERGY POTENTIAL OF THE CONTIGUOUS UNITED STATES

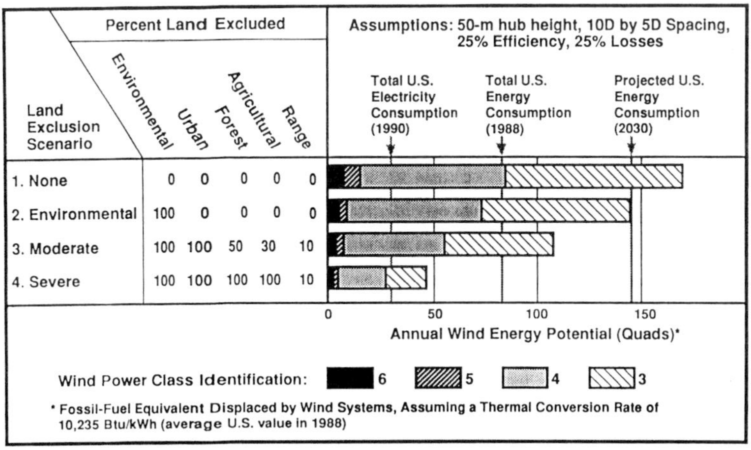

FIGURE 4: PERCENTAGE OF THE TOTAL 1990 ELECTRICITY CONSUMPTION OF THE CONTIGUOUS UNITED STATES THAT EACH STATE COULD PRODUCE (ASSUMPTIONS: AREAS DESIGNATED CLASS 4 AND ABOVE, HUB HEIGHT OF 30 M, MODERATE LAND EXLUSION SCENARIO)

Specifications: Wind resource ≥ Class 4 at 30m (≥ 320 W/m²), 30m hub height, 10D x 5D Spacing, 25% Efficiency, 25% Losses

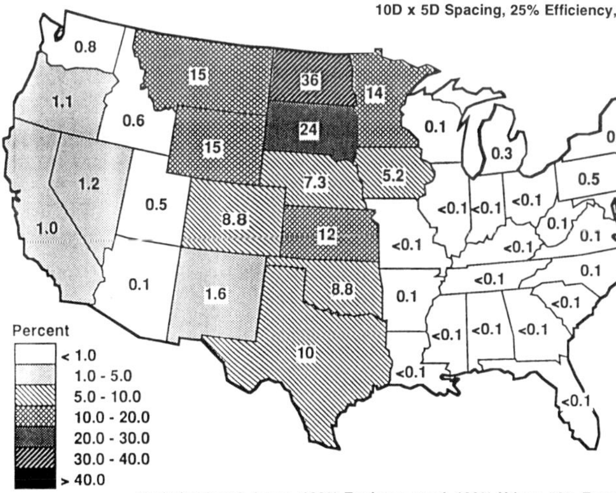

Excluded Land Area: 100% Environmental, 100% Urban, 50% Forest, 30% Agricultural, 10% Range

Wind Energy Technology Development

Susan M. Hock
Wind Energy Program Manager
National Renewable Energy Laboratory
Golden, Colorado, U.S.A.

Background:

Ms. Hock has been involved in wind energy research since 1979. She has a broad background in systems analysis for several solar technologies, as well as wind turbine testing experience, data reduction/analysis, systems performance, economics estimating, and program planning.

Abstract:

The U.S. Department of Energy is sponsoring the development of technologically advanced, high-efficiency wind turbines to provide cost-effective bulk electric power.

No paper. Oral presentation only.

Wind: The Renewable of Immediate Opportunity

Charles Linderman
Director of Fossil Fuels & Renewable Programs
Edison Electric Institute
Washington, D.C., U.S.A.

Background:

In addition to his work at Edison Electric Institute, Mr. Linderman serves as the Treasurer of the Utility Photovoltaic Group and as a member of its Board of Directors. He has been involved in major federal activities concerning energy and the environment, and organized and directed, through the Fossil and Synthetic Fuels Committee of the Institute, an analysis of natural gas demand for electric generation before 2000.

Abstract:

A discussion of near-term opportunities for wind energy in the U.S., a potential program to utilize wind to reduce requirements for agricultural subsidies, and the role of wind in assisting the electric industry on competitive issues.

No paper. Oral presentation only.

CHAPTER 5 -- UTILITY ENERGY STRATEGIES -- PART 1

Environmental Implications of Electric Energy Supply

Thomas D. Morron
Vice President, Customer Services and Marketing
Edison Electric Institute
Washington, D.C., U.S.A.

Background:

In addition to his work at Edison Electric Institute, Mr. Morron serves on the boards of directors of the Electric Vehicle Association of the Americas, the National Food and Energy Council, and the International District Heating and Cooling Association, among others. He is respected nationally as an expert on energy and environmental issues, and represents the utility industry in matters of end-use efficiency on various U.N. and U.S. government panels.

ENVIRONMENTAL IMPLICATIONS OF ELECTRIC ENERGY SUPPLY

Thomas D. Morron
Vice President, Customer Service & Marketing
Edison Electric Institute
701 Pennsylvania Avenue, NW
Washington, D.C. 20004-2696

Good morning, ladies and gentlemen.

First, I would like to thank our gracious hosts for convening this meeting. It is appropriate and timely that representatives of this great industry should meet to discuss what we are doing, and can do, to reduce emissions in the post-Rio era.

It is appropriate because the form of energy we produce and distribute, electricity, is uniquely suited to improving the environment. I say "uniquely suited" because electricity is the cleanest form of energy at the point of use, it is the most flexible and adaptable form of energy, it offers dramatic efficiency gains through continual improvements in applications technologies, and it can be produced from a variety of sources.

As you know, U.S. electricity is produced from a variety of sources. In 1991, our fuel mix was 55 percent coal, 22 percent nuclear, 10 percent hydro, 9 percent gas, 4 percent oil, and less than one-half of 1 percent renewables.

Both for environmental and economic reasons, our use of natural gas for generation has been growing rapidly. We expect at least 33 percent of the additional demand for electricity in this decade to be met by gas. Some estimates run as high as 50 percent.

And yet, as attractive as gas is, we are encountering some difficulties in achieving a smooth interface with the gas suppliers. These problems deal primarily with deliverability and price volatility.

Indeed, every energy source for generation has its own advantages and disadvantages. And that fact causes U.S. electric utilities to favor a diversified fuel mix.

We have learned from experience that government attempts to promote one fuel over another often create more problems than they resolve. Before the first OPEC oil embargo, for example, our government encouraged us to switch from coal to oil; after the embargo, the government wanted us to switch from oil to gas. When gas shortages developed, we were told to stop using gas. Now, we are being encouraged to use more gas. The official signals on nuclear energy have been equally changeable.

Because U.S. electric utilities take very seriously their obligation to provide a reliable, economical electricity supply, we are -- and will remain -- reluctant to put too many eggs in one fuel basket.

We believe that this emphasis on diversity actually contributes to a cleaner environment, in addition to its other benefits. I realize that statement will be questioned. Please bear with me while I explain.

The emphasis of this conference program is on the environmental effects of Producing electricity. That is only one-half of the equation. To achieve a properly balanced perspective, we also must consider the environmental benefits that using electricity can provide.

For the United States, EEI expects electricity demand to grow at an average of 1.9 percent a year for the rest of this decade. Much of this growth is structural -- users increasingly are substituting electricity for fossil fuels in end-use applications. As this happens, the proportion of primary fuels used to make electricity rises. For example, the fraction of total primary energy consumption used to generate electricity in the U.S. has grown from 27 percent in 1974 to almost 38 percent in 1990. By 2010, it will be about 44 percent.

This substitution effect has very significant environmental benefits. The obvious benefit is that an emission-free form of energy is substituted for burning a fossil fuel at the end-use point. The less obvious, but more important, benefit is that end- use electric technologies often are so efficient that changing to them saves energy, even after allowing for the energy losses in generation and transmission. And if less energy is used, fewer emissions are produced, including emissions of greenhouse gases.

Here are a few specific examples: Producing steel with an electric arc uses only half of the energy of a blast furnace. Melting glass with electricity instead of gas uses 65 percent less energy. Drying paint with infrared electric heat uses 90 percent less energy. And for space heating, a heat pump delivers two or three times more energy than it consumes. There are many more examples. And their effects add up.

For example, since 1973, the U.S. has increased its production of goods and services by 47 percent, while using less than 10 percent more energy from all sources combined. We did it by using 61 percent more electricity. At the same time, the rate of carbon dioxide production dropped from four pounds per dollar of GNP in 1973 to a little more than two pounds per dollar in 1990. Let me repeat that very important point: as the U.S. economy has become more electricity-intensive, the rate of carbon dioxide production has gone down.

As you will understand, U.S. electric utilities are proud of that record and want to see such progress continue. What we fear, however, is that attempts to change the fuel mix on environmental grounds could raise the price of electricity. And raising the price of electricity might discourage switching from other fuels to electricity at the end-use point. In other words, with the best of intentions, we could inadvertently reverse that long trend toward using energy more efficiently by using more electricity.

With that background, perhaps you can understand why the U.S. government and U.S. electric utilities have not favored policies that could discourage switching to electricity.

Of course, environmental progress is possible -- and desirable -- on many fronts. U.S. electric utilities support efforts to use fossil fuels in generation more efficiently and with fewer undesirable environmental effects; we favor expanding the role of nuclear power, and we are following a plan to do that; we favor development and use of electric vehicles; and we favor aggressive development of renewable generating technologies, to bring down their costs. I will discuss each of these approaches in turn.

Our efforts to reduce the undesirable environmental effects of fossil fuel use focus mainly on coal, for obvious reasons: it is 55 percent of our fuel mix, natural gas is relatively clean already, and U.S. electric utilities use very little oil.

U.S. utilities are working with the U.S. Department of Energy's Clean Coal Technology Program to demonstrate that improved coal-burning plants can be operated reliably at commercial scale and under realistic conditions. Clean coal technologies are proving to be vastly superior to flue gas "scrubbers," the most familiar approach to controlling power plant emissions. And the potential of these technologies can help to address global environmental concerns, as other coal-rich countries begin to draw more heavily on their reserves.

Clean coal technologies range from the basic, such as improved coal cleaning, to the exotic, such as magneto hydrodynamics. Among the more promising are fluidized-bed combustion and coal gasification combined-cycle systems. Using these technologies, up to 99 percent of potential sulfur dioxide emissions can be eliminated. Further, several of the technologies also reduce the formation of nitrogen oxides.

With methods like fluidized-bed combustion and coal gasification, coal-fired generating efficiencies may approach 60 percent by the year 2020. And when fully realized, advanced coal gasification plants will co-produce electricity and such valuable chemicals as hydrogen, methanol, and gasoline, while nearly eliminating gaseous emissions and solid wastes.

Of course, the U.S. already has made significant progress in controlling fossil fuel emissions, under some very strong emissions control laws.

The basic law regulating air quality is the Clean Air Act of 1970, which was amended in 1977, and again in 1990.

Under this law, several major categories of clean air regulation apply to electric utility operations. These are: New Source Performance Standards, Prevention of Significant Deterioration review, the National Ambient Air Quality Standards program for reviewing sources in "non-attainment" areas, and the visibility protection program for major existing sources.

New Source Performance Standards for new facilities such as power plants are among the most stringent sulfur dioxide and nitrogen oxide limits in the world.

Prevention of Significant Deterioration is an additional level of regulation. It ensures that even in areas where clean air standards are being met, new or substantially modified sources are not allowed to cause air quality degradation.

Six pollutants now have National Ambient Air Quality Standards: Particulate matter, sulfur dioxide, nitrogen dioxide, carbon monoxide, ozone, and lead.

These laws already have had a positive effect on our air quality. The large contribution made by electric utilities can be illustrated with just two striking facts: First, according to the U.S. Environmental Protection Agency, utility sulfur dioxide emissions decreased 19 percent between 1973 and 1989. And second, this decrease took place even though utility coal use increased 96 percent during the same period.

The 1990 amendments further tightened the emissions limits. They require electric utilities to reduce sulfur dioxide emissions by almost 50 percent from 1985 levels by the year 2000. Starting in 2000, total sulfur dioxide emissions will be capped. This means there can be no net increase even if electricity demand continues to grow -- and it will. The emissions cap affects the use of oil as well as coal. Significant reductions in emissions of nitrogen oxide also must be achieved.

Electric utilities will spend $4 to $5 billion a year to comply with the 1990 regulations.

To moderate the impact of these regulations on the hardest-hit companies, and to provide an incentive for exceeding the requirements, the 1990 amendments introduced a system of emissions allowances. The allowances are being awarded to all existing utility power plants. They can be bought and sold by utilities to help them meet emissions requirements. That is, a utility that exceeds the requirements may sell its unused allowances to a utility that is finding compliance more difficult. In this system, the power of the free market is used to meet the national targets in the most cost-effective manner.

Given this record of progress, we believe the U.S. is demonstrating that the world's valuable fossil fuel resources can be used in an environmentally responsible manner. But fossil fuel emissions controls are not the only environmental option we are pursuing.

To expand the role of nuclear energy in power generation, the Administration, Congress, and the nuclear industry have been cooperating closely.

The report on the National Energy Strategy, completed by the Department of Energy (DOE) in the spring of 1991, recommends a doubling of U.S. nuclear generating capacity by the year 2030. To achieve this goal, it identifies a need for plant standardization and for improvements in the licensing process.

As a follow on, DOE is co-funding design certification programs for standard plants and for doing the first-of-a-kind engineering. This second step is intended to achieve near complete designs that will provide a firm basis for construction cost and time schedules before an order is placed.

To improve the nuclear plant licensing process, the Nuclear Regulatory Commission has defined a new licensing approach for standard plants, and Congress has provided guidance to strengthen and improve it in a major new Energy Policy Act that became law just last month.

The new Energy Act also establishes a government corporation to run uranium enrichment programs.

In parallel with the government efforts, the nuclear industry has developed a Strategic Plan for Building New Nuclear Power Plants. This detailed plan has an overall goal of getting new nuclear plants into operation around the turn of the century. It addresses plant operation, standardization, design certification, licensing, financing, state regulatory treatment, fuel supply, waste disposal, and public information. The industry has been pursuing this plan for two years.

One area of disappointment for us is the lack of satisfactory progress on nuclear waste disposal. Congress assigned the responsibility to DOE for developing a deep underground geologic repository for high-level radioactive waste. Currently, the Yucca Mountain site in Nevada is being studied to determine if it is suitable for a repository. This process has been slowed by objections from the State of Nevada, delays in issuing environmental permits, and limits on the use of appropriated monies from the Nuclear Waste Fund. The latest estimate of when a repository will be available is 2010.

U.S. utilities are willing to have the government dispose of spent fuel without reprocessing. This is an economic decision that recognizes the availability of uranium at a reasonable cost for the next several decades. Because the cost of disposal of the highly radioactive elements separated by reprocessing is about the same as the cost of disposing of spent fuel, the U.S. is forgoing the reprocessing step for the time being.

At the other end of the wire, the U.S. commitment to electric vehicles is growing in strength almost daily.

As many of you know, the State of California in 1990 adopted a law requiring that a specified percentage of vehicles sold there in the future be zero emission vehicles. The requirement is 2 percent for 1998 through 2000, 5 percent for 2001 and 2002, and 10 percent thereafter. This law has since been adopted by New York and Massachusetts. Currently, it is under consideration in six other states.

Further, all three U.S. automobile manufacturers have EV programs under way, and have joined in a coalition with the U.S. government and electric utilities to develop improved battery technology.

In addition to these earlier developments, the new Energy Policy Act has numerous EV incentives. Here are the highlights:

- A multi-year, $485 million EV and battery research and development program within DOE to be conducted in cooperation with the electric utility, automotive, battery, and other industries.

- A 10-year, $50 million EV demonstration program.

- A five-year, $40 million program to enter into joint ventures with industry to demonstrate the infrastructure required to operate and maintain EVs.

- A five-year, $50 million program to assist state governments in developing and implementing incentive programs for alternative fuels and alternative fueled vehicles.

- A 10 percent federal tax credit on the purchase price of an EV, up to $4,000 per vehicle.

- A $100,000 federal tax deduction for clean-fuel refueling facilities, including EV facilities.

- And clean-fuel federal tax deductions with maximums ranging from $2,000 to $50,000 per vehicle, depending on vehicle size.

- The Act also establishes percentage requirements for adding alternative-fueled vehicles to government fleets and the fleets of fuel providers. These requirements are phased in from 1993 through 2006.

As you can see, the U.S. has laid the groundwork for dramatic progress in the development and use of electric vehicles.

Now let's look at the outlook for renewable technologies in the U.S. Although their present contribution to our fuel mix is modest, they eventually will play a more important role. Toward that end, the Energy Policy Act provides additional federal support. A key element is a production incentive entailing a direct federal payment of up to 1.5 cents per kilowatt hour for electricity from such sources. The Act also authorizes joint ventures between the federal government and the private sector to develop renewable technologies and applications, and promotes the export of renewable technologies to developing countries.

Hydroelectric generation currently is the leading renewable energy source, supplying about 10 percent of U.S. electricity. It is safe, clean, efficient, and relatively inexpensive. However, its use in the U.S. is expected to grow only modestly in the years ahead, because most sites suitable for large-scale hydro plants already have been developed. Technological advances, however, are making it possible to extract ever greater amounts of energy from existing water resources, thus increasing electricity output without building new dams.

Nevertheless, there is a risk that hydro energy production in the U.S. may actually decline because of the difficulty of relicensing existing facilities. Most hydro facilities must be federally licensed, and many of the existing licenses will expire over the next few years. Of 12 hydro projects already relicensed by the federal government in 1987 and 1988, eight were required to lower their output to accommodate other environmentally compatible uses of their waterways.

Wind power is among the more fully developed of the renewable technologies. By 1989, total world capacity had reached an estimated 2,797 megawatts, with 76 percent of it in California. However, interest is growing among utilities in other parts of the U.S., and wind projects have been announced in Washington, Minnesota, and New England.

Geothermal energy has great promise, because it is dispatchable and can be competitive in price with conventional fuels. But capital costs for geothermal plants are high, and there have been disappointments. Some of the first steam reservoirs to be tapped are declining rapidly. Research now taking place attempts to learn more about the characteristics of geothermal reservoirs.

There is great interest today in harnessing solar energy as a source of electricity. As early as 1983, U.S. electric utilities were involved in some 870 solar energy projects.

In 1990, 354 megawatts of solar thermal electricity -- more than 90 percent of the world's solar generating capacity -- was being produced by one company in Southern California and sold to utilities there. The cost of electricity made with the trough-style collectors used in this system has fallen from 24 cents per kilowatt hour in 1984 to about 8 cents for units entering service in 1989.

In 1991, a consortium of U.S. utilities announced plans to test a 10-megawatt bulb-type collector using molten nitrate salt as the heat transfer medium. This system is expected to be more competitive with fossil fuels than trough collectors using oil to transfer heat, and it may even allow solar heat to be stored for use after sundown.

Despite its many attractive features, solar photovoltaic electricity is not competitive in today's bulk power market. Nevertheless, the industry is building a base of experience with it. A 1989 survey by the Electric Power Research Institute found 219 photovoltaic plants feeding electricity to utility grids in the U.S. Their total capacity, however, is only

11.6 megawatts, and just three plants account for 9.2 megawatts. But research has been bringing down the cost, and experts predict that this progress will continue.

One U.S. utility, Idaho Power, recently announced a plan to design, own, and maintain photovoltaic systems installed at remote sites that otherwise would require expensive line extensions to provide service. Customers using this service will pay Idaho Power a fixed monthly facilities charge, rather than being metered and billed for the energy consumed.

To help lower the cost of photovoltaic power, U.S. utilities have formed a coalition to undertake an 18-month program to integrate photovoltaics into utility operations. Forty utilities are participating, along with representatives from state governments and the federal government.

Of course, new energy sources and new generating technologies are only part of the story. U.S. electric utilities are pursuing two strategies to defer the need for new power plants. One is to help electricity consumers use energy more efficiently. Improving the insulation of homes and encouraging the use of efficient appliances are examples of this approach. The second is to level out demand by shifting some of the peak load use to other times of the day.

Collectively, these strategies are known as "demand-side management," or DSM. Utilities, energy planners, and utility regulators throughout the U.S. are embracing DSM programs. This year, the U.S. electricity industry is spending more than $2.5 billion on about 2,000 of these programs. To date, they have deferred the need for almost 25,000 megawatts of generating capacity. By the year 2000, the amount of deferred capacity is expected to reach 45,000 megawatts.

The new Energy Policy Act will encourage energy efficiency investments by exempting from federal taxes conservation rebates offered to residential customers by electric utilities. Rebates for commercial and industrial customers will receive a partial exemption.

By practicing demand-side management, U.S. utilities are deferring the need for the capital to build new generating plants, and saving the fuel that would have been consumed, and the emissions that might have been produced.

In conclusion, I see electric utilities on the positive side of the environmental issue -- viewed as being part of the solution, rather than a cause of the problem. We look forward to increasing our cooperation with electric utilities in other countries, so that we may continue to progress in this important area.

Government & Electric Utility: Can Environmental Goals Be Mutually Beneficial?

Ronald Drewnowski
Manager, Project Development
Public Service Electric & Gas of New Jersey
80 Park Plaza, PO Box 570
Newark, New Jersey, 07101, U.S.A.

Background:

Formerly Manager of Nuclear Systems Engineering and Environmental Engineering, Mr. Drewnowski is currently working to develop supply-side strategies at Public Service Electric and Gas Company, New Jersey.

This paper was prepared jointly with:

Walter P. Drabinski
President
Vantage Consulting, Inc.
11 West Avenue, PO Box 40
Wayne, Pennsylvania, 19087, U.S.A.

ABSTRACT

The topic of this discussion focuses on the relationship of utilities and regulators in today's competitive environment. We at Public Service Electric and Gas believe that the achievement of environmental goals may be consistent with a utility's vision and long term competitive objectives. As an example, we illustrate our own relationship with the New Jersey Department of Environmental Protection and Energy (NJDEPE). Specifically, Public Service Electric and Gas has established a relationship with its regulators that recognizes the mutual benefits to be derived from meeting environmental goals. While developing plans to meet future energy needs and to comply with federal and state mandates spelled out in the 1990 Clean Air Act and the 1992 Energy Policy Act, the Company has taken a number of bold steps. The result is a plan that provides cost effective, reliable and environmentally sound service to our customers. There are five elements to PSE&G's strategy: Demand Side Management to minimize peak load growth; conservation; utilization of non-utility generators; the addition of emission reduction technologies; and repowering of existing power plants to improve efficiency and environmental performance while utilizing valuable resources.

The results of this program are impressive: peak load growth is projected to be minimal for the next ten years; new sources of power supplied by non-utility generators will be on the system by the year 2000; and in addition to the application of advanced add-on technologies, over 1,500 MW of repowered generation is currently planned by the year 2000.

INTRODUCTION

The relationship between regulators in the State and Federal Government and the utility industry is changing today, with cooperation becoming a necessity as economic, competitive and environmental pressures increase. During the last three years a number of significant regulatory and political changes have occurred that will modify how utilities operate and compete in the future. These changes include: 1990 Clean Air Act Amendments; the Energy Policy Act of 1992; FERC 636 Rules; and worldwide concerns about global warming. These changes are forcing utilities to address not only more stringent environmental goals, but how they will compete in a more open industry. Utility planners expect the advent of Electric Wholesale Generators (EWGs), open transmission, non-traditional rate making, issues of stranded assets and other changes in the electric energy system to force utilities to act in a more competitive manner.

Public Service Electric and Gas ranks as the third largest combined electric & gas utility in the US. It provides service in the central part of the state, often called the Amtrack corridor, serving most of the states largest cities. (Figure 1) The 13 generating stations it operates currently includes 15 steam fired units which utilize coal, gas and oil, 3 nuclear units and 143 combustion turbines. The combustion turbine fleet is the largest in the US and reflects the need for significant amounts of peaking power in the region. As can be seen from the map, many of the generating stations are located very close to densely populated urban areas.

This map (Figure 2) illustrates the problems of Ozone attainment on the east coast. The state of New Jersey has been classified as a severe non-attainment area for ozone and this is the primary pollutant being addressed by the state and PSE&G.

PSE&G ENVIRONMENTAL GOALS

A large part of PSE&G's strategic reassessment is the recognition that there must be greater cooperation between government agencies and utilities if these changes are to be mutually beneficial. It is generally recognized that the implementation of detailed regulations without input from the parties being regulated is not effective. For that reason, PSE&G and the NJDEPE have developed a comprehensive program that we believe optimizes the value to PSE&G's customers and shareholders while achieving mutual environmental goals to protect those same customers.

In conjunction with meeting the specific environmental regulatory requirements, PSE&G has decided to identify corporate environmental targets. (Figure 4) These targets or goals are significant for two reasons: First, they interpret the complex regulatory compliance requirements into understandable goals that are measurable and which employees, customers and stockholders can understand. Secondly, these goals commit PSE&G to go beyond current regulatory reductions. In other words, we are willing to set concrete goals for environmental compliance based on what we believe is needed, not on what we are compelled to do. The two major air emission reduction goals we have established are for NOx and CO2.

NOx has been identified as a precursor to ozone and needs to be reduced significantly to bring New Jersey in to compliance with federal laws. (Figure 4) Stationary sources such as power plants are estimated to emit about 40% of the total NOx emissions in the state.

The proposed State Implementation Plan calls for New Jersey utilities to utilize front end combustion controls which will result in a 30% to 40% overall reduction in stationary source emissions. PSE&G believes it can reduce NOx emissions even more than required in a cost effective manner, enabling the state to avoid more costly efforts in other segments of the state. (Figure 5) Therefore targets of 60% reduction by 1995 and 80% reduction by 2000 have been established. The addition of Bergen Phase I repowering alone will reduce annual NOx emissions by over 10,000 tons per year.

A second concern in the release of Greenhouse gases such as CO2. The recent Rio Conference highlighted the need for nations throughout the world to minimize emissions of greenhouse gases. PSE&G, although not mandated by any legislation to do so has established a goal of limiting CO2 emissions to the 1990 level by the year 2000. The 34% reduction in heat rate of a combined cycle units opposed to a steam unit helps to stabilize CO2 emissions.

Maintaining a healthy economy and achieving environmental goals are key objectives of all parties. The federal government has taken steps in their statutory programs to control the flow of federal funds to states to assure the achievement of environmental goals. State governments are responsible for implementing many aspects of these programs and stand to lose significant sums of money if they do not achieve proscribed levels of success.

Utilities in many cases are trapped by opposing forces which on one hand ask for a cleaner environment and at the same time require cost effective service. Therein lies the crux of the issue. Regulators want environmental improvements. Utilities agree, but must remain competitive if they are to avoid a potential death spiral. A company must be concerned that increased energy prices resulting from environmental controls may send large customers elsewhere. This could happen in a number of ways, including use of self-generation, a switch to other suppliers through open access, or relocation to other states or countries. This loss of customers in turn increases prices even more.

Can environmental goals be mutually beneficial to all parties? We believe the answer is yes. PSE&G has learned that by working together with regulators, comprehensive plans can be developed that achieve the environmental goals the regulators desire along with the level of service the utility is expected to provide. The key to this relationship is in recognizing that the service today's customer desires includes more than reliable, low cost service. Today customers expect a utility to be socially responsible as well as cost effective and reliable. As utilities recognize this objective it is easier to understand how cooperation with regulators can benefit all stakeholders.

ENVIRONMENTAL AND GOVERNMENT POLICY INITIATIVES

In order to put in perspective the changes taking place between utilities and regulators, it is helpful to see how regulators view the changing regulatory environment. At a recent presentation to PSE&G's senior management, the Chief Economist of the New Jersey Board of Regulatory Commissioners discussed the evolution of regulatory jurisdiction from a closed to an open system. (Figure 6)

The closed system as shown in this slide presents the State Regulatory Commission at the center of the process. This can be likened to a pre-Copernican view in which one believes the world revolves around oneself. In this closed system the regulators establish standards of performance within a regulatory jurisdiction and is typical command and control approach to regulation. This system is contrasted by an open, enlightened system where standards of performance are established in relation to market benchmarks. In this system, the customer is shown at the hub of the process.

The closed view (Figure 7) from the utilities perspective would show the utility at the center of the universe, with customers and regulators being portrayed as branches of the process. What is most important about this portrayal of interaction is that there is a balance between the best interest of the customer and other groups to which a utility is responsible. In a sense this view pre-supposes a customers understanding that social values are necessary and that there is a compact with utility investors regarding a fair return.

ENVIRONMENTAL EXTERNALITIES

In developing strategic plans environmental costs must be considered. During recent years a number of academic, scientific and regulatory bodies have tried to put a cost or value on the impact of pollution to mankind and the rest of the planet. These studies have looked at health impacts, remediation costs using statistical or artistic means of developing dollars. Estimates for SOx and NOx run from a few hundred dollars per ton to many thousands of dollars per ton. How are the values for environmental externalities accounted for in a utility's economic decision making system?

The value of environmental or societal benefits are identified and accounted for in a PSE&G's Integrated Resource Planning Process. (Figure 8) The IRP process provides a framework for evaluating the entire spectrum of energy alternatives in a manner that balances conservation and demand side management with the full range of generating alternatives. Multiple scenarios with a range of environmental externality costs and other alternatives are run to assure a set of results that are comprehensive.

REGULATIONS DRIVING UTILITY CHANGES

The current opportunities for addressing environmental goals in a mutually beneficial way are provided or one might say demanded in response to the 1990 Clean Air Act Amendments and the 1992 Energy Policy Act. In fact these two legislative initiatives are causing utilities to rethink the way they do business.

1990 Clean Air Act Amendments

The 1990 Clean Air Act Amendments are having a profound impact on the planning and operation of electric utilities. (Figure 9) In total there are 11 Titles in the CAAA as shown on this slide. Of the 11 specific titles, 5 have an immediate impact on power generation aspects of utilities.

Title I - Non-Attainment - This title addresses the need to reduce ground level ozone. In New Jersey, the State Department of Environmental Protection and Energy (DEPE) is in the process of developing a State Implementation Plan (SIP) to achieve compliance. For electric utilities this means reducing NOx by significant amounts by 1995 with further reductions likely in future years. There has been a high level of cooperation between all the utilities in the state, the DEPE and environmental advocates to achieve the required reductions in a manner that is both cost effective and has the desired impact.

Title III - Hazardous Air Pollutants (HAPS) - This title will identify the level of HAPs power plant emit and determine if these are of concern. Utilities are working with both the state and federal regulators to determine the exact nature of these pollutants and means of reducing them to acceptable levels if needed. Depending on the final outcome, the cost of compliance could be enormous.

Title IV - Acid Deposition Control - The acid rain reduction program is the most visible of the Titles and initially has the highest costs associated with it. Throughout the nation utilities, state and federal regulators are working to implement this legislation in the most cost effective manner.

Title V - Air Permits Emission Fees - Utilities will have to obtain permits for all "grandfathered" power plants and will have to pay a fee for every ton of pollutant emitted from all facilities. This will also constrain the total emissions from a facility and may require the installation of additional control devices.

Title VII - Enforcement - Increased penalties is obviously a concern to senior management at utilities as they recognize that there will be severe costs to be paid for non-compliance.

ENERGY POLICY ACT OF 1992

The Energy Policy Act of 1992 has a direct impact on the competitive production of electricity and related environmental issues. (Figure 10) Many of the 30 Titles in this Act will affect utilities, however I would like to mention the three that impact energy competition and electric production environmental issues. From a competition standpoint, Title VII - Electricity will have a major impact on utility strategy. One major change is the amending of the Public Utility Holding Company Act (PUHCA). Changes include the creation of exempt wholesale generators (EWG's) allowing independents and utility affiliates to own and operate facilities anywhere. With state approval utilities may remove facilities from rate base, create hybrid facilities, and deal with affiliates. The Title also removes any PUHCA constraints from foreign utility systems investments and allows EWG's to be brokers. The Act also amends the Federal Power Act, giving FERC more authority to order wheeling and in essence provides open access to utility transmission. Wheeling can be ordered provided it is in public interest and it does not unreasonably impair reliability. Cost based transmission rates would be established, and there would be limitations on retail wheeling and sham transactions. Utilities will now be required to file annual transmission capability reports.

Global Climate Changes are addressed in Title XVI of the act along with the broad issue of Energy and the Environment in Title XXI. These Titles may have additional impacts on emissions controls as they are further formulated.

The net result of this is legislation is an increase in the competition among utilities and other generation sources for existing customers. The success of a utility to stay competitive and avoid the death spiral mentioned earlier depends to some degree on its ability to meet its environmental goals in a cost competitive manner.

PSE&G'S STRATEGIC APPROACH

What type of strategic approach can be utilized to achieve a win-win result regarding environmental goals? At PSE&G, we have developed a long term strategy (Figure 11) that first identifies the needs of each of the stakeholders and then selects the optimum responses from the options available. One of the first steps we took in developing our strategy was to determine the State of New Jersey's environmental objectives.

In New Jersey the state has clearly defined environmental objectives. These include balancing the need to comply with Federal Regulations while continuing to encourage economic growth. To achieve this end, cost effective solutions for reducing environmental concerns must be identified and implemented.

PSE&G had numerous discussions with State DEPE officials, environmental groups and customers regarding environmental goals and impacts. We provided input into the regulatory process and ultimately developed a series of strategies aimed at meeting everyone's needs in a cost effective manner. Using a comprehensive, Integrated Resource Planning process four elements to the plan were formulated. These include an aggressive Demand Side Management (DSM) program, conservation, the procurement of energy and capacity from Non-Utility Generators, and the application of emission reduction technologies or repowering of a number of units.

The results of this program are impressive. PSE&G expects its peak capacity requirement to remain constant at approximately 10,000 megawatts through the end of the century as a result of the application of a broad range of Demand Side Management and conservation programs. In addition, PSE&G has solicited bids from independent sources of power generation. Contracts and offers for approximately 1,000 megawatts of non-utility generation power are possible between now and the end of the century. These contracts are a combination of energy and capacity and are being developed in a manner that assures both low cost and operational flexibility.

REPOWERING AND TECHNOLOGY ADDITIONS

The third and key element of PSE&G's environmental plan is the application of technologies to existing power plants and the repowering of key units. PSE&G is in the process of developing approximately 1500 megawatts of repowered combined cycle generation at two of its existing generating stations. (Figure 12) One project, Burlington 10 is a 245 MW refurbishment of a 1970's vintage combined cycle unit. After 10 years of operation as a simple cycle combustion turbine unit, its renovation as a modern combined cycle was recently completed.

A larger more prominent project is the Bergen Phase 1 and 2 repowering. This project utilizes the existing Bergen 1 and 2 unit steam turbines which are 285 megawatts each along with new state of the art combustion turbines and heat recovery steam generators (HRSG) to provide highly efficient, environmentally clean generating sources. When completed these two units will each be almost 650 megawatts in size. How was this decision made and why are there mutual benefits to both the state and PSE&G? In 1989 the company was faced with spending a significant sum of money to renovate the aging boilers and support systems of these two units. The cost of these renovations would have been in the hundreds of millions of dollars and when complete the units would not have any improvements in cycle efficiency or their environmental signature. At the time the exact requirement and cost for emissions control were uncertain, however management expected they would ultimately need to spend millions to comply with the impending Clean Air Act Amendments. Therefore a comprehensive analysis of available alternatives was prepared that addressed both the cost effective refurbishment of this unit and the need to improve environmental performance. A decision was reached in early 1990 that the repowering of these two units as gas fired combined cycle was the appropriate approach.

What is a repowered combined cycle unit? A repowered unit utilizes an existing steam turbine and generator. The original boiler is removed and a heat recovery steam generator is added to take exhaust heat from a gas fired combustion turbine. The advantages of this technology are many.

The cycle efficiency is improved by almost one-third. The heat rate of the original conventional boiler/steam turbine is reduced from about 12,000 Btu/Kwhr to 7,900 Btu/Kwhr. This effectively increases the energy conversion efficiency from 28% to 43%.

Emissions of key pollutants are reduced significantly as well with a combined cycle repowered unit. with NOx being reduced by 96%, SOx reduced by 84%, and CO reduced by 63%.

There are major economic savings as well. By repowering these existing sites, PSE&G's customers will be saving approximately 20% of what a previously undeveloped or "green field" unit would cost. Furthermore since the units emissions controls are state-of-art add-on technologies are unlikely to be needed in the near future. In addition to the obvious savings in fuel attributed to the efficiency increase, the fixed and variable operating costs are much lower than for conventional steam units.

One additional benefit of the repowered unit is flexibility in alternate fuel types. The initial units will use natural gas as its primary fuel with oil as a backup. Natural gas is a clean, available fuel that currently results in low cost operation of this unit. An option ultimately available at these units is the ability to burn synthetic gas derived from coal, ore emulsion, sludge or other fuel sources. Should natural gas prices rise significantly or natural gas become unavailable, a modern gasification system can be added to ensure an alternate source of fuel.

A number of other technologies are also being developed and applied where cost effective. Testing at our facilities occurs only after briefing and obtaining approval from the NJDEPE. These include (Figure 15):

- A water injection system using Light Oil Emulsion (LOE) for combustion turbines. This system developed in conjunction with NALCO Fuel Tech has been undergoing testing and refinement for over a year at our Edison Generating Station. Initial results indicate that reductions in NOx of 45% to 60% are possible.

- High pressure water injection technology for combustion turbines is also being developed and tested. Expectations are that NOx can be reduced significantly without the need for extensive auxiliary systems and chemicals as with LOE.

- A first of its kind test of a Selective Non-Catalytic Reduction (SNCR) system on a coal fired, wet-bottom boiler is being conducted during the summer of 1993. This project is being done in conjunction with the Electric Power Research Institute. The development of this technology will permit NOx reductions on units where Low NOx burners and other front end technologies are not applicable.

- Research and development of dry low NOx technology for PSE&G's fleet of 126 FT-4 aero-derivative combustion turbines is being funded. The manufacturer, Turbo Power Marine (TPM) expects to complete development of technology that will result in major reductions of NOx emissions in these 1970 vintage machines by the mid-1990's.

- Low NOx Burners with Over Fire Air are being installed on large coal fired units when applicable. This front tend technology is able to reduce emission rates by up to 70% in some applications.

SUMMARY OF BENEFITS

In summary, we believe that in New Jersey both the regulating agency and PSE&G have come to recognize the value of open communications in resolving complex economic and environmental problems. The drive towards greatly improved environmental performance is occurring at a time when electric utilities are just beginning to grasp the implications of deregulation, competition, and changing price structure. In the cited example, the Bergen repowering project not only will have state-of-the-art environmental performance but operating costs for the plants will be greatly reduced. This contrasts with an option where technologies might have been added to the existing plant but would not achieve emission reductions as great as the repowering. Additionally, operating costs for the latter option would increase since the emission reduction technologies would not improve performance while they would contribute to the operating costs.

ACKNOWLEDGMENTS

PSE&G would like to thank William O'Sullivan and his associates at the NJDEPE for their cooperation, availability, and assistance in developing its NOx reduction program. The authors would like to credit the numerous people at PSE&G who have contributed to the evolution of the many programs mentioned within this paper.

ABOUT THE AUTHORS

Ronald Drewnowski is the Manager , Project Development at PSE&G. Previously he was Manager of Environmental Engineering and Nuclear Mechanical Engineer Manager . Mr. Drewnowski holds a BE degree in Mechanical Engineering and a ME degree in Mechanical Engineering from the Stevens Institute of Technology. He is licensed a professional engineer in New Jersey and Pennsylvania.

Walter P. Drabinski is the President of Vantage Consulting, Inc. a management consulting firm that specializes in utility strategy and operations. He holds a BS in Electrical Engineering from SUNY at Buffalo and an MBA from The Wharton School, University of Pennsylvania. During the last two years, Vantage has assisted PSE&G in developing long term environmental strategies and regulatory compliance plans.

FIGURE 1 PSE&G GENERATING STATIONS

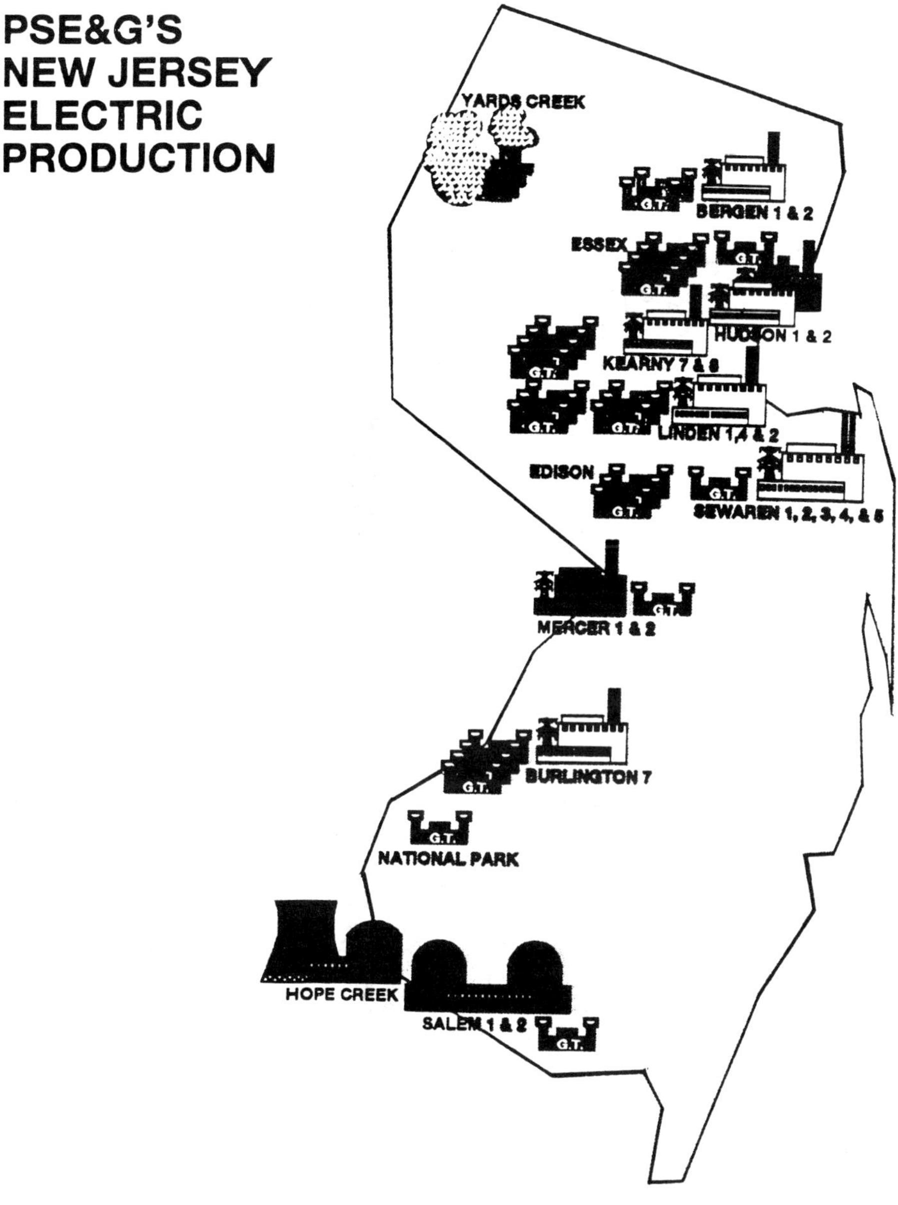

FIGURE 2 OZONE NON-ATTAINMENT AREAS

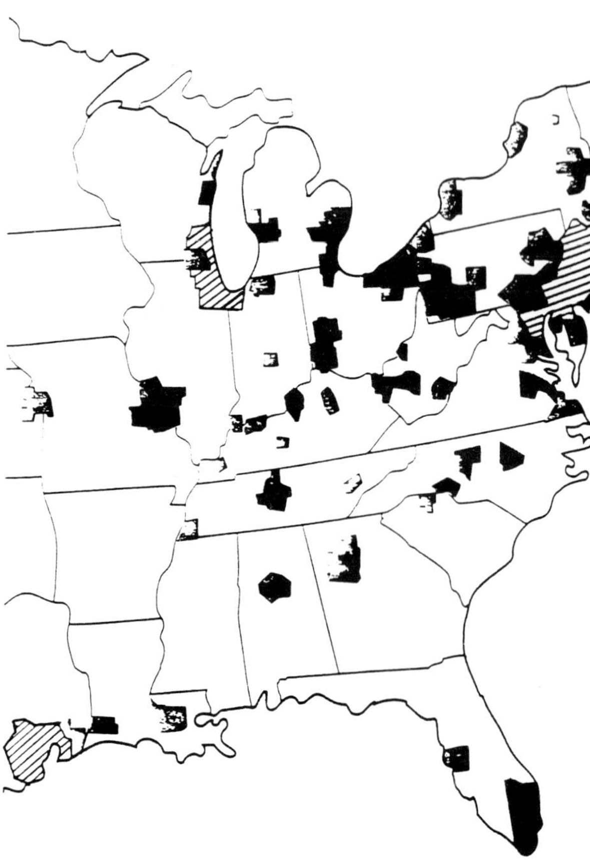

Category	Major Source	Major Modification	Offset Ratio
	TPY VOC or NO_x		
Marginal	100	40	1.1 to 1
Moderate	100	40	1.15 to 1
Serious	50	25	1.2 to 1
Severe	25	25	1.3 to 1
* Extreme	10	Any Increase	1.5 to 1

* *The extreme classification applies only to the Los Angeles South Coast Air Basin and a portion of San Bernadino County. See 56 FR 56726.*

Boundaries are approximate. see 56 FR 56694.

FIGURE 3

ENVIRONMENTAL PERFORMANCE TARGETS	
• Stationary Source Air Emissions	**Reduce NOx emissions by 60% by the year 1995 and 80% by the year 2000.**
	Stabilize CO2 emissions at 1990 levels by the year 2000.
• Mobile Source Emissions	Achieve a 60% alternative fuel fleet by the year 2000.
	Reduce miles traveled on the job as well as stationary operation.
• Recycling	Recycle 75% of all non-hazardous solid waste by the year 1995.
• Hazardous Waste Reduction	Reduce the generation of hazardous waste from our operations by 30% by the year 1995.
• Energy Conservation	Develop and promote electric demand side management activities.
• Water Discharges	Reduce the discharge of pollutants to surface and ground water.

FIGURE 4

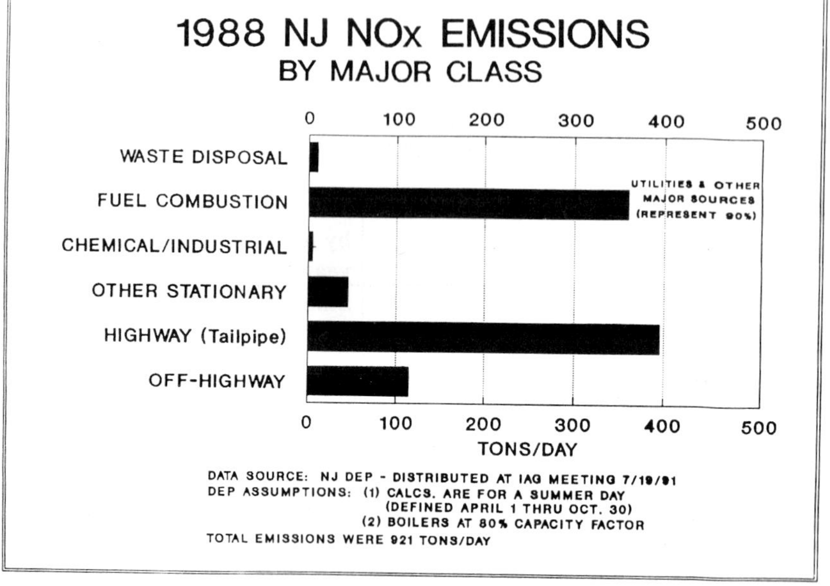

FIGURE 5

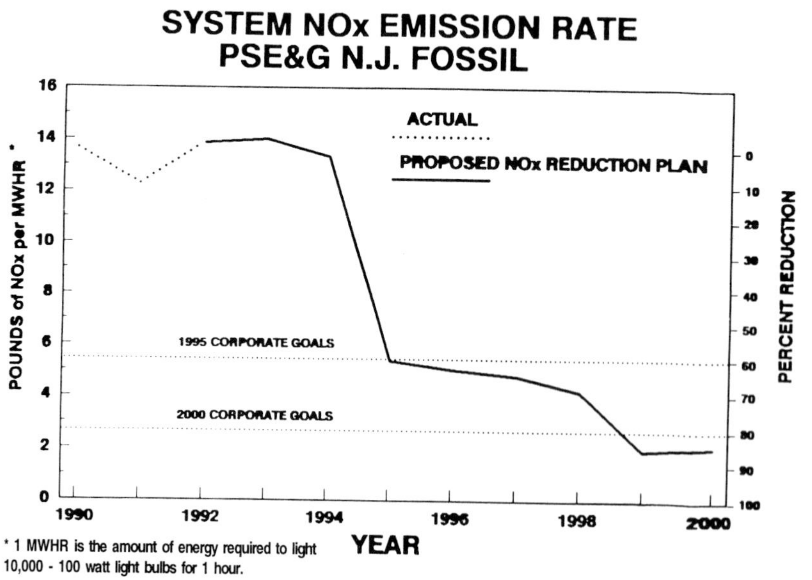

FIGURE 8

PSE&G's INTEGRATED RESOURCE PLANNING PROCESS

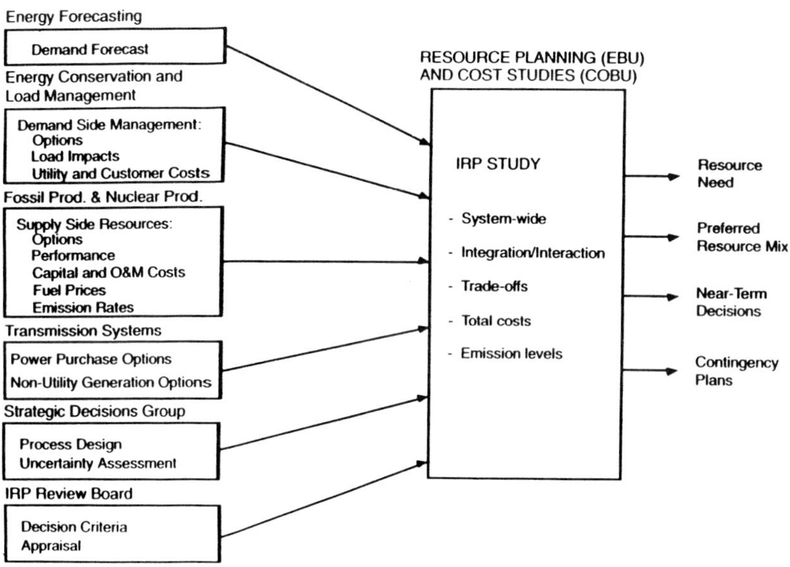

FIGURE 9

1990 Clean Air Act Amendments

TITLE I - *Nonattainment Areas*

TITLE II - *Mobile Source Emissions*

TITLE III - *Hazardous Air Pollutants*

TITLE IV - *Acid Deposition Control*

TITLE V - *Operating Permits*

TITLE VI - *Stratospheric Ozone Protection*

TITLE VII - *Enforcement*

TITLE VIII - *Miscellaneous*

TITLE IX - *Research*

TITLE X - *Disadvantaged Businesses*

TITLE XI - *Job Assistance*

FIGURE 10

Energy Policy Act of 1992

Titles Affecting PSE&G Energy Strategy

- Title VII - Electricity
- Title XVI - Global Climate Change
- Title XXI - Energy and Environment

FIGURE 11

PSE&G Energy Strategy

- Demand Side Management
- External Generation Sources - NUG'S
- Reduction of Emission from Current Sources
  - Repowering Technology
  - Control Technologies

FIGURE 12

PSE&G Repowering Projects

- **Burlington 10 - Refurbish - 246 mw - 1993**
- **Bergen Phase 1 - Repower - 650 mw - 1995**
- **Bergen Phase 2 - Repower - 650 mw - 1998**

FIGURE 13 TYPICAL CONVERSION TO REPOWERED UNIT

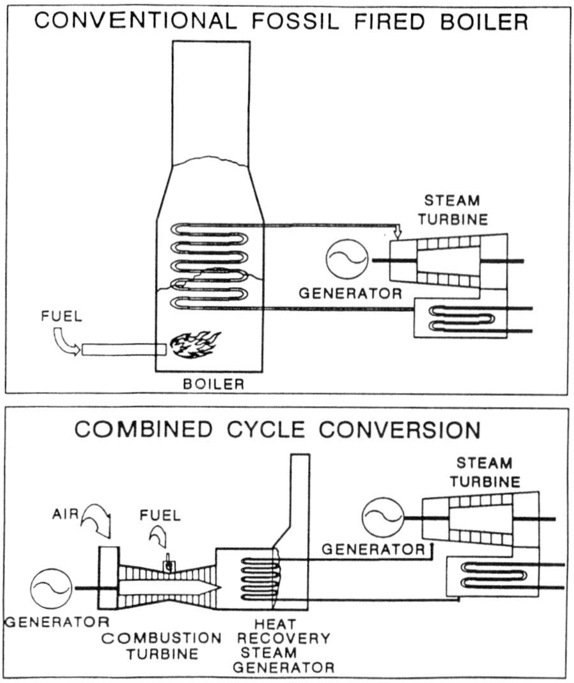

FIGURE 15

NOx Reduction Technologies

Technology	Application	NOx Reduction
• Light Oil Emulsion	Combustion Turbine	45% - 60%
• High Pressure Water Injection	Combustion Turbine	50% - 70%
• Dry Low NOx Technology Adoption to FT-4	Combustion Turbine	60% - 80%
• Selective Non-Catalytic Reduction	Coal Units	30% - 60%

Beyond Regulation: Electric Utility Strategies for a Greener, More Competitive World

Bruce N. Stram
Vice President -- Corporate Planning and Analysis
Enron Corporation
Houston, Texas, U.S.A.

Background:

Dr. Stram is Vice President of Corporate Planning for Enron. He has also worked with Natomas North America, the U.S. Government Interstate Commerce Commission, Department of Energy, Civil Aeronautics Board, and Department of Transportation. After receiving his Ph.D., Dr. Stram joined the research faculty at the University of Maryland, working on environmental and policy issues.

Abstract:

Evolving trends toward greater competition and more stringent environmental regulation, and consequent utility strategies.

No paper. Oral presentation only.

Strategic Utility Planning for the Twenty-First Century

Thomas J. Foley
Manager
Battelle Portland Operation, Pacific Northwest Laboratories
Portland, Oregon, U.S.A.

Background:

Mr. Foley has been with Pacific Northwest Laboratories for the past ten years. Previous to that, he was at the Northwest Power Planning Council, a planning agency for the Northwest. Mr. Foley received his academic training in mathematics and economics.

STRATEGIC UTILITY PLANNING: NEW TOOLS ARE NEEDED

Thomas J. Foley
Pacific Northwest Laboratory
Box 999
Richland, WA 99359 USA

ABSTRACT

Utility planning has historically been assisted by computer models of the system. The models and the myriad of inputs to the models are used to select resources that yielded the lowest "expected" costs of serving loads over the planning horizon. Forecasts of load growth, including all of the independent variables that drive load, fuel costs, expected operating performance of generating plants and everything else about the future was assumed to be known.

Reliance on past planning tools--which focussed on single point solutions--drove decisions toward large generating plants, because of the perceived economies of scale. In retrospect, too little regard has been given to the difficulty (read, impossibility) of forecasting the future. Loads have not always grown as expected, generating plants have not always operated as expected, and the large sizes and long construction lead-times of some generating options have left many utilities with economic problems, even bankruptcy. As a result, in part, of these experiences, utility planning has recently moved away from reliance on large plants to smaller, more dispersed plants.

Unfortunately, most utility planners are still using the same models they used in the past. We have moved away from reliance on large plants, but have kept the models that helped drive us there in the first place. We have dealt with the symptoms, but not the fundamental problem. Continued use of the old models is analogous to using weapons of the last war to fight the next war. The country that does so usually lives to regret it.

The future cannot be known. Therefore, planning should contain a healthy dose of risk management. If one considers worst case conditions about environmental concerns, for example, it becomes clear that simple adaptation to unforeseen events will not be an effective means of protection. It is important to actively manage and reduce risks. Risk reduction can be achieved through higher expenditures on resources with fewer future risks, research and development on technologies of the future, and so forth. Individually, we do this all the time in the development of our investment portfolios, in the purchase of insurance against future catastrophic events, and in denying ourselves near-term pleasures, in order to assure better health in old age.

This paper will explore some of the shortcomings of the old, existing tools, and what can be done to improve the planning function within utilities. Shortcomings include the inability to adequately address future uncertainties as well as the inability to model all of the benefits that accrue to conservation and renewable resources. Although the paper focusses on electric utilities, the discussion is appropriate for any utility providing services to retail customers.

INTRODUCTION

Many of the ideas in this paper were formulated prior to the adoption of the Energy Policy Act of 1992 (EPACT), which amended the Public Utilities Company Holding Act (PUHCA). The changes made in EPACT make the future even less certain for utilities, and move the utilities in the direction of needing more sophistication in planning. As a result of EPACT and further anticipated changes to PUHCA, some utilities may cease to look like utilities as we know them today and instead focus on one or more niche markets. However, to survive at all in the utility business of the future, good planning will be paramount. Given the state of flux in the business of generating and delivering power to customers, I believe that this discussion is timely.

The Future is Unknown

It was often said in the early 1980s that utility planning was riskier than it had ever been before. I believe it had always been risky, but that we had not been hurt, because load was growing so rapidly that most planning problems were mitigated. In fact, the fundamental problems were always there. We assumed that we knew what the future held in store and we planned that way. We planned to supply electricity in ways we would never use to plan our retirement portfolios, for example. We didn't diversify in the selection of generation plants. We didn't consider the risks inherent in our actions and how rapidly the future can change from the one envisioned. We always chose the action with the lowest expected costs if everything went as planned, and did not pay enough attention to the down-side risks of our actions.

If we have learned anything in utility planning, it is that we cannot know what the future will be. Partially completed generating plants under construction in the 1970s are a concrete legacy of planning and building resources to an unknown future without hedging against unknowns. Even in retrospect, it cannot be said that planners should have foreseen higher costs, higher interest rates, and more regulation. To the contrary, what can be said is that uncertainty is the norm, and we should accept that in our planning and take steps to enable us to adapt to whatever the future holds.

Of course, it is not only utility planning that is fraught with the difficult task of planning under uncertainty. We all face uncertainty in our own individual planning and in our daily businesses. It is the reason that we buy insurance, we buy options as businessmen, and we diversify our investment portfolios.

In utility planning, it is easy to forget just how uncertain the planning environment is, because in planning, we tend to use models that are quite precise. We get trapped into believing the precise numbers that are the results of our modelling efforts. In addition, regulation which focusses on estimated avoided costs, rate impacts, and the like exacerbate the tendency. But a brief examination of history helps shock us into an understanding of how rapidly and how radically situations can change.

Consider the following examples:

1) In 1900 the U.S. Government was going to close the U.S. Patents Office because, "Everything had been invented."

2) Until the mid-seventies, nearly all automobiles had carburetors. Where are the carburetor makers today?

3) In 1948 Thomas Watson, then CEO of IBM, testified before Congress that he could not foresee the need for more than five computers of the capacity of the one just completed by his company. Today, virtually all of us have more power in our laptop computers. The computer Mr. Watson spoke of filled a good-sized room.

4) In the 1990s the Soviet Union, one of two super powers in the world, ceased to exist and virtually all of Eastern Europe changed its form of government.

5) In 1903 the Wright brothers flew. In 1963, 60 years later, we landed a man on the moon.

6) Tris Speaker, a hall of fame baseball player said, "The Babe will never make an everyday ballplayer."

7) Not one of the top ten vacuum tube manufacturers were in the top ten of transistor manufacturers.

The significance of recalling these events and many others like them should be clear. Conditions that exist today can change rapidly and radically. EPACT puts an exclamation mark at the end of the sentence. Planners should take action now to enable them to adapt to an environment that will probably be totally different than the one we plan in today.

PLANNING FOR UNCERTAINTY

Here I will discuss uncertainties that are not well addressed in planning. Then I will examine conservation as a resource that is not modelled well in the current tools. I will also show that conservation implemented by non-utilities can be a threat to utility's native loads, and discuss how utilities might protect themselves.

Many of the planning models identify <u>the</u> set of resources that "minimize" either the net present value costs or the rate impacts of meeting future loads. In either case, a single forecast is used and the model internally has perfect vision of the future. This approach to planning is analogous to accounting, not decision making. Planning should be about identifying those actions that are important to meet near-term and long-term objectives while managing the risk attendant with future uncertainty.

In current utility planning, the cost of meeting known load is determined with 100% probability. We know that costs will most probably be either higher or lower than estimated because of all we have discussed to this point. But this estimated number (either rate impact, societal cost, or other metric) is used in the regulatory process as if it were sacrosanct. It has been the subject of countless intellectual discussions about the appropriateness of using the rate impact measure (RIM) versus the societal cost measure as the basis of resource selection, especially conservation resources. These calculated <u>future</u> costs can dictate many of the actions taken and inappropriately constrain the use of renewables and conservation.

We have experienced large variations from forecasted values, even when we have been estimating performance of technology. We typically think of forecasting errors in the 10% to 20% range. But in fact, we see much greater variances from our forecasts than this. The changes can come from myriad directions. For example, nuclear plants were designed to be operated by 100 people. These same plants now have over 1,000 people operating them. Much of the change has been dictated by regulatory agencies, but the point is that change can come from anywhere.

Consider what we really know when we do planning. For each set of decisions, we can estimate a cost or rate, but we know that the actual outcome will be different, and that from our planning perspective, the outcome could have a large range of costs with different probabilities of occurring. Now consider another set of actions. They also might have a wide range of outcomes. What we have now are two different distributions of cost and probability that overlap each other. Figure 1 depicts this idea and might reflect for some real decisions the extent of our knowledge about outcomes.

Information like that shown in Figure 1 can display the risks inherent in a decision for one set of actions over another. Is one too risky, even though it is lower cost over a large part of the range? Are there actions to preserve low costs by moving the entire distribution? A decision has to made. A model that can provide this kind of information can help.

FIGURE 1. OVERLAPPING DISTRIBUTIONS OF RESULTS OF ACTIONS

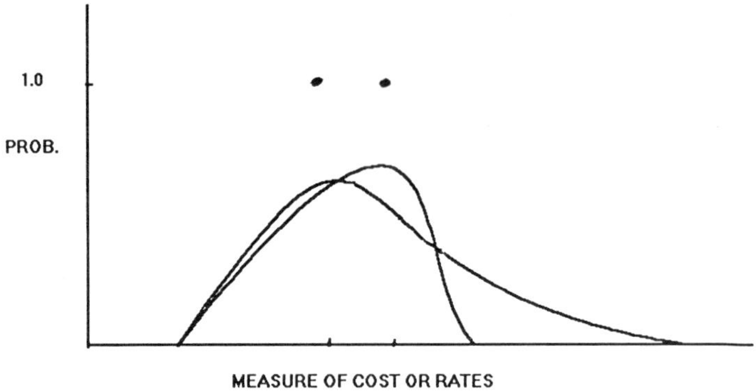

MODEL IMPROVEMENTS

As we have discussed, dealing with future uncertainty has to be a fundamental part of planning. There are Monte Carlo modelling approaches that specifically address the uncertainty of all key driving variables. Multiple runs of the model with estimates drawn from specific ranges of the key variables will yield information like that shown in Figure 1.

At a minimum, the following possibilities should be in the forefront of the minds of utility planners:

- Rapidly changing load growth.
- Changing environmental regulations.
- Revolutionary technological change.
- Rapidly increasing fuel costs.
- Resource performance and costs.
- Others?

Load Growth. Over the past 20 years, loads have been stagnant in some periods and have soared in others. Annual differences of 7% have been experienced. A growth rate of 3.5% over 20 years would double the resource requirements of a system, while a 1% growth rate would only require additions equal to about 20% of the existing system. Forecasting over a broader expected range and examining all of the ways one can be wrong instead of the one (optimal) way to be right should be a fundamental objective of utility planing. Forecasting is an important part of this change.

Other Uncertainties. Load growth is only one of several examples of how the future utility planning environment can change. As the earth's population swells to 10 billion[1] in the next 50 years, environmental issues will become more global than they are today. It is possible that world agreements, like the Montreal Protocol to reduce ozone depleting gasses, for example, could result in our inability to generate power from fossil fuels.

Perhaps the most pressing environmental concern is global warming. The issue of CO_2 emissions and its effect on global warming is being fiercely debated. Often, the debaters take one side of this issue or another; either global warming is a potential problem or it is not. But, what if the probability of global warming being a major environmental problem is only 5%, for example. Should we take action? Given the dramatic consequences predicted by some, I would say yes. No one of us would get in an airplane or on an operating table if the probability of personal damage were as high as 5%. And, we pay a lot to ensure against accidents with a lot less than 5% probability of happening.

Earlier, I discussed technological change. Change can occur quickly, and utilities would be well advised to be part of the change instead of reacting to it. In today's new utility world with the fear of capital being "stranded" by competition with cheaper technology, utility planners would be well advised to carefully watch and consider what technological change can do to business.

History has taught us how quickly fuel prices and availability changes. Under federal law, utilities were prevented from using natural gas as a significant fuel for generating electricity. Now, virtually every utility is considering gas-fired generating resources as the resource of choice. Already we can see the price of natural gas rising. Despite the "gas bubble or sausage," it would not be surprising if the price of natural gas continues to rise, driven by market forces and near-term constraints in the delivery system.

Earlier nuclear power and, at various times, coal power were considered the clear choices to power baseload power plants. Experience with the operating expenses and operating performance of nuclear technology and fear of the ramifications of global warming has cooled the urge to build more of these "conventional" power plants. Rather than adopting a single technology, it is paramount for risk mitigation to consider and hold open options for a diverse set of resources, including fossil, renewables, and conservation resources.

---

[1]World population, at 2 billion 40 years ago, is now at about 5 billion and is estimated to be at 10 billion in the next 40 years. In one human lifetime, population, which took forever to reach 2 billion people, will have soared to 10 billion. The potential amount of pollution generated by 10 billion people will be considerably higher than we see today, unless mitigative action is taken now.

Conservation Resources

Because conservation is the new kid on the block, current planning models do not accommodate conservation very well. The models were created before conservation was an important resource. Thus, many of the benefits of conservation to utilities' economic health are not appropriately accounted for in the planning function. Those benefits that usually are considered include reduced line losses, reduced plant derating for peaking plants on hot summer days, lower transmission and distribution costs associated with conservation (i.e., none), and the fact that conservation requires no reserve margins. Attempts have been made to include the environmental benefits that derive from conservation relative to generation. However, the benefits are at once difficult to estimate and are often revised in the political process to be "acceptable."

Still not considered by most planners are benefits accruing to conservation because:

- The smaller unit size makes it easier to avoid under- or over-building.

- Smaller unit size also allows for a longer "construction period" to acquire a given amount of capacity and energy, because there is little "lumpiness" with this resource. In contrast, large resources have to be built in advance of need with growth counted on to work away the surplus.

- Efficiency standards for new uses yield more resource when it is needed than when it is not needed. In a high growth mode, more new uses of power are purchased and the efficiency gains, which are the same per unit, are higher in absolute terms than in a low growth mode.

- Efficient building shells have capacity benefits that increase relative to less efficient shells as the indoor-outdoor temperature differential gets larger. Thus, capacity benefits calculated under average conditions underestimate the real benefits that accrue to conservation measures.

- A certain amount of conservation, beyond what appears to be cost-effective, may be appropriate as an integral part of a diverse resource mix in order to decrease the risks discussed in this paper.

Until utilities adopt models and planning techniques that enable them to account for these benefits, generation will be built when it is not justified from an economic and risk management perspective.

OTHER CONSIDERATIONS IN PLANNING

Earlier, I addressed technological change. Using conservation as an example, I want to explore a possible future condition that could affect the future solvency of utilities. I will also suggest actions that can be taken to mitigate the potential damages and perhaps to improve on the economic health of utilities.

The situation I see evolving and put forth for discussion is one in which historic competitors such as electric and natural gas utilities will see each other as a minor irritant relative to the effect conservation by do-it-yourselfers or Energy Service Companies (ESCOs) can have on utilities' loads. Gas and electric utilities compete, sometimes bitterly, but loss of load to each other is constrained by earlier investment decisions made by consumers, consumers' lack of capital, uncertainty about how another fuel would affect bills, availability of gas pipelines, and so forth.

To date, the adoption of conservation has been constrained by the reluctance of some utilities to consider conservation as a resource, uncertainty in performance, availability of measures, lack of capital, and so forth. Will this continue to be the case?

Consider an environment in which utilities' acquisition of conservation lags, for whatever reason, while technological progress puts inexpensive, high quality, energy efficient technology into customers' hands. Customers can deploy these new, cheaper, and higher quality conservation options through their own initiative or spurred by marketing efforts of ESCOs acting as a catalyst, possibly with ESCOs providing the investment capital. We know that technology can develop quickly. One only has to look at computer technology to see how quality can increase even while costs decrease rapidly. Parenthetically, computers are and will continue to an important element of many conservation options.

Individual commercial buildings in Portland, Oregon, have achieved 45% energy savings using better lighting and energy management systems. The lights and management systems are off-the-shelf technology. Assume technology improves for all end-uses. Utilities' loads could conceivably fall by 50% in a short period, if all measures were implemented. I suspect that this outcome would hurt any utility, especially any that have surplus power. Utilities can guard against this possibility by being an active participant in providing efficiency improvements to their customers. Is the scenario of low-cost, easily implemented conservation measures likely? Can utilities afford to ignore this possibility? Even if the likelihood is far less than certain? I suggest that they cannot and that utilities should take action to protect themselves against such a possible outcome.

A couple of final words on conservation as a key resource in planning. First, other competitors will appear in the future that do not currently exist. Consider retail wheeling, which would allow any utility to compete for utilities' native loads. Utilities will probably compete for the largest, often most inefficient loads. If utilities can work with their customers to make them as efficient as is feasible, it is more likely that a trusting relationship will be built and, possibly more importantly, the now smaller load will not be as attractive a target for competing utilities. On the other hand, large, possibly inefficient loads, will be vulnerable to outside competition, especially if subsequent efficiency improvements are offered by the competing utility. Again, the possibility of this future threat should be pushing utilities to implement aggressive conservation programs in their respective service territory and they should use these programs to build a relationship of trust with their existing and future customers.

Second, utilities are appropriately concerned about capital being made economically obsolete (stranded) by newer technology or in periods of low fuel prices for alternative generation. If one could define an embedded resource, one that could not be made obsolete, it would be important to utilities. But conservation and some renewables may be precisely this kind of resource. That is, imbedded resources like some conservation and renewables that are an integral part of a building will not be shut down like economically obsolete generating plants. The value would be reduced, because they would be saving lower cost power, but they would not be torn out. They would continue to be used and useful. These two thoughts are material for additional papers.

SUGGESTED REGULATORY CHANGE

In order to encourage better risk management, I would recommend one regulatory change. Regulators typically do not allow utilities to spend more for power than the "avoided cost," which is an outcome of the planning effort. However, if we think again about insurance and other hedges against future catastrophes, there is always a cost associated with them. Regulators and utilities ought to think about the benefits of resources such as renewables or conservation that have no associated fuel costs as a way to avoid future risk, even though they represent higher cost power today.

CONCLUSION

The future is highly uncertain and failure to account for this uncertainty can be disastrous. I have suggested that utilities adopt new planning techniques, in order to better address these uncertainties. Many suggestions have been made in this paper relative to the risk mitigation properties of conservation. The suggestions have appeared before and some of them have been incorporated since the early 1980's into the planning done in the Northwest by the Northwest Power Planning Council.

A new concern expressed in this paper is the potential risk to utilities if they do not embrace conservation and take control of it in their service territories. The risk is from conservation done by customers themselves or through efforts of ESCOs. Utilities can adopt aggressive acquisition programs to protect themselves from ESCOs and other agents of conservation deployment. Alternatively, utilities can work with their commissions to adopt tariffs that make utilities indifferent to who causes efficiency improvements to end-uses they serve. One suggestion is to work towards a tariff that decouples profit from sales by basing utilities' revenues on the number of customers it serves, instead of the number of KWhs it sells.

Finally, there are other reasons why utilities want the loads they serve to be as efficient as possible. In parts of the world, utilities are now free to offer power to loads outside of their historic service territory. It appears as though it is only a matter of time before open access to customers will be a reality in the U.S. The Federal Electric Reliability Council (FERC) continues to consider rules to make open access a reality. To the extent that loads are very efficient, and as a result smaller, they should present a less desirable target to utilities from another service territory. Further, in this new environment, the risk of stranded capital, which would probably be deemed to be "not used and useful," is very real. Conservation and some renewables can be embedded in end-using structures, and would minimize, if not eliminate the fear of the investment becoming economically obsolete, and therefore stranded.

CHAPTER 6 -- UTILITY ENERGY STRATEGIES -- PART 2

Planning Now for the Post-2000 Period

Eric Toolson
Manager, Resource Planning & Evaluation
Sacramento Municipal Utility District
Sacramento, California, U.S.A.

Background:

Mr. Toolson has been involved in resource planning issues for the last 15 years and is currently the Manager of Resource Planning and Evaluation at the Sacramento Municipal Utility District.

PLANNING NOW FOR THE POST-2000 PERIOD

Eric Toolson
Manager, Resource Planning & Evaluation
Sacramento Municipal Utility District
P.O. Box 15830, MS99
Sacramento, CA 95852-1830

The topic I am going to speak on is "Planning Now for the Post-2000 Period."

To start with, what we want to do is look at what this post-2000 period might represent from the utilities' perspective. First of all, we expect increased competition. PURPA, passed in 1978, allowed increased competition on the generation side. We're also looking at increased competition on the transmission side such as open access and retail wheeling. We're also looking at increased competition on the demand side. Vendors, other than the utility itself, will be promoting various demand-side technologies.

Second, there will be greater emphasis on customer value. Instead of providing just a single product, there will be a range of products offered to maximize customer value. Let me give you an example. The Sacramento Municipal Utility District has peak load at about 2,000 MW. We serve approximately 500,000 customers in the Sacramento region. One of those customers is the State of California, and they have the computer facility going in downtown. We're in the process of negotiating with them what we call a "premium power package." That is a package that would have dual distribution feeds. It would have power quality equipment on the front end before it enters into the facility. It might even have a backup facility. Their original concept was a diesel generator. Our recommendation is a small fuel cell. It's this type of customizing that will occur more and more frequently as we enter into the post-2000 period. It won't just be a sale of kilowatt hours and kilowatts.

Environmental regulation, as we all know, will be more stringent in the post-2000 period. It will be harder and harder to generate from supply-side resources, and environmental factors will play an increasingly prominent factor in the licensing of such sites.

Finally, as in any area, there are some significant uncertainties that need to be addressed. Let's look at some of those areas of uncertainty. Fuel cost and availability -- for the last twenty years, the fuel cost has been very difficult to predict, and any forecast of fuel cost in the post-2000 period is really not much more than pure speculation with a very broad band of possible outcomes. In addition, availability of the fuel is in question for a lot of resources. Let me give you another example. SMUD has a geothermal resource in the Geysers Area which is the largest geothermal field in the world. We've had a facility there since the mid-70's. We expect that over the next twenty years the steam from that facility will decline so that the output from that plant will be reduced approximately 50%. That's just one of the issues that we are facing.

Another example of fuel uncertainty is biomass. We are looking at purchasing biomass in the county adjacent to us. That biomass plant is connected with a timber mill, in a co-generation type of mode. However, as the U.S. Forestry Department re-evaluates their timber policy, there's a question of whether the wood waste is going to be available. We have a series of biomass plants throughout central California that are hauling waste in from as far as 200 miles and up to $100 a ton because they can't find it elsewhere. So, there are some very significant issues and questions, not only of fuel cost, but also its availability.

The second example of uncertainty is economic load growth. We're in the midst of an economic recession. All of you have felt it. We've particularly felt it in the Sacramento area. Not only is the state government having a hard time balancing its budget, but there are three military plants, two of which have already been scheduled for closing. The third is currently under debate. Significant issues are how long the recession will last; when it rebounds; how far will it rebound; will the economy be restored to its pre-1990 conditions; or will it only be a partial restoration. These economic factors drive the load growth. There are other factors that drive the load growth as well, but the economic factors are among the main drivers. This contributes to quite a bit of uncertainty regarding our load growth in the future. Are we going to see 1%, 2% or 3% load growth? Are we going to see much self generation? Are we going to see a lot of conservation?

The third issue of uncertainty is transmission access and retail wheeling. The independent power producers would like to have open access to the utilities transmission lines and provide competition. That sounds like a good goal. We're supportive of that. The question is how much access? At what cost? What effect will it have on the reliability? What access will they have to which utility customers? These are large unresolved issues.

Fourth, when we talk about advanced and renewable (A&RT) resources, and as you know SMUD is very actively involved in that area, what are those costs? What is their performance going to look like in the post-2000 period? Photovoltaic might cost $10,000/kW now. Will it drop to $4,000 as projected? Will it drop more? Will it drop less? Fuels cells, how available will they be? What will their cost and performance be? Those are all uncertainties as we plan for the post-2000 period.

Against this background of uncertainty, SMUD has developed a certain emphasis in its resource planning area. When we look at a resource, these are the parameters that we are trying to satisfy. Of course, cost is important. The resource needs to be competitive with both supply- and demand-side resources. As important as cost is its variability. Are resource costs fairly predictable, since they can cause considerable price risk in the forecast?

Second, we need to be environmentally sensitive, not only to air; we need to be sensitive to water and land use as well. All these factors are taken into consideration.

Third, we need to add diversity. Not only fuel diversity, but we need to have technology diversity. For those of you that are familiar with SMUD's history, until 1989 we had a 900 MW nuclear plant. Well, 900 MW is a lot when you only have a 2,000 MW peak load. Despite where you are on the argument, if nuclear power is safe or not, the issue is that 900 MW out of a single unit on a single shaft is too much for 2,000 MW system. In the planning world, when you have a 900 MW contingency you need 900 MW of reserves. So the technology issue is important. If we are going to serve 2,000 MW, we don't want to do it with two plants, we want to do it with 20 plants. We want to do it with plants that use different fuels. We don't want to do it out of plants that operate on just natural gas or any other single fuel, because that doesn't offer any diversity.

Fourth, operability. These plants need to be operated so we can match load with generation. We are in the process of becoming a control area and then second-by-second you need to match the load with the generating performance. A lot of the renewables are not as operationally flexible as some of the natural gas type units. With renewable resources such as wind and solar, you take the output when you can. You can couple it with a storage facility and that helps, but otherwise you still need to have operational flexibility in your system.

Then finally, reliability. Resources need to be reliable. They need to be there when you count on them on a hot summer afternoon. They need to come up and help meet load.

Let's talk a little bit about how we translate that into our resource selection process. Conservation is obviously our resource of first choice. Not only because typically it's least in cost, but also because of its environmental benefits and because of its local economic attributes as well.

Second, distributed resources. Previously, utilities had only large central plant resources. The generating unit was typically located remotely from the load center. Huge 500-2,000 MW units cost a fair amount in terms of transmission and distribution to get that power to the customer. Now, utilities are looking at taking smaller units of generation, placing them directly at the customer location. It has several benefits. There's transmission and distribution savings that can be incurred, both capital and in terms of losses. Also, distributed generation improves reliability. You don't have to worry about T&D outages. Finally, it helps the diversity of the system. Instead of having one 1,000 MW unit, you could have 100 1-MW units.

So, distributed generation is a very important concept that we're looking at for the post-2000 period. SMUD has a program to put PV on rooftops throughout the customer service area. This is a form of distributed generation.

Renewable resources offer us both technology and fuel diversity. In addition, we emphasize non-polluting resources as we look out in the future because of the obvious environmental benefits.

The question is why should we plan for post-2000 resources? Why don't we wait until 2000? If we waited, we would certainly reduce a lot of our risk. We would know what our load growth is. We would know a little bit about the fuel prices. We would have some idea of what the technologies are.

However, there are a lot of reasons for planning. One is the long lead times. Do you remember the slide that Tom Foley put up a few minutes ago about the lead time for a coal plant being 12 years? You need to start thinking about some of the supply-side resources that have a long lead time.

Second, we want to promote small commercialization. SMUD's serious about advancing renewable technologies. The advances in renewable technologies, however, for the most part, are not at the point where they are commercial and cost effective. They are not as competitive as traditional resources, unless you have them in remote locations and other specialized instances.

What can we do? SMUD has a program where we will be investing in these resources year by year in a sustained orderly development effort to help bring down the cost, so that these resources can play a major role in our post-2000 generation mix. Those sort of plans need to be played out five or ten years before you hope these products meet the market.

Finally, we develop the infrastructure on demand-side resources that are required. Whether the delivery mechanism is through trade allies or in-house auditors, this infrastructure takes time to develop.

To give you an idea of what it takes to license and construct a plant these days, SMUD is looking at licensing a 250 MW pump storage project. Our best estimate is it should take us 12 years. So if we start in 1993, that resource wouldn't be available until 2005. That gives you some idea of the licensing, the studying, and the construction process that's involved in building a new facility.

As we talk about the role of demand-side management, we need to look at what are the causes of reduction of energy consumption. What role does the utility have versus what role do the market forces and regulation have in conserving energy consumption? There are several factors that cause a reduction of energy consumption or energy conservation. The easiest way to promote conservation is to send the proper market signals regarding the cost of power. Lead people react to what the true cost of power is. Price elasticity simply says that if your price goes up, your consumption goes down.

Second, if the conservation measure is currently cost effective, one way to implement these savings is through regulation or standards. An example of this is the refrigerator. New federal efficiency standards that take effect in 1993 require a 19 ft^3 refrigerator with a freezer on the top and automatic defrost to have an annual consumption of less than 700 kWh/year. To give you a point of comparison, SMUD has a program where they will go out and retire refrigerators that are old and inefficient. Through our monitoring and evaluation on this program, we have found that the refrigerators we have brought in that are approximately 15 years old have an annual consumption of almost 2500 kWh/year. So you can see the difference between the older inefficient refrigerators of 2500 kWh/year and the new refrigerators at 700 kWh/year.

Then as a last resort, utility programs can be enacted to encourage greater efficiency (i.e., the energy efficiency after all the market forces and regulatory impacts have occurred). At that point it is appropriate for the utility to enter into the market and to try to correct the market inefficiencies.
This is a graph that I took from an EPRI report. I thought it was very informative. It talks about the causes of energy reduction. What's interesting about it is that it displays the magnitude of various causes of energy reduction over the years. What we're doing here is looking at a thirty-year period from 1970-2000. You can see this first line. It shows the load growth that was common in the 1970's. Everybody was saying we are going to run out of capacity, utilities forecasting 6-7% annual load growth. What happened? Prices started going up, people conserved. This whole response is the post-73 impact of market forces and standards; mostly market forces. Prices went up, people voluntarily conserved on their own. That's the best way to conserve. Get people proper pricing, let them respond to the signals in the free market. You can get an idea of the magnitude of that conservation. Instead of looking at a load growth of 6-7% per year, it was reduced to 2-3%.

The second category is the post-1998 impacts and market forces and standards. Standards like the refrigerator standard that we've talked about. In California, there are building standards. They require so much insulation. Those are regulatory standards that cause a significant amount of reduction in energy consumption. You can see it's not nearly as much that occurred the first time. That's what you'd expect. You give people proper pricing, and conservation is incurred.

Finally, this last part is the maximum technical potential. In the year 2000, if we were to implement every possible technically viable energy efficiency measure, estimates are that you could reduce the load from 25% to 45%. These are very soft numbers. In reality a lot of those potential measures are not economic and there are also significant market barriers. So when we look at the realistic projection as to what a utility can cause to happen above market standards, then you are looking at this other block here. We're still talking about 100,000 gW hours by the year 2000. So it's a fairly significant amount.

Let's talk about where that energy conservation is going to come from. We're going to look at it by sector and by end uses. If we look at it by sector first, and then end uses second, 42% will come from the commercial sector, industrial 32% and the lowest is residential. That's still very significant at 26%.

Residential includes both single- and multi-family units. On the industrial side, 22% comes from motors. Instead of having somebody go out and rewind an inefficient motor, have them replace it, have them go to a higher standard and become more efficient. On the residential side there's a host of measures focused on space heating, water heating, cooling, and lighting. Those are the end uses. Those are the areas that utilities are trying to impact in terms of increased energy efficiency in the year 2000 and beyond.

The question is how much of our energy can we save by the year 2010? In 1990, we estimated 6% of the energy can be saved. Two years later that almost doubled to 11%. There were several reasons for this change, a couple of them being that we had little experience in conservation in 1990. Now we have a better handle on what it can do. There have been some estimates up to 20% and that's using very aggressive type of strategies. So the range is somewhere in between there. It's not going to meet all the needs, but it's going to meet a significant amount of the needs.

Utility spending on energy efficiency has increased from $900 million in 1989 to $1.2 billion today. Some forecast that as much as $100 to $150 billion will be spent annually by the year 2010. That's just a huge amount. To give you some perspective, SMUD alone is spending $50 million in 1993 on energy conservation. That's $50 million out of the total revenue requirement of approximately $800 million. So, my math on the stand is not good, but lets say about 7%. That's a fairly significant expenditure. Fifty million going into various rebates, audits, direct installs; all measures to promote conservation.

What's the world going to look like after the year 2000? Let me just share a couple of ideas. These aren't my ideas. These are ideas from different authors.

One is competition will be introduced on the demand side just like it has been on the supply side. On supply-side you don't just buy kilowatt-hours from the electric utility anymore, you buy it from qualifying facilities, you buy it from independent power producers. Some utilities generate themselves. Many people propose that the same is going to be true in demand side. There will be a lot of different vendors of demand-side measures. The utility would only be one of them. The utility may decide that it can't offer those programs efficiently and pulls theirs out of the market altogether.

SMUD is currently contracting with what are called ESCO's (Energy Service Companies) to come in and reach markets that SMUD has had a hard time reaching. For instance, small commercial/industrial customers. People that are just going month-to-month and have a little mom-and-pop store or a restaurant. Utility bills aren't their highest concern. Those are hard markets to reach. We are using an independent company to reach those. That will be more prominent in the post-2000 period.

Another idea--the customers will drive the markets. The markets will be established. They give the example of the compact fluorescent lightbulb. Right now utilities are putting a fair amount of money into that. These rebates should cause customer acceptance. People will become familiar with it. It will cause economies to scale as people purchase more of them. They will start legitimizing the products so that the trade allies have these on their shelves and start dealing with them on a regular basis. The customers and the retailers become comfortable with the technology.

Third idea, distributed generation will be used to reduce T&D expenditures. We've talked about this a little bit. PG&E (Pacific Gas & Electric), the nation's largest utility, has started a program where they found that if they implemented a high level of conservation measures in a T&D constrained area they could avoid the capital cost associated with the transmission/distribution up to $10 and $25 million dollars. That's a lot of money to justify conservation. As we head out into the future, utilities expect most of their expenditures to be in the area of distributed generation. Therefore, demand side will have an increased value by reducing the need for expensive, and hard to license upgrades to the T&D system.

Fourth, now we feel that rebates aren't compatible with retail wheeling. There's been a lot of concern about how competitive the utility is going to be. In the future with all the different forces in the market, some feel that the market will have reached enough maturity that it won't need any great utility involvement in the post-2000 period.

And fifth is that a lot of the energy efficiency will be occurring just through mandated standards like the refrigerator program. Something that is clearly cost effective. Customers are required to do it. Therefore the utility doesn't have to get involved. Then you don't have all the program administrative costs when a utility tries to implement something that is really clearly cost effective.

Let me go to the last slide and sum this up for SMUD. This is what our resource plan looks like in the next twenty years. This is a dynamic plan and it changes frequently. The top level is conservation. We expect to meet all load growth with conservation over the period from 1993 to 2010.

The second area is future advanced and renewables. This is the area we are talking about that we need to invest in now to make sure that resources such as solar, thermal, photovoltaic, fuel cell, advanced geothermal, advanced CT's are available. We expect that contribution to be about 400 MW by the year 2000. Our current renewables include hydroelectric. You can see that we are unique in that about 1,000 MW of our current system is renewable. We have 130 MW of geothermal. We've committed to purchase 50 MW of wind and 50 MW of biomass in the next four years. We also have the largest existing photovoltaic plant in the nation.

The other categories are not too critical. There are some short term purchases and storage. You can see that our fossil generation remains fairly consistent through 2010. Future fossil is about 25%.

So, given all those factors, that is what we at SMUD translate into our future resource plan.

Expectations for Post-2000 Era

- Increased Competition
- Greater Emphasis On Customer Value
- More Stringent Environmental Regulation
- Formidable Uncertainties

Areas of Uncertainty

- Fuel Cost/Availability
- Economic/Load Growth
- Transmission Access/Retail Wheeling
- Cost/Availability of Advanced and Renewable (A&RT) Technologies

SMUD Resource Planning Emphasis

- Cost & Variability
- Environmental Impact
- Diversity
- Operability
- Reliability

SMUD Resource Preferences:

- Energy Efficiency
- Distributed
- Renewable
- Non-Polluting

Why Plan for Post-2000 Resources Now

- Long Lead Times
- Promote Commercialization
- Develop Infrastructure

TYPICAL HYDROELECTRIC PROJECT SCHEDULE REQUIREMENTS

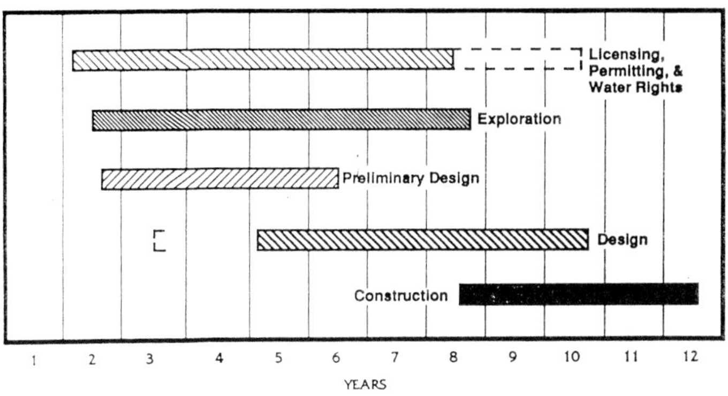

Causes of Reduction In Energy Consumption

- Price Elasticity
- Regulatory Standards
- Utility Energy Efficiency Programs

Energy Consumption Impacts in U.S.*
(1970 - 2000)

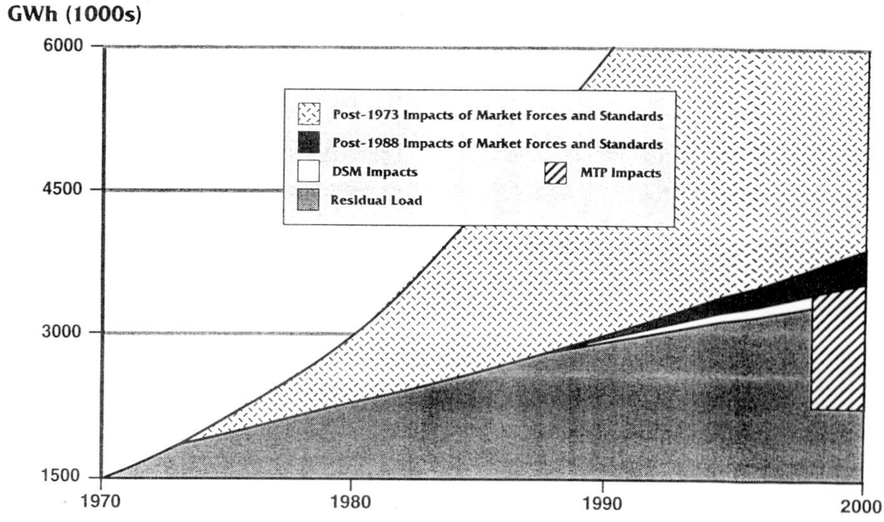

* EPRI, September, 1990

DSM Impacts in 2000 by Sector and End-Use*

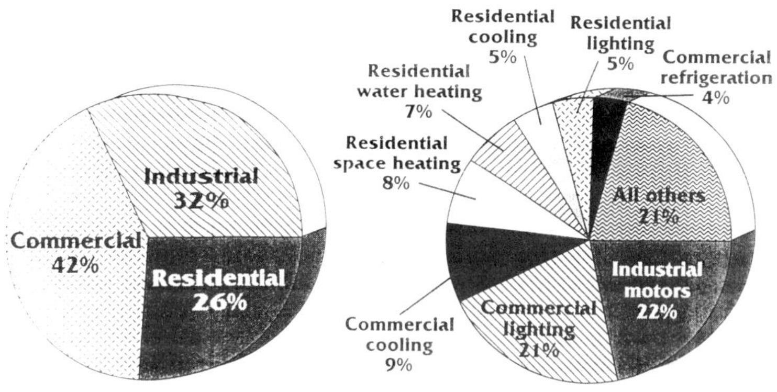

* EPRI, December 1992

Maximum Achievable Energy Efficiency Achievable by 2010

- 1990 EPRI — 6%
- 1992 EPRI — 11%
- Oak Ridge — up to 20%

SMUD'S RESOURCE MIX

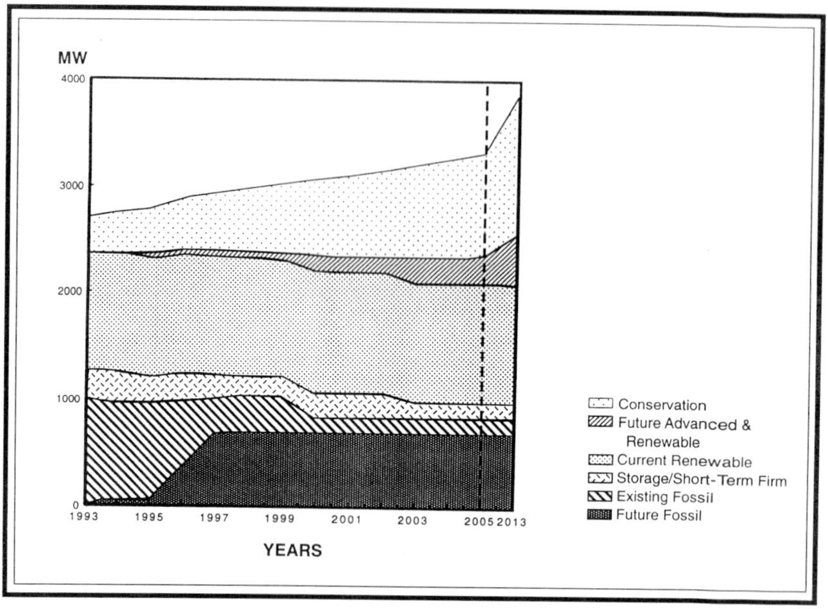

The Three Faces of IRP

Edward A. Reid, Jr.
Director, Market Development
Columbia Gas Distribution Companies
Columbus, Ohio, U.S.A.

Background:

Mr. Reid's responsibilities at Columbia Gas have included managing research programs to develop lower emission gas equipment and new gas end-use applications, HVAC's, NGV's, and fuel cells; and, currently, directing development of new market opportunities, including natural gas vehicles, gas cooling, cogeneration and power generation.

THE THREE FACES OF IRP

Edward A. Reid, Jr.
Columbia Gas Distribution Companies
200 Civic Center Drive
Columbus, OH 43215, USA

ABSTRACT

Integrated Resource Planning, as it is currently practiced in most jurisdictions, is none of the above. It is disaggregated, dealing separately with the individual fuels which compete for end-use markets. It is site based, ignoring the energy consumed in the production, transformation, transmission, and distribution of the energy to the point of end use. It is reactive to the current situation, rather than focused on a long range plan for future energy consumption and demand. This paper explores the importance of a comprehensive approach to IRP which takes into account energy, environmental and business issues.

INTRODUCTION

The subject of my presentation today is Integrated Resource Planning. This talk was originally titled <u>The Three Faces of IRP</u>. One interpretation of that title would suggest that we're looking at comedy, tragedy and reality. However, in context, I believe those three faces are most appropriately electric IRP, gas IRP and a truly integrated, truly resource based planning process which Commissioner Kretschmer of Illinois has referred to as "comprehensive IRP".

By comprehensive IRP, I am referring to a truly integrated, truly resource based planning process that provides simultaneous consideration of all retail energy forms and deals from both an energy conservation and an environmental perspective with optimal fuel selection for each end use. This comprehensive approach to IRP would be based on the conservation of resource energy, rather than retail energy forms. That is, it would take into account the full energy trajectory efficiency from the mine mouth or well head through the end use, with a sensitivity to the environmental impacts of the resource energy form. In this comprehensive IRP perspective, we would also be discussing true planning, with a concentration on appropriate future development and full consideration of all the available options to supply existing and new end uses.

The key element of this comprehensive integrated resource planning process is the issue of optimal fuel selection. Optimal from the standpoint of conservation of resource energy, reduction of criteria pollutant emissions and anticipation of the need to maintain the criteria pollutant emission caps under the Clean Air Act Amendments of 1990. This concept also includes anticipating the impact of the air toxics regulations to be implemented by EPA under the 1990 Clean Air Act Amendments, beginning in 1994. This comprehensive approach to integrated resource planning leads us to a potential future direction with a focus on all energy sources, at the resource level, and all energy end uses in the context of resource efficiency, emissions reduction, energy security and economic competitiveness resulting in a truly integrated, truly resource based energy plan.

This graphic shows the energy trajectory efficiencies from source to site for natural gas, fuel oil and electricity generated by existing electric generating stations using a variety of fuels.

From this energy trajectory efficiency analysis, we conclude that the natural gas industry delivers resource energy to the customer three and one-half times more efficiently than the electric utility industry.

Shifting our attention to end use efficiencies, we can compare the resource energy efficiency of commercially available gas and electric space heating systems. In the case of gas space heating systems, the annual fuel utilization efficiency (AFUE) of the gas furnace is multiplied by the resource energy efficiency (REE) with which natural gas is delivered from the well head to the customer's meter to calculate a resource-based annual fuel utilization efficiency (RAFUE) which combines the effects of the resource efficiency and the appliance efficiency. For the gas heat pump and for electric space heating systems, the heating seasonal performance factor (HSPF) of the appliance is multiplied by the resource energy efficiency (REE) of the energy delivery system to calculate a resource heating seasonal performance factor (RHSPF). Note that the resource annual fuel utilization efficiency of the NAECA minimum efficiency gas furnace is approximately equal to the resource heating seasonal performance factor of the most efficient electric heat pump commercially available.

We can make similar comparisons for gas and electric water heating systems. In all cases, the energy factor (EF) of the water heater is multiplied by the resource energy efficiency (REE) of the energy delivery system to calculate a resource energy factor (REF) for each appliance. Note that the resource energy efficiency of the minimum efficiency gas water heater is twice the resource energy efficiency of the NAECA minimum electric water heater.

From these analyses, we can conclude that natural gas end use equipment satisfies the customer's end use energy needs more efficiently than electric end use equipment for a range of selected applications including: residential and commercial space heating, water heating, cooking and laundry drying; for commercial and industrial space conditioning; and, for a variety of commercial and industrial engine-driven end use equipment.

The other issue which must be taken into account in a truly integrated, truly resource based energy planning process is the emissions of a variety of pollutants. This slide compares the relative pollutant emissions resulting from the combustion of natural gas and 1.3 percent sulfur coal, which is the national average coal sulfur content. The third column in the slide shows the ratio of emissions of coal versus natural gas for sulfur oxides, nitrogen oxides, carbon dioxide and particulates.

Approximately 55 percent of all of the electricity generated in the United States each year is generated from the combustion of coal. We have used coal for the emissions comparisons in this presentation because most of the end uses discussed in this presentation would be expected to be provided with electricity from intermediate load coal plants. Most electric utilities base load their hydroelectric and nuclear power plants, because they have the lowest marginal generation cost. Intermediate loads are typically served by plants burning either coal or fuel oil, with peaking requirements provided by simple cycle gas or oil turbines.

Dealing with pollutant emissions in an integrated resource planning context requires placing some value on the emissions. The environmental externalities costs of emissions from energy production and use are the topic of significant debate within the regulatory community and the utility industry. There are a broad range of costs imputed to the environmental externalities associated with various pollutant emissions, as shown in this slide. The values used in the remainder of this discussion are the values established by the New York State Public Service Commission, which tend to be at the lower end of the range of costs imputed to the various pollutant emissions. The externality cost values at the low end of the range are typically the values determined by the Electric Power Research Institute. The values at the high end of the range tend to be the values for emissions released in the Los Angeles air basin.

Applying the New York State Public Service Commission environmental externalities costs to various end use energy applications takes us one further step in our discussion of optimal fuel selection. This slide compares the environmental externalities costs of space heating using a NAECA minimum efficiency gas furnace and a NAECA minimum efficiency electric heat pump, based on the New York State Public Service Commission environmental externality cost estimates. Note that the operation of the NAECA minimum efficiency electric heat pump results in 23 times higher environmental externalities costs than the operation of a NAECA minimum efficiency gas furnace. In the case of an electric resistance furnace, the ratio would be 46:1 in favor of the natural gas furnace.

This slide shows the same information for water heating equipment, comparing NAECA minimum efficiency gas and electric water heaters. In this case, use of the electric water heater results in 50 times the environmental externalities costs associated with the operation of the gas water heater.

This slide compares the environmental externalities costs and the resource energy cost associated with the operation of existing and commercially available new gas and electric space heating equipment. The values for the electric space heating equipment are highlighted in the box on the graph.

This slide and those that follow make a very important point about the relative value of electric integrated resource planning, gas integrated resource planning and a truly comprehensive integrated resource planning process. In electric IRP, you deal only with opportunities contained in the box on this slide. While it is obvious that electric IRP offers substantial opportunities to reduce both environmental externalities costs and resource energy cost, it does not offer the opportunity to achieve the minimum available resource energy cost or environmental externalities cost.

On the other hand, dealing strictly with gas IRP presents a somewhat more limited range of opportunity to reduce resource energy and environmental externalities cost, but ignores the potential opportunities available through fuel optimization by upgrading customers using electric equipment to current technology gas equipment.

Similar comparisons can be made for water heating. Note that the lower point in the red box in this slide shows a heat pump electric water heater. While the heat pump electric water heater has almost the same resource energy cost as a minimum efficiency gas water heater, it still has approximately 25 times the environmental externalities cost.

This slide shows a similar comparison for commercial cooling equipment, comparing the various electric motor-driven and gas engine-driven mechanical refrigeration systems with gas absorption equipment. Note that, with the exception of double-effect gas absorption air conditioners, the range of resource energy cost for gas and electric commercial cooling equipment are very similar. However, the environmental externalities costs of using electric cooling equipment are five to six times the externalities cost of operating the gas cooling equipment.

This final graph compares existing and current technology central station electric power generation (shown within the triangle) with current technology gas cogeneration systems. The dot at the peak of the triangle is an existing coal-based electric generating station. The points at the base of the triangle are current technology, simple cycle and combined cycle gas turbine power plants. The primary difference between the combined cycle gas plant shown at the left side of the base of the triangle and the gas cogeneration systems is the 50 percent of the recoverable thermal energy from the cogeneration plants assumed to be recovered and used at the site.

In summary, the efficiencies of natural gas end use appliances tend to be somewhat lower at the site than the efficiencies of current technology electric end use appliances. However, the energy trajectory efficiency of electricity delivery is substantially lower than the energy trajectory efficiency of the natural gas delivery system. As a result, the natural gas end uses are more resource efficient. In addition, the natural gas end uses result in substantially lower emissions and associated externalities costs; and, therefore, in lower total societal costs. In addition, for the future, natural gas cogeneration is preferable to the construction of new central station electric generation.

The issue of fuel switching (or, more appropriately, fuel optimization) has already been recognized for application in combination utilities. However, with straight electric and gas companies it is a complicated and contentious issue. Columbia believes that the environmental benefits of a fuel optimization approach to integrated resource planning are compelling. In addition, we believe state commissions must take into account the fact that competing incentives for both gas and electric end use equipment are confusing to the customer and are not economical from an integrated resource plan cost perspective.

Environmental externalities costs are also a very contentious issue. There is a wide range of costs imputed to externalities. We recognize that compliance with the Clean Air Act Amendments of 1990 will result in reductions of SO_x emissions by approximately 50 percent and of NO_x emissions by approximately 20 percent after the year 2000. However, we must also remember that the remaining 50 percent of the SO_x emissions and the remaining 80 percent of the NO_x emissions will persist. Their emissions will continue to impose costs on the environment. The only way to reduce CO_2 emissions is to move in the direction of higher resource energy efficiency and to select fuels with lower carbon contents.

Other future issues to be dealt with from the environmental externalities cost perspective are the costs of controlling air toxics emissions under the Clean Air Act Amendments of 1990, beginning in 1994; and, the longer term issue of high level nuclear waste storage.

In summary, truly integrated, truly resource based planning has great potential, while both electric and gas IRP have more limited potential. Contemporaneous review of gas and electric IRP programs in an overall comprehensive IRP assessment is essential. Another essential concept in comprehensive integrated resource planning is that incentive funding must track the benefits resulting from the IRP program. In the case of fuel optimization, most of these benefits will flow to the electric utility, as the increased direct end use of natural gas displaces existing coal-based electric generation. Finally, we must not lose site of economic competitiveness in the development of IRP programs.

This discussion leads to a series of inescapable conclusions. If there is to be electric integrated resource planning, then there must be gas integrated resource planning, because all of our customers in the gas industry are also customers of the electric utility industry. If there are to be both electric and gas integrated resource planning, then there must be a truly comprehensive approach to IRP, because it is not possible to plan effectively in a vacuum. If we plan effectively, we can succeed in the IRP context in reducing both resource energy consumption and the emissions of criteria pollutants while providing substantial benefits to utility customers and the nation. Finally, we believe we must succeed. The price of failure is intolerable.

Integrated Resource Planning

Preferred Comprehensive Perspective

- INTEGRATED:
  - all retail energy forms
  - simultaneous consideration
  - end use optimization
- RESOURCE:
  - full trajectory efficiency
    - minemouth/wellhead through end use
  - environmental sensitivity
- PLANNING:
  - concentration on appropriate future development
  - full consideration of all available options

Integrated Resource Planning

Potential Future Direction

- FOCUS:
  - all energy sources
  - all energy end uses
- CONTEXT:
  - resource efficiency
  - acid rain precursors
  - global warming gases
  - air toxic emissions
  - radioactive waste
  - energy security
  - economic competitiveness
- RESULT:
  - integrated, resource-based energy plan

Energy Trajectory Efficiencies

Source to Site

	Extraction	Processing	Transportation	Conversion	Distribution	Cumulative
Natural Gas	96.8%	97.6%	97.3%	*	99.2%	**91%**
Fuel Oil	96.8%	90.2%	98.4%	*	99.8%	**86%**
Electric						
Coal	99.4%	90.0%	97.5%	33.4%	92.0%	27%
Oil	96.8%	90.2%	98.4%	32.5%	92.0%	26%
Gas	96.8%	97.6%	97.3%	31.8%	92.0%	27%
Nuclear	94.6%	57.0%	100.0%	31.7%	92.0%	16%
Hydro	100%	100.0%	100.0%	29.1%	92.0%	27%
Other	100%	100.0%	100.0%	17.3%	92.0%	16%
Weighted	*	*	*	*	*	**25%**

Energy Trajectory Efficiency

Conclusion

- The natural gas industry delivers resource energy to the customer 3.6 times as efficiently as the electric utility industry.

Full Trajectory Resource Energy Efficiency

Gas and Electric Space Heating Systems

GAS SPACE HEATING	AFUE * REE = RAFUE
NAECA minimum GF	0.78 * 0.91 = 0.71
Condensing GF	0.94 * 0.91 = 0.85
	HSPF * REE = RHSPF
Gas Engine Heat Pump	1.40 * 0.91 = 1.27
ELECTRIC SPACE HEATING	HSPF * REE = RHSPF
Resistance	1.00 * 0.27 = 0.27
NAECA minimum EHP	2.00 * 0.27 = 0.54
High efficiency EHP	2.60 * 0.27 = 0.70

Full Trajectory Resource Energy Efficiency

Gas and Electric Water Heating Systems

	EF * REE = REF
GAS WATER HEATER	
NAECA minimum	0.55 * 0.91 = 0.50
Subchamber	0.62 * 0.91 = 0.56
Subchamber - powered	0.72 * 0.91 = 0.65
Subchamber - condensing	0.85 * 0.91 = 0.77
ELECTRIC WATER HEATER	0.90 * 0.27 = 0.24

Environmental Externalities Costs

Range of Imputed Values

Pollutant	Cost Range	NYSPSC
	$ / ton	$ / ton
SOX	$ 630 - 27,436	$ 840
NOX	$ 101 - 27,676	$ 1,840
CO2	$ 1 - 35	$ 1
Particulates	$2,020 - 4,000	$ 2,020

Full Trajectory Resource Energy Efficiency

Conclusion

- Natural gas end use equipment satisfies customers' end use energy needs more efficiently than electric end use equipment for selected applications including:
  - residential & commercial space heating
  - residential and commercial water heating
  - residential and commercial cooking
  - residential and commercial laundry drying
  - commercial and industrial space conditioning
  - commercial and industrial engine -driven equipment
  - electric power generation

Environmental Externalities Costs
Space Heating

Pollutant	NAECA Gas Furnace	NAECA EHP
	$ / yr.	$ / yr.
SOX	0.02	106.6
NOX	6.6	88
CO2	3.27	7.79
Particulates	0.8	121.2
TOTAL	10.69	343.59

Ratio = 32:1

Integrated Resource Planning

Space Heating Comparison

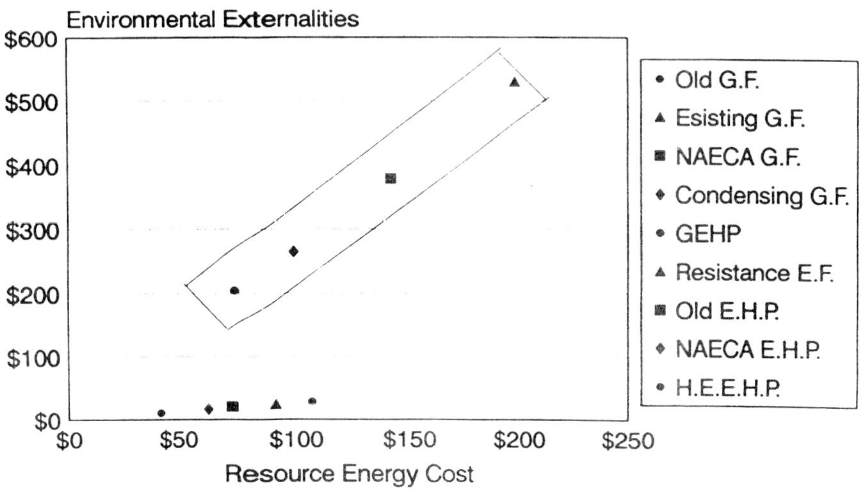

Integrated Resource Planning

Water Heating Comparison

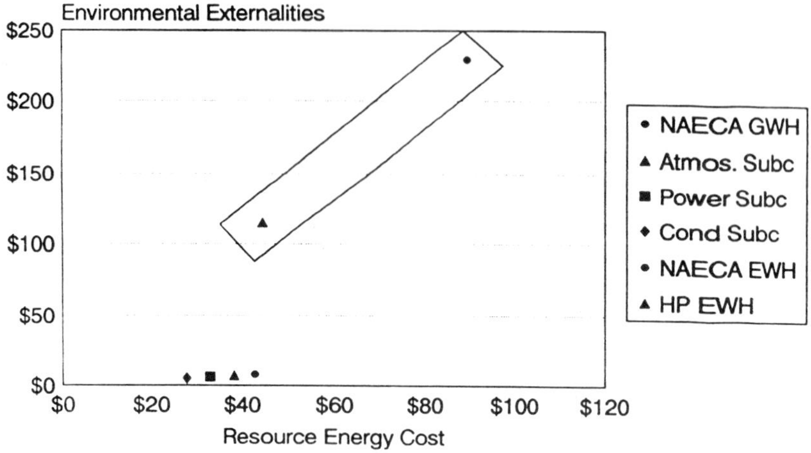

Integrated Resource Planning

Commercial Cooling Comparison

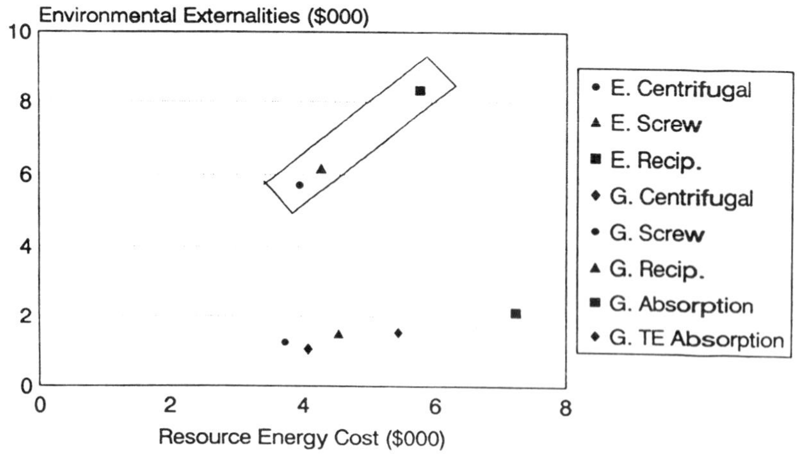

200 RT, 2000 hr. / yr., No heat recovery

Integrated Resource Planning

Cogeneration comparison

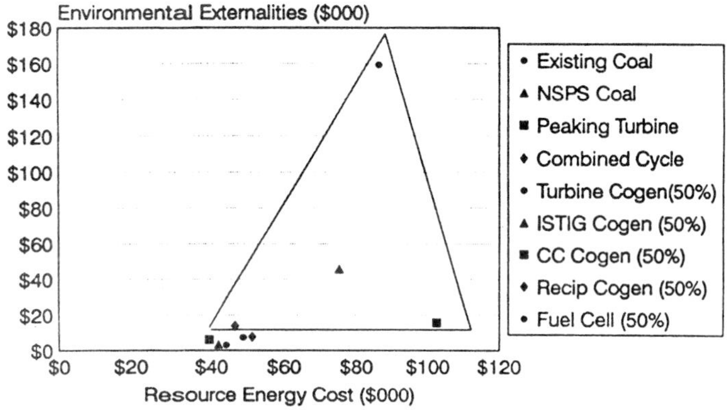

1 MW Generating Capacity @ 65% LF

Environmental Externalities Costs

Future Issues

- Wide range of imputed costs
- CAAA will reduce SOX by 50%
- CAAA will reduce NOX by 20%
- Increased efficiency reduces CO2
- CAAA air toxic emission limits 1994
- High level nuclear waste storage

Integrated Resource Planning

Summary

- Natural gas site efficiencies lower
- Electric trajectory efficiency lower
- Natural gas more resource efficient
- Natural gas end use emissions lower
- Natural gas externalities costs lower
- Natural gas total societal costs lower
- Natural gas cogeneration preferable

Integrated Resource Planning

Fuel Switching

- Recognized for combination utilities
- Complicated and contentious
- Environmental benefits compelling
- Competing incentives confusing
- Competing incentives not economical

Integrated Resource Planning

Opportunities and Challenges

- IRP has great potential
- Electric IRP has limited potential
- Gas IRP also has limited potential
- Contemporaneous review essential
- Incentive funding must track benefits
- Economic competitiveness crucial

Integrated Resource Planning

Inescapable Conclusions

- If Electric IRP
  - then Gas IRP
    - our customers are their customers
- If Electric and Gas IRP
  - then (comprehensive) IRP
    - can't plan effectively in vacuum
- If we plan effectively
  - then we must succeed
    - the price of failure is intolerable

Opportunites and Prospects for Demand-Side Efficiency Improvements

Michael A. Kuliasha
Associate Director / Energy Division
Oak Ridge National Laboratory
Oak Ridge, Tennessee, U.S.A.

Background:

Dr. Kuliasha is currently Associate Director of the Energy Division of Oak Ridge National Lab, with primary interests in research and development, relating to energy end-use and the environment, including conservation technology, transportation systems, advanced computing, environmental assessment, and economic analysis.

OPPORTUNITIES AND PROSPECTS FOR DEMAND-SIDE EFFICIENCY IMPROVEMENTS

Michael A. Kuliasha
Oak Ridge National Laboratory
P.O. Box 2008, MS-6189
Oak Ridge, Tennessee 37831-6189, USA

ABSTRACT

Substantial progress has been made over the last 20 years in improving energy efficiency in all sectors of the U.S. economy. Although there remains a large potential for further efficiency gains, progress in improving energy efficiency has slowed recently. A combination of low energy prices, environmental challenges, and life-style changes have caused energy consumption to resume rising. Both new policies and technologies will be necessary to achieve cost-effective levels of energy efficiency. This paper describes some of the promising new demand-side technologies that are currently being implemented, nearing commercialization, or in advanced stages of development. The topics discussed include finding replacements for chlorofluorocarbons (CFCs), new building equipment and envelope technologies, lessons learned about conservation program implementation, and the role of utilities in promoting the efficient use of energy.

INTRODUCTION

Investing in energy efficiency can be a good business decision that can save money, improve comfort and functionality, and contribute to an improved environment. A number of studies have compared the economics of improving energy efficiency with the cost of new energy supplies to meet the growing demand for energy services.[1,2] A common conclusion is that in many cases improving energy efficiency is several times more cost effective than developing new energy supplies. The energy saved from improving energy efficiency can be used to supply new demand, and more efficient products decrease the rate of growth in demand. The lower capital requirements of increasing energy efficiency as compared to developing new energy supplies is particularly important in the capital-constrained economies of developing countries. Consequently, improving energy efficiency can improve the environment, liberate investment capital, and reduce the need for developing new energy supplies and their environmental impacts.

A certain level of efficiency improvement will likely be achieved without any major policy initiatives. Figure 1 shows the results of a study commissioned for the United States Department of Energy's (DOE) National Energy Strategy.[3] The study examined three possible scenarios for future energy use in the U.S. for the period 1988 to 2010. The frozen efficiency scenario projects energy consumption given no further improvements in energy efficiency. As such, it is a direct measure of the growth in demand for energy services which are projected to grow from 85 EJ to 121 EJ by the year 2010. The second scenario looks at where the United States is headed if current government policies and programs remain in effect. Under this scenario, energy consumption will grow to 108 EJ in 2010, an 11% improvement over the base energy demand. The third scenario examined energy consumption if currently available, cost-effective efficiency improvements were adopted by all consumers. This scenario projects annual consumption of only 93 EJ in 2010, which represents a 23% savings in energy consumption.

However, there are serious questions as to whether even the 11% efficiency improvement projected based on historical long-term trends will be achieved. In the last several years, a combination of low energy prices, environmental challenges, and life-style changes have caused energy consumption to rise substantially. In the industrial sector, energy productivity is stagnating. Transportation energy use is up as fleet average fuel economy has stabilized and vehicle miles traveled continues to increase. Commercial building energy use is rising substantially due to office automation.

Both new policies and technologies will be necessary to achieve cost-effective levels of energy efficiency. This paper describes some of the promising new demand-side technologies that are currently being implemented, nearing commercialization, or in advanced stages of development. The topics discussed include finding replacements for CFCs, new building equipment and envelope technologies, lessons learned about conservation program implementation, and the role of utilities in promoting the efficient use of energy.

REPLACING CFCs

The fully halogenated CFCs have been identified as contributing to reductions in stratospheric ozone. The United States and 33 other countries have ratified the Montreal Protocol mandating a gradual reduction in the production of certain compounds containing chlorine and bromine, including CFC-11, CFC-12, CFC-113, and CFC-114. All of the major producers of these chemicals in the U.S. have announced that they will cease production of the fully halogenated CFCs as soon as alternative refrigerants and blowing agents are available.

The transition away from fully halogenated CFCs could have a negative impact on energy end-use because these CFCs have high thermal efficiencies in both refrigeration systems and foam insulations. An assessment of the energy impacts of substitute refrigerants found that the major energy impacts will occur in those applications that rely almost exclusively on CFC-blown foam insulation. These include household refrigerators, freezers, water heaters, roofs of commercial buildings, and insulated building foundations.[4]

Both the direct and indirect effects of CFC alternatives should be considered when analyzing the environmental consequences of replacements. Direct effects include the ozone depletion and greenhouse warming caused by the refrigerants, solvents, and blowing agents that escape into the atmosphere. Indirect effects are the energy use and resulting CO_2 emissions attributable to the application. Even small decreases in energy efficiency are unacceptable for CFC replacements from a global warming perspective.[5]

The refrigeration, air conditioning, and insulation industries have been working on improving the energy efficiency of refrigeration systems and insulations that use CFC alternatives. In 1988, CFC alternatives resulted in a 10 to 15% reduction in energy efficiency in many applications. In refrigeration systems new formulations, synthetic oils, and redesigned equipment have resulted in recent test results showing essentially equal energy efficiency for refrigerator/freezers using CFCs and those using either hydrofluorocarbon (HFC) 134a or a mix of hydrochlorofluorocarbons (HCFCs) consisting of HCFC-124, HCFC-22, and HFC-152a.

The results of recent laboratory and field tests show that polyisocyanurate insulation board blown with CFC alternatives do not currently perform as well as insulations blown with CFCs. Figure 2 shows the performance of polyisocyanurate foam laminate boardstock blown with CFC-11, HCFC-123, and HCFC-141. A cooperative industry/government project on alternative blowing agents concluded that foams blown with alternative blowing agents exhibited long-term thermal conductivity values 7 to 15% above those for CFC-11 foams under similar conditions.[6] Given the relative importance of indirect environmental effects discussed previously, additional work is needed to improve the energy efficiency of insulations blown with non-CFCs.

A current issue in the insulation industry is the effect of aging on the insulating value of foam insulations. A large part of the high thermal resistance of polyisocyanurate and polyurethane is due to the low thermal conductivity of the blowing agents trapped in the foam cells. However, over time, air diffuses into the foam and the blowing agents diffuse out. This aging process degrades the thermal resistance of the insulation. Test results show that the thermal resistance of some foam insulations can decrease more than 25% during two years.The effects of aging resulting in decreased thermal performance is also obvious in Figure 2.

The challenge of redesigning equipment and products to eliminate the use of CFCs can also spark innovations that result in improved products with higher efficiencies than their CFC counterparts. For example, a joint industry/government project is currently underway to design an advanced refrigerator/freezer that uses half the energy of the best model sold in 1991. The project is considering advanced refrigeration systems and vacuum insulations that are being pursued at least in part because of the necessity of finding alternatives to CFCs. Also, important enabling technologies have been developed such as a leak detector that is sensitive to both CFC and non-CFC refrigerants and a ratiometer that can easily and quickly determine the composition of a refrigerant mixture.

NEW BUILDING EQUIPMENT TECHNOLOGIES

The efficiency of conventional building heating and cooling equipment has continued to improve, primarily through the application of more efficient burners, better compressors, improved heat exchangers, and microprocessor controls. The next big step in HVAC efficiency improvement has just begun with the increasing availability of variable-speed systems. These systems match building load by varying speed rather than on-off cycling resulting in higher efficiency through reduced cycling losses and by improved heat exchange at the part-load conditions that make up the majority of building operation.

If successful, several technologies currently under development may have a profound effect on the economics of building heating and cooling. One such technology is advanced absorption systems. Absorption systems are widely used in some world markets for building cooling. A typical double-effect absorption chiller has a coefficient of performance (COP) of around 0.95. A new absorption cycle, termed the triple-effect cycle, has been invented at Oak Ridge National Laboratory that offers the potential of a 50 to 60% performance improvement over conventional double-effect machines.[7] The triple-effect cycle consists of two conventional single-effect absorption systems using different working fluids that are coupled so that both the condenser and absorber reject heat from the high-temperature loop are used to produce additional refrigerant in the low temperature loop with both evaporators providing useful cooling. Analytical predictions are that the new cycle could result in practical absorption machines with COPs in excess of 1.5. The technology has been licensed and is under development at the Trane Company and Apache Corporation for large building cooling and cogeneration applications respectively.

In addition to absorption chillers, thermally activated heat pumps that burn fuel to produce both heating and cooling using a refrigeration cycle are also under development. These systems are particularly attractive in locations with a substantial heating load, because their heating COP can be more than double that of an efficient furnace while providing cooling at a primary energy efficiency (considering electricity conversion and transmission losses) on par with electric systems. Heat pump systems that use an internal combustion engine converted to run on natural gas instead of an electric motor driving a conventional compressor are already on the market. Small engine-driven heat pumps are currently being field tested. Absorption systems that heat pump in the heating mode as well as provide cooling are also nearing commercialization.

Lighting consumes a significant portion of the electricity consumed in buildings. White light can theoretically be produced at almost 350 lm/W. By comparison, a fluorescent lamp operating with a conventional ballast has an efficiency of 63 lm/W, a high-intensity discharge lamp produces about 50 lm/W, while an incandescent lamp has an efficiency of about 16 lm/W. An obvious energy efficiency option is to minimize the use of incandescent lamps. Compact fluorescent lamps can be substituted in many of the applications that would normally use an incandescent lamp. The efficiency of conventional fluorescent lamps can be improved by the use of a solid state ballast, a high-frequency (20,000 Hz) ballast that can replace conventional magnetic ballasts. A cool-white bulb operating with a solid-state ballast has an efficiency of 80 lm/W and can reduce lighting energy needs by 25%. Improved lighting fixtures are also available with reflective surfaces that provide more useful light per watt. A further benefit from improving lighting efficiency is that it reduces the load on the building cooling system.

RECENT ADVANCES IN BUILDING ENVELOPE SYSTEMS AND MATERIALS

New products and issues regarding old products and design practices continue to dominate the building envelope systems and materials industry.

Recent test results have shown that loose-fill fiberglass insulations can suffer from convective cells when large temperature gradients are present across the insulation. The effect of these convective cells is that the insulating value of the material can be less than half the rated value as determined under more moderate conditions. Further testing has shown that a thin blanket placed on top of the loose-fill insulation will disrupt the convection currents and restore full insulating value to the system. Some states have already incorporated these findings into their building codes.

Low-emissivity (low-E) windows are readily available where one of the inner surfaces of a double-glazed window is coated with a thin film of a transparent low-emissivity material which reflects infrared radiation. Although low-E windows were originally designed to reduce heat loss in winter, later developments in low-E coating technology have yielded modified coatings that block most of the solar near-infrared radiation, while transmitting a high level of visible radiation. These coatings allow daylighting while minimizing excessive solar gain.

Windows with the same high thermal resistivity as walls are a technical possibility. Double-glazed windows with a hard vacuum between the two panes are under development. Another option is to fill the gap between the two panes with a high-tech insulator called aerogel. Aerogel consists of sparse skeleton of tiny glass particles that are transparent to visible light but that scatter infrared radiation. Recent advances in thin-film coatings for windows provides a means of reducing heat gain significantly without proportionally reducing daylight transmittance. Compared to dark tinted glass, these spectrally selective products can transmit up to four times as much daylight with equal or better solar heat control.[8]

Another interesting window development is switchable windows. These windows have an electrochromic coating that changes the opacity of the window depending on the applied voltage. This new control option for glazing could be used to reduce glare, minimize heat gains while allowing daylighting, or to control heat losses.

CONSERVATION PROGRAM IMPLEMENTATION

Because new construction adds less than 1% per year to the building inventory, improving the energy efficiency of existing buildings offers the greatest near-term potential for energy savings. For example, government weatherization assistance programs weatherized 198,000 single-family and small multi-family dwellings in 1989. The annual energy savings per dwelling is estimated to be 16.4 MBtu at the site. Over the 20-year projected life of the weatherization measures, these homes will save the equivalent of 12 million barrels of oil, or approximately the amount of oil added to the Strategic Petroleum Reserve in 1992.[9]

Efficiency gains of 30-40% are possible and can be cost-effective with the proper selection of retrofit measures. The performance of new retrofit measures such as radiant barriers in attics and "outsulation" (insulation applied to exterior walls of concrete block) have been quantified in field tests and added to the repertoire of available measures. Recent field tests of outsulation in Arizona measured a 10% reduction in electricity consumption and a 0.5 kW reduction in cooling demand.

Much has been learned in the last 10 years about the effective implementation of conservation programs. A number of studies have clearly demonstrated the benefits of using an audit-directed approach for selecting retrofit measures. That is, the energy savings of various efficiency measures are analyzed including the cost of the measure and interactions between various measures. Table 1 compares the results of four different field tests in Wisconsin, New York, and North Carolina. Recent field tests show that an advanced audit-directed procedure can almost triple the overall cost-effectiveness of low-income weatherization assistance programs as compared to the priority system formerly used.[10]

The cost-effectiveness of retrofit measures is highly dependent on the original condition of the building and climate. A detailed study of small commercial buildings in the U.S. found that relamping to lower light levels, adding manual switching for zoned lighting, and temperature setback/setup control were cost effective in all locations.[11] Additional insulation was generally cost effective in colder climates. The average payback for groups of cost-effective measures was about 2.5 years.

In residential buildings, the most effective weatherization programs were found to: occur in moderate and cold climates; weatherize high energy users; use an integrated audit that considers both envelope and equipment measures and the interactions between measures; use measured equipment efficiencies in calculating paybacks; include distribution system diagnostics; and use a blower door to guide infiltration control.

THE ROLE OF UTILITIES IN PROMOTING ENERGY EFFICIENCY

With so many options available for improving energy efficiency and with favorable economics in many applications, one has to wonder why more efficient systems are not the norm. The primary factors limiting the widespread utilization of existing cost-effective technologies are institutional rather than technological. These barriers have been characterized as both structural and behavioral and include such problems as distortions in energy prices, access to capital, regulatory policies, codes and standards, infrastructure limitations, information gaps, and perceived risk.[12]

One problem has to do with the differing economic ground rules that are applied to energy consuming investments compared to energy supply investments. A building owner often makes decisions on simple payback with three years being a common rule-of-thumb. Thus investments to improve energy efficiency must generate a 33% return on equity. Electric utilities compare options over a 30 or 40 year life using life-cycle costs and typically earn a 10% return on equity. Thus there is a built-in bias in the economy in favor of supply options.

A very promising development for overcoming this institutional barrier is the ongoing change in the role of utilities. Many utilities and utility commissions are beginning to view utilities as energy service providers rather than just energy providers. That is, energy itself is not an end product. It is the light, shaft power, heating, cooling, and other services energy provides that makes it such a valuable commodity. If utilities extend their planning and investment decisions all the way to the end-use and if regulatory agencies allow utilities to earn an appropriate return on end-use investments, then the playing field is leveled between supply and end-use efficiency options.

While current world energy prices are unusually low limiting the current cost effectiveness of some efficiency measures, it is important for practitioners and policy makers to bear in mind that the replacement rate for many end-uses is slow and that products purchased today will continue to consume energy over the life of the product. Future energy costs as well as environmental concerns should play a factor in investment decisions and policies. Investments in energy efficiency made today will continue to generate a return as well as contribute to an improved global environment.

ACKNOWLEDGEMENTS

This work was sponsored by the U.S. Department of Energy under contract number DE-AC05-84OR21400 with Martin Marietta Energy Systems.

REFERENCES

1. Arthur H. Rosenfeld and David Hafemeister, "Energy-Efficient Buildings", *Scientific American*, April 1988.

2. J. Goldemberg et. al., *Energy for a Sustainable World*, Wiley Eastern, Ltd., New Delhi, 1988.

3. R. S. Carlsmith et al., *Energy Efficiency: How Far Can We Go?*, ORNL/TM-11441, Oak Ridge National Laboratory, Oak Ridge, Tennessee, January 1990.

4. S. K. Fischer and F. A. Creswick, "How Will CFC Bans Affect Energy Use", *ASHRAE Journal*, Vol. 30, No. 11, November 1988.

5. S. K. Fischer, P. J. Hughs, and P. D. Fairchild, *Energy and Global Warming Impacts of CFC Alternative Technologies*, ORNL/M-1869, Oak Ridge National Laboratory, Oak Ridge, Tennessee, December 1991.

6. J. E. Christian et. al., *The Technical Viability of Alternative Blowing Agents in Polyisocyanurate Roof Insulation: A Cooperative Industry / Government Project*, ONRL/CON-367, Oak Ridge National Laboratory, Oak Ridge, Tennessee, June 1993.

7. R. C. DeVault, *Triple Effect Absorption Chiller Utilizing Two Refrigeration Circuits*, U.S. Patent 4,732,008, March 1988.

8. B. J. Davids, "Taking the Heat Out of Sunlight - New Advances in Glazing Technology for Commercial Buildings", *ACEEE 1990 Summer Study on Energy Efficiency in Buildings*, American Council for an Energy-Efficient Economy, Washington, DC, 1990.

9. M. A. Brown et. al., *National Impacts of the Weatherization Assistance Program in Single-Family and Small Multifamily Dwellings*, ORNL/CON-326, Oak Ridge National Laboratory, Oak Ridge, Tennessee, June 1993.

10. T. R. Sharp, "Advancing Weatherization Performance: Measured Results from the North Carolina Field Test of an Advanced Measure Selection Technique", *Proceedings of the Sixth International Energy Program Evaluation Conference*, Chicago, August 25-27, 1993.

11. R. J. Kedl and T. K. Stovall, *Cost-Effectiveness of Single and Multiple CACS Conservation Actions in Small Commercial Buildings*, Oak Ridge National Laboratory, ORNL/CON-176, August 1985.

12. E. A. Hirst, "Individual and Institutional Behavior Related to Energy Efficiency in Buildings", *Journal of Environmental Systems* 16(1), pp. 57-74, 1986-87.

TABLE 1: COMPARISON OF
CONSERVATION PROGRAM FIELD TESTS

	Average Cost per House ($)	Heating Energy Savings (%)	Annual Energy Savings per $100 (MBtu/yr/$100)
1983 WISC Study	2,250	10	0.5
1985 WISC Field Test	1,603	19	1.3
1988 NY Field Test	1,690	25	1.5
1991 NC Field Test	1,400	33	1.2

Cost includes $300/house for administration and audit

FIGURE 1: CONSERVATION SCENARIOS

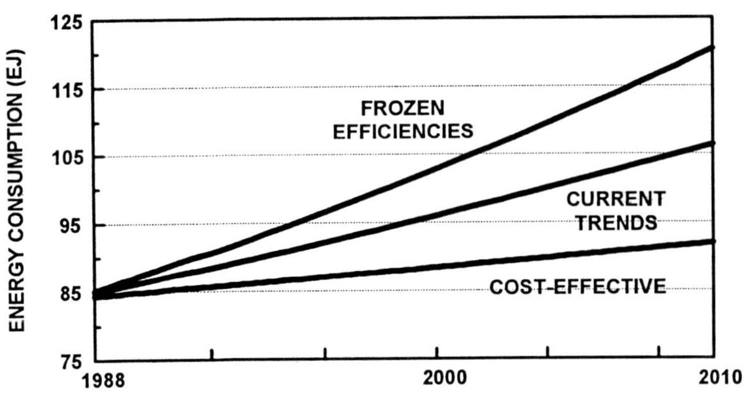

FIGURE 2: THERMAL PERFORMANCE OF FOAM INSULATIONS

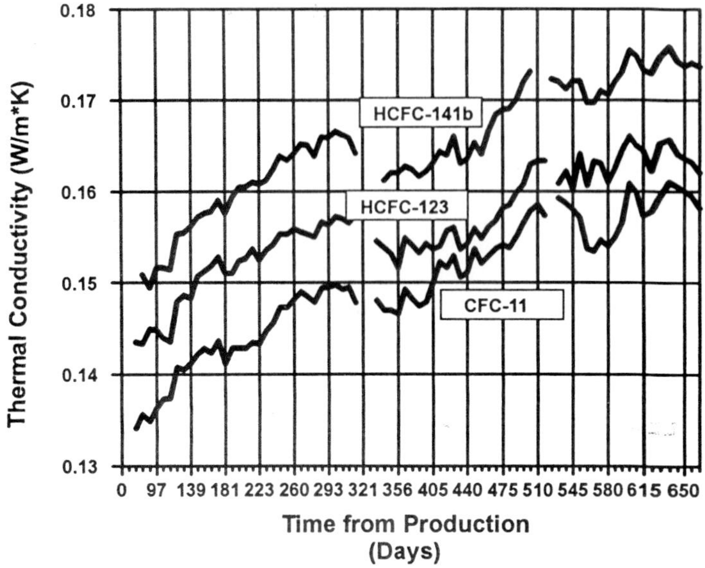

CHAPTER 7 -- ZERO EMISSION VEHICLES

Electric Vehicles -- Opportunity & Challenge

Lawrence G. O'Connell
Manager, Technical Support
Electric Power Research Institute
Palo Alto, California, U.S.A.

Background:

Mr. O'Connell's work at EPRI is currently Manager, Technical Support, Customer Systems Division, with emphasis on Electric Transportation. Prior to this he was Manager of the EPRI Transportation Program. He managed Transportation Systems Research at Lawrence Livermore National Lab and was Resource Manager for the Separator Project. He has worked with the U.S. Atomic Energy Commission, Pacific Gas & Electric and the California Department of Transportation.

ELECTRIC TRANSPORTATION: OPPORTUNITY AND CHALLENGE

Larry O'Connell
Electric Power Research Institute
3412 Hillview Avenue
Palo Alto, California 94303 USA

ABSTRACT

In the relatively short period since their invention, petroleum-powered vehicles have become the mainstay of our ground transportation system, providing cost-effective and expansive mobility of people, goods, and ideas. To offset some of the problems caused by overuse of petroleum in transportation--such as deterioration of air quality and the balance of trade--federal, state, and local governments are encouraging the development and use of alternative-fuel vehicles. Among these, electric vehicles show high potential as a means of improving urban air quality and reducing dependence on imported petroleum. However, today's electric vehicles are suitable only in selected niche applications. A number of development efforts are focusing on overcoming the technical and market barriers to this promising technology.

INTRODUCTION

Rapid, mechanistic transportation is a relatively modern phenomenon. In fact, for most of human history, walking has been the predominant mode of land transport. In slightly more than two centuries, however, vehicles powered by petroleum--now the dominant transportation fuel--and other fuels have become one of the most cost-effective means for moving people and goods over long distances. The mobility they provide affords extensive benefits: greater use of land and greater movement of ideas and goods--and thus the opening of global perspectives and markets.

It is crucial to keep these benefits in view as headlines remind us regularly of the toll petroleum-powered vehicles extract on society. Although these problems must be solved, sacrificing mobility is not a viable economic option.

PROBLEMS ASSOCIATED WITH PETROLEUM-POWERED VEHICLES

The problems caused by transportation are real and serious. For example, EPA studies have estimated that highway vehicles produce 27% of all nonmethane organic gas (NMOG) emissions, 28% of nitrogen oxide (NOx) emissions, 50% of carbon monoxide (CO) emissions, and 24% of carbon dioxide (CO_2) emissions [1 and 2]. All of these emissions are potentially dangerous: NMOG and NOx combine in sunlight to form smog, CO is a serious health hazard, and CO_2 is the principal gas associated with global warming. Despite a steady tightening of federal tailpipe standards over the last 20 years, air quality remains a critical issue in most large U.S. cities.

Further, about 62% of all petroleum consumed in the United States today goes to the transportation sector. Figure 1 shows the evolution of this trend [3]. An examination of the balance of payments brings the implications of this oil addiction into sharp focus: In 1991, imported oil accounted for $51 billion of the total trade deficit of $65.4 billion [4].

ELECTRIC VEHICLES: PART OF THE SOLUTION

Due to their unique characteristics, electric vehicles (EVs) can be part of a solution to these transportation-related problems. For example, because EVs produce no emissions while operating, they remove pollutants from crowded urban areas, thereby significantly reducing health risks. Further, even accounting for the power plant emissions required for EV battery recharging, EVs are much cleaner than their gasoline counterparts [5].

Increased use of electricity in transportation could also help cure America's oil dependence. Only about 4% of U.S. electricity is generated from petroleum. The remainder is produced from a variety of domestic energy sources. Coal accounts for more than 50% of U.S. electricity generation, nuclear for just over 20%, while hydro, natural gas, and other fuels make up the difference. Due to this fuel diversity, EVs use much less oil than gasoline vehicles. For instance, an electric car based on 1995 technology recharged during a utility's peak period would require only two barrels of petroleum over 100,000 miles of operation. In contrast, a gasoline car would need 94 barrels to cover the same distance. (These figures assume an average efficiency for future EVs of 4 m/kWh and an average efficiency for gasoline vehicles of 28 mpg.)

These potential benefits clearly justify the pursuit of EV technology advances. In fact, federal, state, and regional initiatives to force fuel diversification in the transportation sector have created a major policy push for EVs.

To date, the biggest boost for EVs has been in California. Facing severe air quality problems, California passed the world's most stringent clean-vehicle regulations in 1990. These rules force the use of increasingly low-emitting vehicles and require automakers to begin producing zero-emission vehicles (ZEVs) for sale in the state. Starting in 1998, 2% of all new passenger cars and light-duty trucks offered for sale--or more than 24,000 vehicles--must have zero emissions, a qualification only EVs meet at the present time. By 2003, 10%,--or about 122,000--of new vehicles sold must be ZEVs. (These figures assume a 0% increase in annual registration from 1990 levels.) Other states, particularly those in the northeast, are beginning to adopt similar rules, creating an even larger potential market for EVs in this decade.

Other efforts bolster the momentum created by the ZEV mandate:

- The 1990 Clean Air Act Amendments created two programs that promote alternative-fuel vehicles. One sets up a pilot alternative-fuel vehicle (AFV) program for California, and the other requires some urban fleets to begin using AFVs. In 1998, 30% of new light-duty vehicles in these fleets will run on alternative fuels, and by 2000, 70% of such vehicles will be AFVs.

- The Energy Policy Act of 1992 creates a tax credit equal to 10% of the EV purchase price (up to $4,000). It also establishes a $485 million EV R&D program, a $50 million EV demonstration program, and a $40 million EV infrastructure demonstration program. Further, it requires alternative-fuel providers (including electric utilities) and federal and state governments to begin buying AFVs for their fleets in 1996.

- More than 30 states have either adopted or are considering legislation to promote EVs. In many cases, states are not simply contemplating individual bills, but establishing comprehensive programs to ensure that significant numbers of AFVs will come into use.

TECHNOLOGY DEVELOPMENT: CHALLENGES AND STATUS

Automotive firms around the world have responded to these mandates and incentives with vigorous EV development programs. However, the challenges are significant. Using current technology, EVs cost more and deliver lower performance than do conventional vehicles. Considerable effort and progress will be necessary before EVs become a major factor in the marketplace. Nevertheless, advances are being made in vehicle and battery development, and several products are appearing that show promise as commercial vehicles for specific market niches.

EPRI's Role in EV Development

In the United States, the electric utility industry, through EPRI, has been active in the EV movement and instrumental in developing two EVs: the G-Van--developed and manufactured by Conceptor Industries with General Motors--and the TEVan--developed with Chrysler. Both these vehicles, shown in Figures 2 and 3, have successfully achieved federal safety certification.

These limited-production electric vans are undergoing multi-year field evaluations in utility fleets across the country. These evaluations will provide valuable information on product performance and limitations, areas for improvement, user acceptance, and charging behavior. The vehicles themselves are proving EV technology under real-world conditions and are introducing many people to EVs. By proving the feasibility of EV technology, these vehicles have contributed to the strong support for EVs in the policy initiatives described above.

Other Vehicle Development Projects

Other projects are resulting in concept and limited-production EVs, as described below:

- Dodge EPIC: This concept vehicle has been presented by Chrysler at autoshows around the country. The EPIC features a two-speed automatic transmission, air conditioning, and power brakes and steering. Equipped with a nickel-iron battery, the van has a top range of 100 miles and a top speed of 65 mph. Chrysler's next production EV after the TEVan will probably be based on the technology in this vehicle.

- GM Impact: GM fired the public imagination with its small electric concept car, the Impact. Designed from the ground up for performance, the Impact is one of the fastest EVs around. With maintenance-free, sealed lead-acid batteries, it achieves a 120-mile range, reaches a top speed of 75 mph, and can outrace the Mazda Miata and Nissan 300ZX in acceleration heats. In fall of 1993, GM plans to produce 50 Impacts for leasing to consumers to two- to four-week trial periods. This pilot program will run for two years.

- Ford Ecostar: Ford is building 82 Ecostar electric minivans that will be tested in utility fleets in the United States, Europe, and Mexico. The field evaluation, to begin this year, will run for 30 months. Equipped with a sodium/sulfur battery and Ford's ac powertrain, the vehicle achieves a 100-mile range and a top speed of 75 mph. Nine of the vehicles produced will be hybrid versions. The hybrid includes a small internal-combustion engine to extend vehicle range to about 250 miles.

- Nissan FEV: In 1992, Nissan unveiled its Future Electric Vehicle (FEV)--the first EV equipped with Nissan's quick-recharge, light-weight nickel-cadmium batteries. According to the company, the prototype EV can be recharged to 40% capacity in 6 minutes and to 100% capacity in under 15 minutes. With a ground-up aerodynamic design, the FEV offers a top speed of 81 mph and a range of 100 miles at about a 45-km/h constant speed. Nissan has not yet announced production plans.

Battery Development

High-performance, low-cost batteries are the key to making EVs acceptable to consumers. To help create a battery that meets all the criteria for a market-ready EV, the Big Three U.S. automakers--GM, Ford, and Chrysler--joined to form the United States Advanced Battery Consortium (USABC). With participation from the U.S. Department of Energy (DOE) and EPRI, representing the utility industry, USABC is developing batteries to meet the 1998 ZEV requirements and to eventually provide EV performance comparable to that of conventional cars. USABC also funds basic battery research through Cooperative Research and Development Agreements (CRADAs) with national laboratories.

USABC has a number of contracts in place:

- Ovonics Battery Co. in Troy, Michigan, received a contract to scale its nickel-metal hydride technology to EV battery-pack size, improve battery performance, and speed batteries to mass production.

- W.R. Grace & Co. and Johnson Controls, Inc., were granted a contract to develop a thin-film lithium polymer bipolar battery. The battery will use a solid-polymer electrolyte that may alleviate problems inherent to liquid electrolytes.

- SAFT America, Inc., was given a contract to develop a lithium (aluminum)-iron disulfide battery. This elevated temperature system achieves high energy and power densities in a relatively small, low-weight battery.

Despite these efforts, USABC activities may not result in a commercial battery by the 1998 ZEV deadline. To fill this gap, EPRI helps develop and test promising near-term batteries. Encouraged by favorable evaluation results for the Electrosource Horizon battery, for example, EPRI has initiated a utility-sponsored project to fund development and pilot production of this advanced lead-acid system.

Looking farther into the future, EPRI has teamed with DOE to perform exploratory research. This work is searching for novel cell components that could become the building blocks of advanced batteries for EVs.

BUILDING THE EV INFRASTRUCTURE

As noted above, EPRI has been active in advancing EV technology. EPRI's focus is currently shifting, however. With automakers taking over the lion's share of vehicle development, EPRI is now able to concentrate on a role better suited to the electric utility industry: providing the electricity infrastructure to facilitate EV charging.

EPRI framed its work in this area by helping establish the Infrastructure Working Council (IWC). By merging the expertise and resources of many organizations that will contribute to infrastructure development--automakers, utilities, Society of Automotive Engineers (SAE), DOE, and others--IWC has created a five-year action plan. This plan has outlined work in three general areas:

- Developing convenient, standardized charging stations and a charging network
- Educating consumers about EV capabilities and benefits
- Addressing utility impacts and optimizing the load management benefits of EVs

By conducting or encouraging these activities, EPRI hopes to smooth the path to wide-scale EV introduction and use. EV owners will be knowledgeable about EV capabilities and assured of quality charging services, and electric utilities will be prepared to realize the benefits of EV commercialization.

Developing Charging Options

The main focus of EPRI's work to date has been developing charging options. Most charging is expected to occur in residential garages, while vehicles are parked overnight. Such "refueling" will be more convenient for consumers than stopping at a gas station. Supplying the electrical service for residential charging is not expected to pose problems for utilities or their customers.

Providing a reliable network of stations for charging away from home will require a somewhat greater effort. This network will fulfill the EV driver's need to extend range with daytime "opportunity" charging--at work, in public parking lots, or even curbside at municipal parking meters. In the future, improvements in battery technology may allow charging at retail fast-charge stations similar to the corner gas station. Through such strategies as issuing RFPs and sponsoring design competitions, EPRI is encouraging work on hardware configurations for charging away from home.

EPRI is also exploring the fast-charge concept. While attractive to customers, the utility industry is concerned with the potential high costs of investing in fast-charge equipment, the effects of fast charging on batteries, and the impacts of fast charging's high electricity-load requirements on the utility system. EPRI's research is therefore seeking to identify the least-cost equipment solution, quantify fast-charging effects on battery cycle life, and develop low-cost on-site energy storage options to mitigate utility system impacts.

Activities in Other Areas

A fundamental infrastructure issue for utilities is ensuring adequate electrical supply for EV charging. This doesn't appear to be a problem in aggregate: A recent study conducted by Argonne National Laboratory estimated that in 2010, more than 31 million EVs could be recharged nationwide without impacting the utility system [6]. However, because individual utilities could face highly localized effects, the industry is examining methods for encouraging customers to do the majority of charging during off-peak hours. If most battery charging occurs at night, EVs could help utilities make more efficient use of generating capacity.

In general, encouraging people to buy EVs and use them effectively will require a wide-scale public education effort. As a first step toward introducing consumers to this new, beneficial technology, EPRI is establishing an EV Research Network. This Network will consist of offices at utilities across the country that will test, evaluate, and demonstrate the latest EV technologies. For example, several Network utility offices are evaluating the Chrysler TEVan, while other offices will be hosting field demonstrations of the new public charging station prototypes. The Research Network will also offer an array of resources, including exhibits, videos, and an on-line library and reference center, to help familiarize visitors with EV technology.

CONCLUSIONS

It is clear that EVs can be part of a nationwide effort to address and correct the effects of a transportation system that is almost wholly dependent on petroleum. But bringing attractive, reasonably priced EVs to market will require extensive efforts in a variety of areas. The vehicle, battery, and infrastructure activities now in place will help achieve the near-term goal of ensuring that some EV and charging options are ready for controlled demonstrations before 1998 to prepare for the ZEV deadline. More important, as this work leads to market-ready products, it will help achieve the broader national goal of creating a transportation system that protects natural resources while providing the mobility necessary for national economic health.

REFERENCES

1. Motor Vehicle Manufacturers Association. *MVMA Motor Vehicle Facts & Figures 1991*. Detroit, Michigan. 1991.

2. The Alliance to Save Energy. "Carbon Dioxide Emission, by Activity," *Alliance Update*, Fall 1991, Volume II, Number 3.

3. U.S. Department of Energy. *Annual Energy Outlook 1992 with Projections to 2010*. DOE/EIA-0383 (92). Washington, D.C.: Energy Information Administration, Office of Integrated Analysis and Forecasting, 1992.

4. International Trade and Commerce. Telephone Communication, Winter 1993.

5. *Electric Van and Gasoline Van Emissions: A Comparison*, TB.CU.177.10.89, Electric Power Research Institute, 1989.

6. *Meeting the Electric Vehicle Generating Capacity in 2010*, Electric Power Research Institute, Forthcoming.

FIGURES

FIGURE 1: OIL ADDICTION

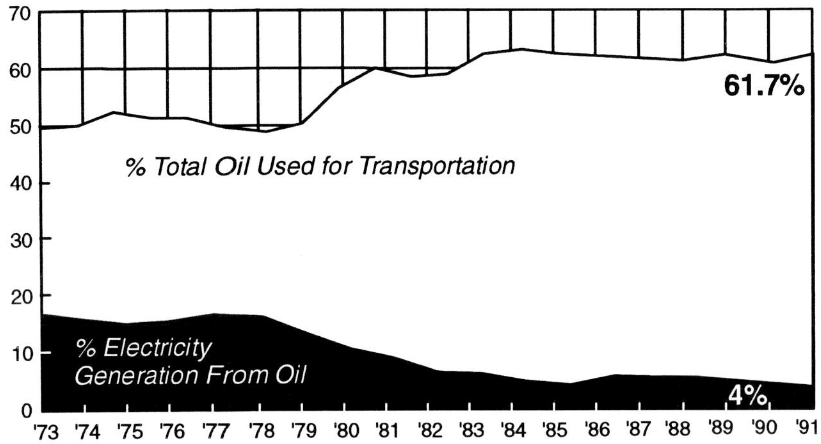

CSLO2381-01 8/04/93

FIGURE 2: THE ELECTRIC G-VAN

FIGURE 3: THE CHRYSLER TEVAN

Solid Polymer Fuel Cells: An Alternative to Electric Batteries in Vehicles -- An Overview

Maria Sanchez
Senior Vice-President
International Academy of Science
Independence, Missouri, U.S.A.

Background:

For the past seven years, Dr. Sanchez has worked at the International Academy of Science in hydrogen fuel cell research, photon irradiation, and computer design and control systems. With a background in heat transfer, thermodynamics and physical chemistry, she has also worked in Technical Systems at AT&T, and at Western Electric.

SOLID POLYMER FUEL CELLS: AN ALTERNATIVE TO ELECTRIC BATTERIES IN VEHICLES--AN OVERVIEW

R.E. Billings and M. Sanchez
International Academy of Science
26900 East Pink Hill Road
Independence, MO 64057-3284

Abstract

As the world becomes increasingly concerned about our environment and the effects various pollutants have on our planet, increasing focus is being placed on clean energy alternatives for satisfying growing energy demands. With legislation in California requiring that 2% of total vehicle sales in 1998 be zero emission vehicles, the automobile industry is in urgent need of a non-polluting energy solution. Although electric batteries offer a clean source of energy, vehicle range is limited and depreciation costs are high. The hydrogen fuel cell is another zero emission alternative offering greater vehicle range and lower depreciation charges. An overview of several recent solid polymer fuel cell projects will be presented.

Introduction

Although the fuel cell was developed in 1839, a major push toward advancement and commercialization of this important technology has only been underway the past fifteen years. Today, more than 200 million U.S. dollars are spent worldwide each year on fuel cell research and development,[1] in search of a solution to meet our growing energy demands while protecting the environment.

Various fuels can be used in different types of fuel cell, but the hydrogen fuel cell is of particular interest from the environmentalist's point of view. In the case of such a fuel cell, hydrogen and air are continuously provided to the cell, with water and energy being electrochemically produced with zero emissions. Fuel cells have no moving parts, make no noise, and utilize hydrogen very efficiently.

The basic chemistry of the hydrogen fuel cell, as depicted in Figure 1, is as follows:

Hydrogen flows into the anode where it dissociates in the presence of a catalyst, forming hydrogen ions (protons) and giving up electrons to the anode.

Anode reaction: $H_2 \rightarrow 2H^+ + 2e^-$

Meanwhile, oxygen is being supplied to the cathode. Hydrogen ions are transported though an electrolyte (an ion-conductive substance), as the liberated electrons flow, by way of an electrical cable, through an external load to the cathode to participate in the water-forming reaction.

Cathode reaction: $2H^+ + 2e^- + ½O_2 \rightarrow H_2O$

The overall fuel cell reaction follows.

Overall reaction: $2H^+ + ½O_2 \rightarrow H_2O$

Each individual cell (anode/cathode sandwich) typically produces close to one volt. Cells are assembled in series and constructed into stacks to acquire the desired voltage.

Over the years, various types of fuel cells have been developed. Oftentimes these fuel cells are categorized by the electrolyte incorporated into their design. Among the most common configurations include the phosphoric acid, molten carbonate, solid oxide, solid polymer electrolyte and alkaline fuel cells. Of these, only the solid polymer fuel cell offers advantages of low operating temperatures and a safe, non-volatile electrolyte--two important considerations for choosing an appropriate fuel cell system for transportation requirements.

The Solid Polymer Fuel Cell

The electrolyte utilized in the solid polymer fuel cell is substantially different than that used in other fuel cell configurations. Under normal conditions, the dry electrolyte looks and feels like a thin sheet of transparent plastic. But, when placed in water, this unusual substance significantly expands and, a highly acidic environment, comparable to that of a ten weight percent sulfuric acid solution[2], is produced in the hydrated membrane.

The idea of using an organic ion-exchange membrane in solid polymer fuel cells was first introduced by Grubb in the 1950's[3]. These early membranes included hydrocarbon-type polymers which were found to be inadequate for fuel cell applications because of their chemical instability[4].

In the early 1960's, shortly after development of the stable and non-reactive substance, Teflon®, the Du Pont Company's Plastics Exploratory Research Group, working on new derivatives of this revolutionary product, stumbled upon a new substance that had the interesting characteristic of being able to interact with the environment.[5] These researchers found that if they attached sulfonic acid groups to the ends of the perfluorinated chains, the resultant product was proton selective--in other words, cations (positively-charged ions) were allowed to pass through this substance, while anions (negatively-charged ions) were rejected.

Soon researchers at Du Pont discovered the importance of this Teflon®-based derivative, for chloralkali applications--hence, a major effort was put forth optimizing this product, later named Nafion®, for this particular industry. Nonetheless, various research efforts have demonstrated the usefulness of Nafion® in other technological fields including electrochemistry. In 1966, General Electric demonstrated the potential of using solid polymer electrolytes in fuel cells when they constructed a unit for use in the U.S. Gemini Space Program.[6] Even today, Nafion® is still the most widely tested ion-exchange membrane used in solid polymer fuel cell research.[7]

The basic chemistry for the solid polymer electrolyte when used in a hydrogen/oxygen fuel cell is demonstrated in Figure 2. The positively-charged hydrogen ions, or protons, generated at the anode, are attracted to the negatively-charged sulfonic acid radicals.

These sulfonic acid groups provide reactive sites for the hydrogen ions which are then allowed passage through the membrane. Transported hydrogen ions chemically combine with oxygen on the other side of the membrane producing water.

Over the years other manufacturers have begun producing similar solid polymer electrolytes, providing attractive alternatives to Nafion® in solid polymer fuel cells. Japanese companies, including Asahi Chemical and Chlorine Engineers, have developed their own ion-exchange membranes. Here in the United

States, a new series of perfluorinated membranes has recently become available from the Dow Chemical Company.[8,9] Like Nafion®, the Dow experimental membrane also has a Teflon®-based backbone, however, the side chain containing the sulfonic acid group is shorter. Like the Du Pont product, the Dow membrane has also produced favorable results in solid polymer fuel cells although this polymer was also developed specifically for the chloralkali industry. One of the major differences in the Dow polymer is that it absorbs approximately half the amount of water absorbed by Nafion®. Nonetheless, the experimental membrane still possesses many of the exceptional ion-exchange properties of Nafion® and several researchers have achieved significant improvements in fuel cell performance utilizing this product.

Traditionally, conventional solid polymer fuel cells have required as much as 4 mg/cm^2 of platinum black on both the cathode and the anode,[10] for a total loading of 8 mg Pt/cm^2. Typically, this catalyst has been hot-pressed directly to the membrane utilizing a Teflon® binder. These prepared membrane are then, often times, bonded to the surface of porous electrodes by using heat and pressure. This process is believed to be necessary to produce the required contact between the electrocatalyst and the membrane surface.[11]

Recent Solid Polymer Fuel Cell Research and Development Projects

Solid polymer fuel cell technology has progressed considerably in recent times: catalyst loading densities have decreased and fuel cell performance has increased[12-18]. An overview of several solid polymer fuel cell research projects conducted between 1990 and the present time shall be discussed below.

International Academy of Science

The Billings Cell, developed at the International Academy of Science, employs an advanced state-of-the-art solid polymer fuel cell design vastly different from that of traditional fuel cells available today. One major difference in the Billings Cell is that it is far more compact than the cells offered by many other fuel cell manufacturers. It is also one of the few available solid polymer fuel cells specifically designed to be able to function bi-directionally--as a fuel cell, hydrogen and oxygen from the air are chemically combined to generate the required energy; alternatively, as an electrolyzer, water and electricity are provided so that hydrogen and oxygen can be produced. A proprietary method of catalyst application provides satisfactory results with total cell loading densities of less than 1 mg/cm^2.

Presently, system life-testing is being conducted and additional research is underway to further minimize catalyst loadings. The overall cell design is being carefully evaluated and further experimentation shall be carried out in order to optimize the complete solid polymer fuel system in terms of performance, reliability, and cost. It is anticipated that the Billings Cell will be ready for commercialization in the near future.

Ballard Power Systems

The anodes and cathodes of the Ballard solid polymer fuel cell[19] are prepared by applying a small amount of platinum black to one surface of a thin sheet of porous carbon paper which has been "wet-proofed" with Teflon®. The electrolyte is then hot-pressed between the anode and cathode, producing a membrane/electrode assembly. This assembly, which is the heart of the fuel cell, measures less than one millimeter in thickness.

On the back side of the anode and cathode are connected channeled graphite flow field plates. The channels are used to supply reactant gases, while the ridges between the channels make electrical contact with the electrodes and conduct current to the external circuit. Product water is rejected from the back of the cathode into the oxidant gas stream where it is carried out of the fuel cell by the excess oxidant flow.

Recent developments by Ballard Power Systems, based on the Dow experimental membrane,[20] have reduced the size of the fuel cell, making possible the use of their cell in motive power applications.

Texas A&M University

Researchers at Texas A&M University are studying ways to improve platinum loading in solid polymer fuel cells. Utilizing a cell fabrication technique developed by Los Alamos National Laboratory, scientists at Texas A&M have been able to reduce cell platinum loading to .90 mg/cm^2 with minor impact on fuel cell performance.[21]

In the reported experiments, a 5-cm^2 fuel cell was operated at 50°C with inlet gas pressures of one atmosphere. Data was collected using both pure oxygen and air as the oxidant in the hydrogen polymer fuel cell. A comparison of fuel cell performance for very high and low platinum loadings is presented in Figure 3. Although a substantial change in cell platinum loadings existed between the two samples, decreasing the catalyst loadings only moderately diminished fuel cell performance when oxygen was used. In the case of air, however, a more notable decrease in cell performance was realized due to mass transfer limitations; or, in other words, fewer reactive sites were available for reaction in the low platinum-loaded sample.

The effect of increasing reactant gas pressures on fuel cell performance for the low platinum-loaded sample is shown in Figure 4. Again, data is reported using pure oxygen and air as oxidants. As is clearly apparent, increasing reactant gas pressures notably improves fuel cell performance, particularly in the case of the hydrogen-air solid polymer fuel cell.

Weizmann Institute of Materials Research and the Polymer Institute

The Weizmann Institute of Materials Research in Israel in conjunction with the U.S.-based Polymer Institute, have been investigating alternatives to frequently used techniques for preparing solid polymer fuel cell electrodes. Their procedure gives results equal if not slightly superior to the conventional method.

Instead of hot-pressing and sintering mixtures of the catalyst containing carbon black with Teflon® as is commonly done in traditional fuel cell electrode preparation, these researchers use a process known as phase inversion. In the described method, a polymer is dissolved in a solvent, carbon black containing the catalyst is mixed with the polymer solution, a film is cast and immersed in a non-solvent, and the dissolved polymer is then precipitated out of its dissolved form in a highly porous state.[22]

Electrodes containing .3 and .6 mg Pt/cm^2, was prepared using the phase inversion process and then bonded to the perfluorinated membrane using a procedure similar to that previously used by the Los Alamos group.[23] Resulting polarization curves, comparing the phase inversion electrodes to .35 mg Pt/cm^2 commercially-available electrodes, are presented in Figure 5. The electrolyte utilized in this experiment was a 3-mil cast Nafion® membrane. Figure 6 shows similar data collected using commercially-available, 7-mil, Nafion® 117.

BCS Technology Inc.

Based on the cell design developed by Ticianelli, Srinivasan, and their coworkers at Texas A&M University,[24,25] researchers at BCS Technology have designed what they refer to as a "unitized regenerative fuel cell"[26]--a solid polymer fuel cell capable of operating in reverse as an electrolyzer.

Utilizing a total cell platinum loading of 2 mg/cm^2, Pt/Ir-on-C and Pt-on-C were evaluated as catalysts for the oxygen electrode in this fuel cell. Pt/Ir-on-C demonstrated a better electrolyzer performance with slightly lower fuel cell performance. This cell was operated alternately in the fuel cell and electrolyzer modes and data was collected, as shown in Figures 7 through 10, using both Nafion® 117 and the Dow experimental membrane. A slight declination in cell performance resulted from cycle to cycle. This deterioration in performance was believed to be caused by electrode retention of reactant gas humidification water.

United Technologies' Hamilton Standard Division

United Technologies has been investigating the use of solid polymer fuel cells for naval applications.[27] These fuel cells need to be able to handle high shock and vibration and have "perfect" sealing of reactant gases. To satisfy these stringent requirements, researchers at the Hamilton Standard Division have developed a "sheet metal" design which they feel offers many advantages over traditional carbon/graphite electrodes.

The United Technologies' design consists of two assemblies: a membrane and electrode assembly and a metal/compression pad assembly, both of which are very flexible. The membrane and electrode assembly utilized in their fuel cell is basically the same as that used in their electrolyzer except that a different catalyst is employed on the oxygen electrode and a special wetproofing film is included on this side of the cell. This assembly has no rigid structural properties. The metal/compression pad assembly consists of expanded sheet metal screen fluid fields, sheet metal separators, individual cell compression pads, and nonmetal frames and gaskets. A separate chamber is incorporated for cell coolant. Like the membrane and electrode assembly, the sheet metal/compression pad assembly is very flexible and produces an exceptionally positive seal for separating the various fluids and maintaining a "perfect" overboard seal. A polarization curve for the "sheet metal" fuel cell system is given as Figure 11.

Los Alamos National Laboratory

Researchers at Los Alamos National Laboratory, concerned with the high cost of platinum catalysts, has been working on lowering platinum loadings in the solid polymer fuel cell. They have developed a new process whereby a thin film catalyst layer is cast from solutions of suspended Pt/C catalyst and solubilized Nafion® ionomer.[28] The direct application apparently provides enhanced bonding at the interface between the membrane and the catalyst layer. Consequently, fuel cells utilizing these ultra-low platinum-loaded membranes demonstrate superior performance when compared to other low-catalyst loaded cells.

The latest process developed at Los Alamos for casting ultra-low platinum, thin film catalyst layers, involves the direct application of a prepared Pt/C/solubilized Nafion® ink onto the surface of a dry solid polymer electrolyte and then baking the electrode/membrane assembly at 150-190°C. The other side is then prepared in a similar fashion. Data collected at various platinum loading densities for hydrogen and oxygen/air, while utilizing the Chlorine Engineers' perfluorinated membrane product, is depicted as Figure 12. Of particular interest are the curves for .12 and .17 mg/cm^2--although the electrode platinum loading was increased by .05 mg/cm^2, little improvement in overall performance was realized.

Fuel cell performance, as affected by reactant gas pressure, is presented in Figure 13. Because of the detrimental effect of an inert nitrogen blanket in the gas diffusion backing of the oxygen electrode, one atmosphere of oxygen performed superior, in the solid polymer fuel cell, to five atmospheres of air at high current densities.

In Figure 14 are presented oxygen and air fuel cell polarization curves for a Dow experimental membrane with electrode loading of .13 mg Pt/cm^2 applied in a similar manner. This cell is capable of delivering 1.3 W/cm^2.

Solid Polymer Fuel Cell Vehicle Demonstration Projects

Several projects aimed at demonstrating the potential of the solid polymer fuel cell in vehicular applications have been and are still being conducted. Below, several of these projects will be discussed.

International Academy of Science--LaserCel1™ Prototype Vehicle

In 1991, the International Academy of Science engaged in a demonstration project with the Pennsylvania Energy Office to build the world's first fuel cell automobile.[29] This prototype, a converted electric Ford Fiesta Postal vehicle shown in Figure 15, derives its cruise power from the LaserCel™ solid polymer fuel cell installed beneath the hood. In the LaserCel1™ prototype, the fuel cell consist of two stacks of cells inside of a single enclosure. The small stack consists of 16 cells, and produces 12 volts to power the vehicle's accessories. The large stack, consisting of 135 cells, outputs 100 volts, and is used to power the drive motor. Both hydrogen and air are pumped to the cell using a specially-designed, load-following, air compressor designed specifically for this fuel cell application.[30] The cell utilizes hydrogen stored on-board in a metal-hydride form--the safest method of hydrogen storage presently available.[31] With a 300-pound hydride tank, the vehicle has almost a 200 mile range. Specific components of this system can be seen in Figures 16.

Currently researchers at the Academy are focussed on optimizing the various components of the fuel cell system, getting all the pieces ready for commercialization. In the meantime, plans for a dual fuel cell powered Eagle are on the slate. This prototype will demonstrate the ability of the fuel cell to provide all the necessary power for vehicle operation.

Ballard Power Systems--The Ballard/BC Bus Demonstration Project

Ballard Power Systems, in conjunction with the Province of British Columbia and the government of Canada, has recently converted a diesel city bus for the Vancouver BC Transit.[32] This 30-ft vehicle is powered by a 105-kW fuel cell based on Ballard technology. The objective of Phase I of this project was to provide a commercial transit bus, fully-powered by a solid polymer fuel cell, providing the same driver acceptance as the corresponding diesel counterpart.[33]

Gaseous hydrogen, stored on-board the bus in DOT-approved glass-wound composite cylinders operating at 3000 psi, provides the necessary fuel required for the 150 km projected vehicle range.[34]

Phase II will involve conversion of a 40-ft transit vehicle using improved hydrogen storage and more compact solid polymer fuel cells.[35]

General Motors/Los Alamos National Laboratory

Los Alamos National Laboratory and General Motors have been working on developing a solid polymer fuel cell for vans and passenger cars.[36] They plan to carry on-board reformers on these vehicles to provide their hydrogen needs.

Conclusions

Recent solid polymer fuel cell research and development is demonstrating the real potential of this technology for vehicular applications. Platinum loading densities are being minimized to acceptable levels and fuel cell systems are being optimized to achieve ideal vehicle range and performance. Several demonstration projects prove the feasibility of this important new technology for meeting the growing world energy demands without compromising the environment.

References

1. J.H. Hirschenhofer, "Latest Progress in Fuel Cell Technology," IEEE AES Systems Magazine, (November 1992).

2. K. Kinoshita, F.R. McLarnon, and E.J. Cairns, *Fuel Cells--A Handbook*, U.S. Department of Energy, Morgantown, West Virginia (May 1988).

3. W.T. Grubb, *Proceedings of the 11th Annual Battery Research and Development Conference*, PSC Publications Committee, Red Bank, NJ (1957).

4. K. Kinoshita, F.R. McLarnon, and E.J. Cairns, *Fuel Cells--A Handbook*, U.S. Department of Energy, Morgantown, West Virginia (May 1988).

5. W.G. Grot, "Discovery and Development of Nafion® Perfluorinated Membranes," The Castner Metal Lecture, paper presented at the Society for the Chemical Industry Third London International Chlorine Symposium, June 5-7, 1985.

6. H.A. Liebhafsky and E.J. Cairns, *Fuel Cells and Fuel Batteries*, John Wiley, New York, NY (1968).

7. K. Kinoshita, F.R. McLarnon, and E.J. Cairns, *Fuel Cells--A Handbook*, U.S. Department of Energy, Morgantown, West Virginia (May 1988).

8. G.A. Eisman, *Proceedings of the Symposium on Diaphragms, Separators, and Ion-Exchange Membranes*, Edited by J.W. Van Zee, R.E. White, K. Kinoshita, and H.S. Burney, The Electrochemical Society, Inc., Pennington, NJ (1986) p. 156.

9. B.R. Ezzell, W.P. Carl, and W.A. Mod, *Industrial Membrane Processes*, AIChE Symposium Series Vol. 82, No. 248, American Institute of Chemical Engineers, New York, NY (1986) p. 45.

10. A.J. Appleby and E.B. Yeager, *Energy*, **11**, 137, (1986).

11. K. Kinoshita, F.R. McLarnon, and E.J. Cairns, *Fuel Cells--A Handbook*, U.S. Department of Energy, Morgantown, West Virginia (May 1988).

12. E.A. Ticianelli, C.R. Derouin, A. Redondo, and S. Srinivasan, *J. Electrochem. Soc.*, **135**, 2209, (1988).

13. E.A. Ticianelli, C.R. Derouin, and S. Srinivasan, *J. Electroanal. Chem.*, **251**, 275, (1988).

14. S. Srinivasan, E.A. Ticianelli, C.R. Derouin, and A. Redondo, *J. Power Sources*, **22**, 359, (1988).

15. K. Prater, *J. Power Sources*, **29**, 239, (1990).

16. M. Bernardi, *Proceedings of the Symposium on Fuel Cells*, Edited by R.E. White and A.J. Appleby, San Francisco, CA (November 1989) p. 51.

17. P.J. Shultz, *Proceedings of the Symposium on Fuel Cells*, Edited by R.E. White and A.J. Appleby, San Francisco, CA (November 1989) p. 87.

18. S. Srinivasan, S. Somasundaram, M. Enayetullah, D. Swan, and A. J. Appleby, *Proceedings of the Symposium on Fuel Cells*, Edited by R.E. White and A.J. Appleby, San Francisco, CA (November 1989) p. 71.

19. B.N. Gorbell and K.B. Prater, *Project Hydrogen '91 Conference Proceedings*, Edited by T.N. Veziroglu and R.E. Billings, Kansas City, MO (September 16-18, 1991) p. 247.

20. K.B. Prater, *J. Power Sources*, **29**, 239, (1990).

21. D.H. Swan, O.A. Velev, I.J. Kakwan, A.C. Ferreira, S. Srinivasan, and A.J. Appleby, "The Proton Exchange Membrane Fuel Cell--A Strong Candidate As a Power Source for Electric Vehicles," *Project Hydrogen '91 Conference Proceedings*, Edited by T.N. Veziroglu and R.E. Billings, Kansas City, MO (September 16-18, 1991) p. 185.

22. J. Manassen and I. Cabasso, "A Polymer Chemist's View of Fuel Cell Electrodes," *Proceedings of the 34th International Power Sources Symposium*, Cherry Hill, NJ (June 25-28, 1990) p. 408.

23. E.A. Ticianelli, C.R. Derouin, A. Redondo, and S. Srinivasan, *J. Electrochem. Soc.*, **135**, 2209, (1988).

24. Ibid.

25. E.A. Ticianelli, C.R. Derouin, and S. Srinivasan, *J. Electroanal. Chem.*, **251**, 275, (1988).

26. H.P. Dhar, "A Unitized Approach to Regenerative Solid Polymer Electrolyte Fuel Cells," *J. Appl. Electrochem.*, **23**, 32 (1993).

27. J.F. McElroy, T.M. Molter, R.N. Sexauer, "SPE® Hydrogen/Oxygen Fuel Cells for Rigorous Naval Applications," *Proceedings of the 34th International Power Sources Symposium*, Cherry Hill, NJ (June 25-28, 1990) p. 403.

28. M.S. Wilson and S. Gottesfeld, *J. Electrochem. Soc.*, **139**, 28, (February 1992).

29. R.E. Billings, *The Hydrogen World View*, American Academy of Science, Independence, MO (1991).

30. R.E. Billings, M. Sanchez, P. Cherry, D.B. Eyre, *Project Hydrogen '91 Conference Proceedings*, Edited by T.N. Veziroglu and R.E. Billings, Kansas City, MO (September 16-18, 1991) p. 257.

31. R.E. Billings, *The Hydrogen World View*, American Academy of Science, Independence, MO (1991).

32. J. Glanz, "Can Fuel Cells Go Where No Device Has Gone Before?," *R & D Magazine*, (May 1993), p. 36.

33. B.N. Gorbell and K.B. Prater, *Project Hydrogen '91 Conference Proceedings*, Edited by T.N. Veziroglu and R.E. Billings, Kansas City, MO (September 16-18, 1991) p. 247.

34. Ibid.

35. Ibid.

36. J. Glanz, "Can Fuel Cells Go Where No Device Has Gone Before?," *R & D Magazine*, (May 1993), p. 36.

Figure 1. The Solid Polymer Electrolyte Fuel Cell

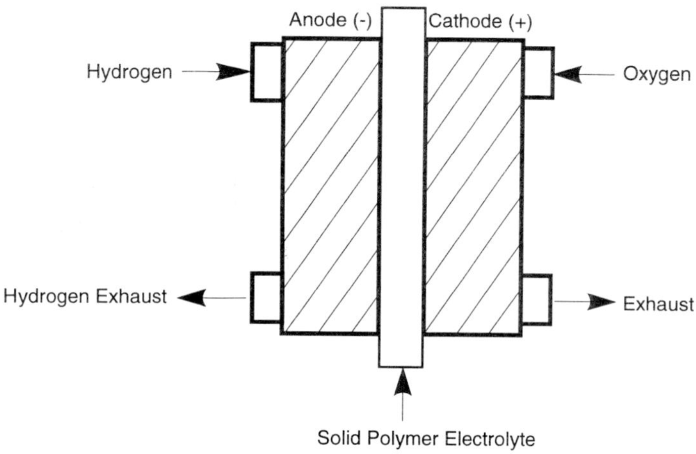

Figure 2. Basic Chemistry of the Solid Polymer Electrolyte

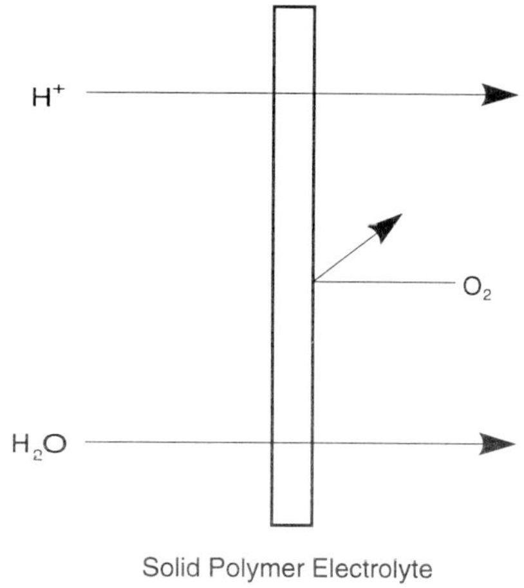

Figure 3. Effect of Platinum Loading on Fuel Cell Performance

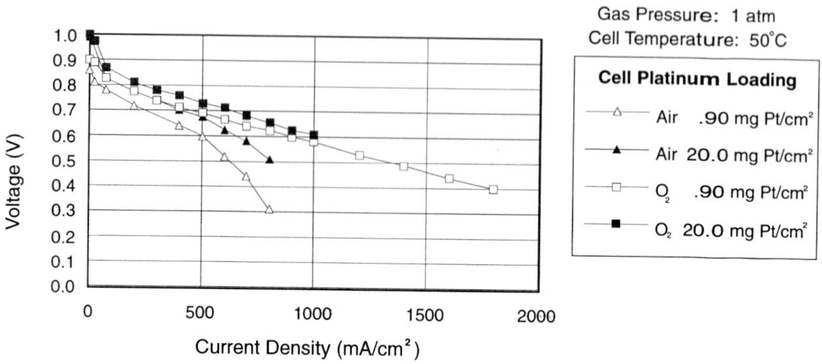

Source: Texas A&M University--1991

Figure 4. Effect of Gas Pressure on Fuel Cell Performance

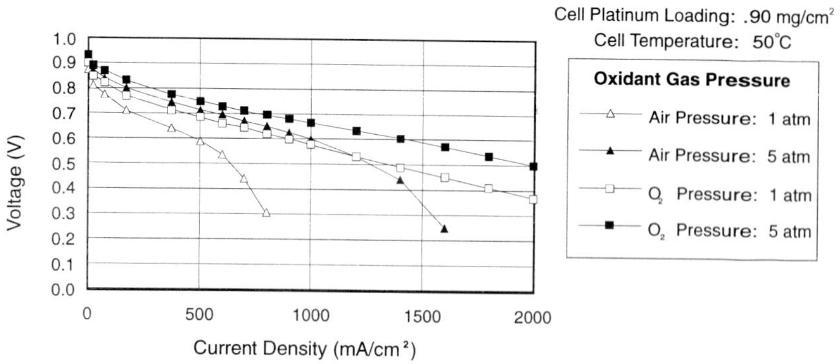

Source: Texas A&M University--1991

Figure 5. Fuel Cell Performance of Phase Inversion Prepared Electrodes Utilizing Cast Nafion

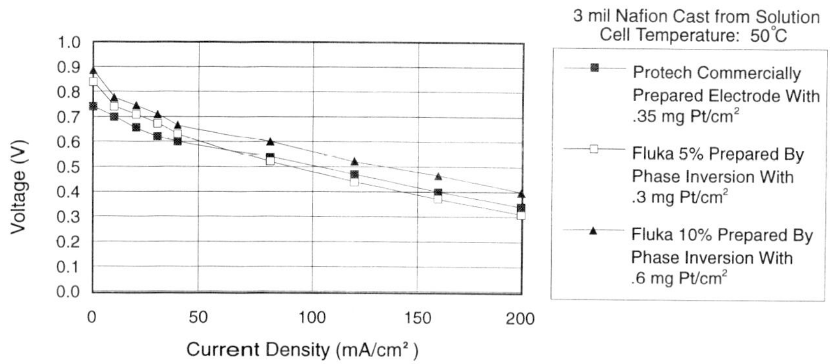

Source: Weizmann Institute of Materials Research and the Polymer Institute--1990

Figure 6. Fuel Cell Performance of Phase Inversion Prepared Electrodes Utilizing Commercially Available Nafion

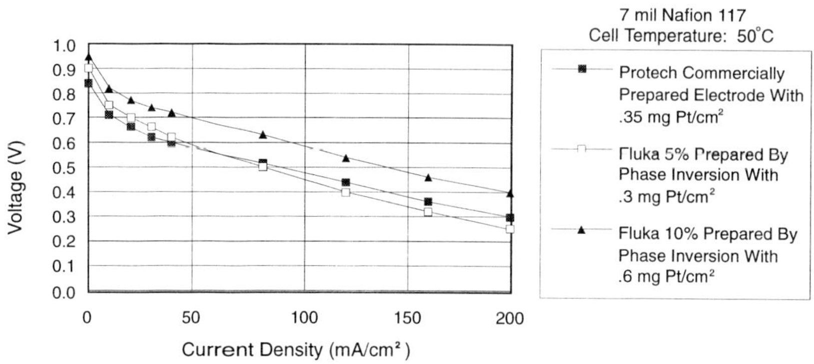

Source: Weizmann Institute of Materials Research and the Polymer Institute--1990

Figure 7. Regenerative Fuel Cell Performance With Nafion Electrolyte and Pt/Ir-On-Carbon as Oxygen Electrode

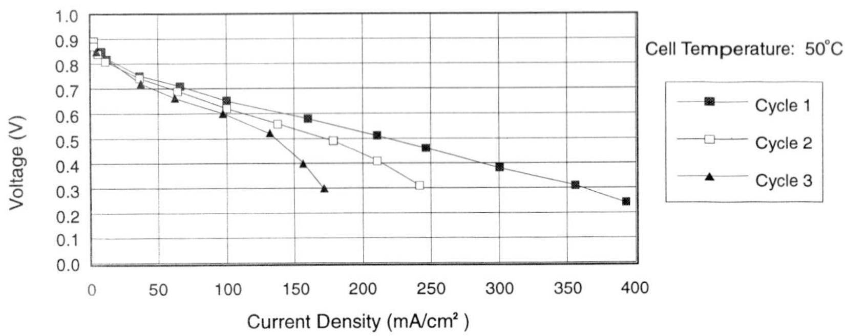

Source: BCS Technology Inc.--1993

Figure 8. Regenerative Electrolyzer Performance With Nafion Electrolyte and Pt/Ir-On-Carbon as Oxygen Electrode

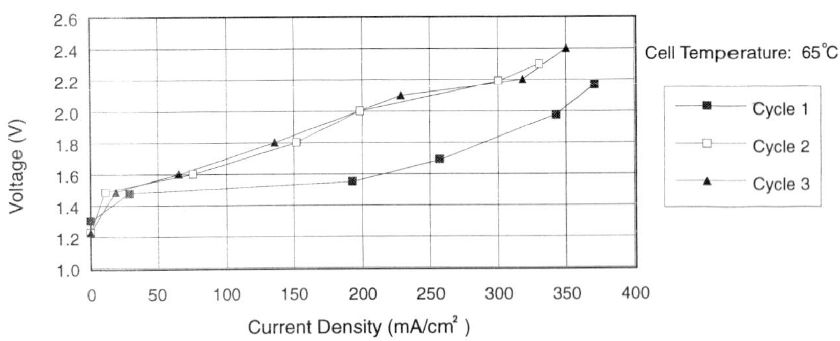

Source: BCS Technology Inc.--1993

Figure 9. Regenerative Fuel Cell Performance With Dow Electrolyte and Pt/Ir-On-Carbon as Oxygen Electrode

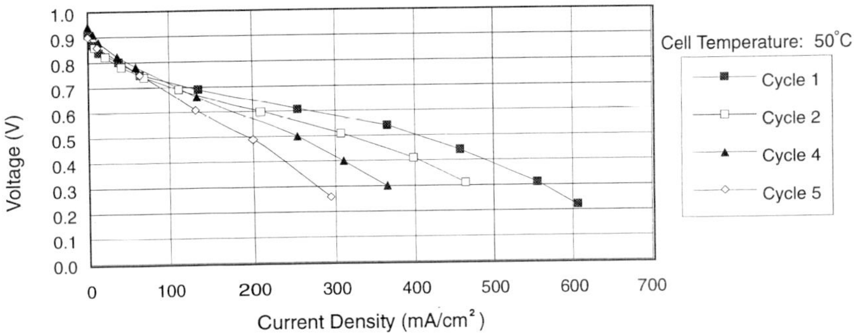

Source: BCS Technology Inc.--1993

Figure 10. Regenerative Electrolyzer Performance With Dow Electrolyte and Pt/Ir-On-Carbon as Oxygen Electrode

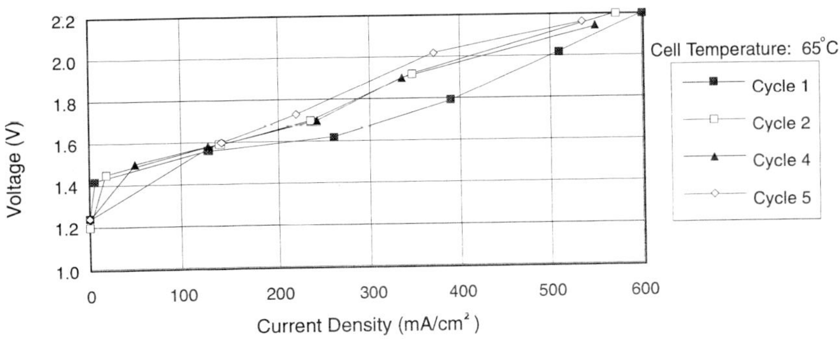

Source: BCS Technology Inc.--1993

Figure 11. Performance Curve for United Technology Fuel Cell

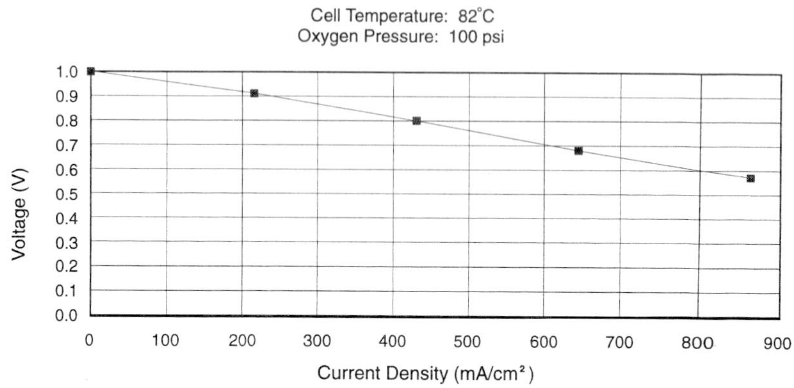

Source: United Technologies' Hamilton Standard Divsion--1990

Figure 12. Effect of Ultra-Low Platinum Loading on Fuel Cell Performance

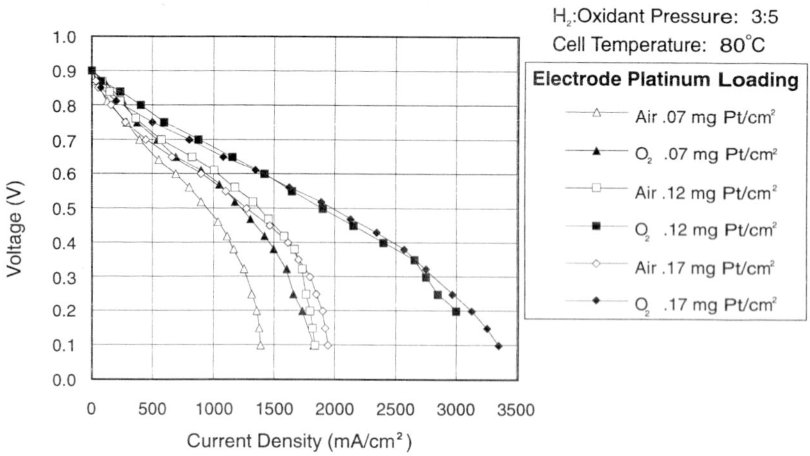

Source: Los Alamos National Laboratory--1992

Figure 13. Effect of Gas Pressure on Ultra-Low Platinum Loaded Fuel Cell Performance

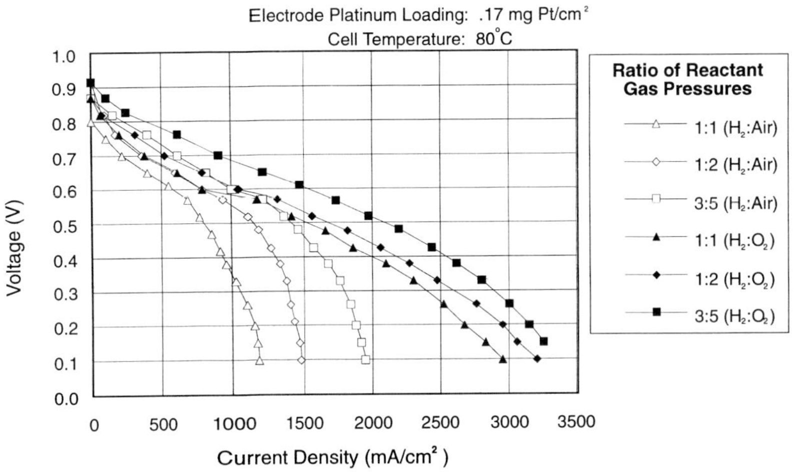

Source: Los Alamos National Laboratory--1992

Figure 14. Polarization Curves for Ultra-Low Platinum Fuel Cells Utilizing the Dow Experimental Membrane

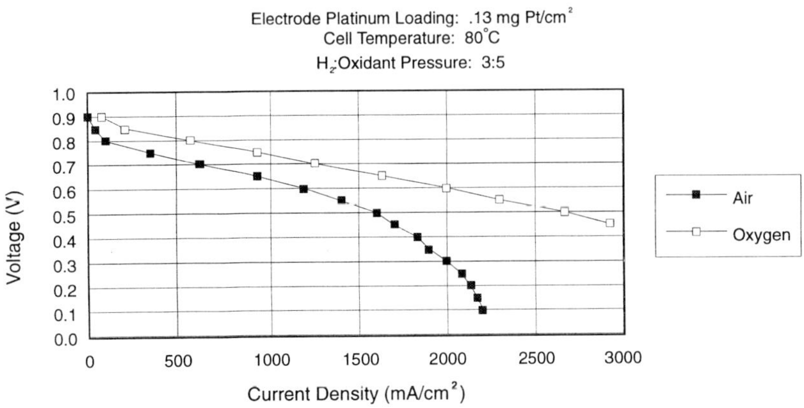

Source: Los Alamos National Laboratory--1992

Figure 15. The LaserCel1™ Fuel Cell Prototype

Source: International Academy of Science--1991

Figure 16. LaserCel1™ Hydrogen System

Load-Following Gas Compressor

LaserCel™ Fuel Cell

Metal Hydride Storage Vessel

Source: International Academy of Science--1991

Minus Emission Vehicles that Clean the Air

Roy McAlister
President
American Hydrogen Association
Tempe, Arizona, U.S.A.

Background:

In addition to his duties at the AHA, Mr. McAlister is Founder and President of Trans Energy Corporation, where he directs the manufacturing division, the research division, and the overall business operations. He has developed and patented numerous energy conversion systems, and has taught engineering at the University of Kansas and Arizona State University.

MINUS EMISSIONS VEHICLES CAN CLEAN THE AIR

Roy E. McAlister
Professional Engineer
AMERICAN HYDROGEN ASSOCIATION
602-921-0433

ABSTRACT

50% of the oil used for transportation in the U.S. is imported. 97% of the transportation system is powered by engines that are dedicated to oil-sourced fuels such as gasoline and Diesel fuel. U.S. domestic oil reserves constitute less than 3% of world oil reserves, but the U.S. consumes about 25% of the world's oil production. In order to sustain leadership in living standards, the U.S. must adopt more efficient energy conversion practices.

Humanity has dominion over all other species of life. This dominion brings the moral responsibility to adopt energy-conversion practices that are based on sustainable resources. It is a moral responsibility to less fortunate emerging nations, future generations, and other species of life that are threatened with extinction due to fossil combustion practices.

Photosynthesis produced all fossil reserves. Overall conversion efficiencies of about 0.09% or less result from making electricity in a conventional fossil-fired steam power plant or driving a gasoline-fueled car when original photosynthesis is taken into account. Solar engine-generators called Solar Stirling Dish Gensets have been developed that are 30-times more efficient than photosynthesis.

Direct-injection and spark-ignition engines called Precision Spark Injection (PSI) engines have been developed that are up to 2.5 times more efficient than most automobile engines. Producing electricity in Solar Stirling Dish Gensets and using advanced engines based on PSI Technology will improve overall propulsion efficiencies to about 15% to 20%. This is over 100 times more efficient than continuing to burn fossil gasoline in present cars.

Direct injection of hydrogen in internal combustion engines clean the air that passes through the engine. Atmospheric contaminants such as carbon monoxide, hydrocarbons, tire particulates, Diesel soot, aldehydes, and peroxyacetylnitrate are reduced by operation of an engine on stratified-charge hydrogen. Vehicles with internal combustion engines using such technology will clean the air as they are driven. This will serve as a stop-gap until hydrogen fuel cells become popular.

INTRODUCTION:

The Industrial Revolution has progressed by exploitation of fossil fuels. 50% of the oil used for transportation in the U.S. is imported. 97% of the transportation system is powered by engines that are dedicated to oil-sourced fuels such as gasoline and Diesel fuel. U.S. domestic oil reserves constitute less than 3% of world oil reserves, but the U.S. consumes about 25% of the world's oil production. Very low energy-conversion efficiencies have been accepted because we disregarded and postponed the need to find a way to sustain the world's economy of producers and consumers.

SOLAR ENERGY PRODUCED FOSSIL BIOMASS

Radiation of light from stars provides the primary energy for building green plant tissues. The Milky Way Galaxy of stars includes our Sun and nine satellite planets. During the last 5 billion years the Milky Way has rotated at the rate of about one revolution every 250 million years. Our solar system has made about 19 trips around a center of rotation in the Milky Way Galaxy. During one of these 250 million year periods a fossil resource bank containing the chemical fuel potential of solar energy in organic tissues developed in the Earth's crust.

In Earth's history this means that on one of the Milky Way's 19 rotations, great amounts of vegetable matter was deposited and covered with sufficient alluvium to seal it from air and prevent oxidation. Petrocarbon deposits of coal, oil, and natural gas developed in the places that the alluvium was sufficiently thick and strong to retain the pressures developed by anaerobic decomposition of the vegetable matter. Our fossil bank account diminishes as we burn the equivalent of about 180 million barrels of oil each day.

Solar energy accounts for nearly all of the energy available for the human economy on Earth. Solar energy produced all of the coal, oil, and natural gas that we have on Earth. Resources such as limestone, diatomatious earth, and corral reefs are products of solar energy conversions by nature.

Most green plants convert less than 1% of the available solar energy to stored energy. Photosynthesis starts with water-splitting reactions. Oxygen from water is released to the atmosphere. Hydrogen is retained and combined with carbon dioxide from the atmosphere to form plant tissues. More than 50% of solar energy is in the infrared spectrum and about 4% is in the ultraviolet region. Plants can only utilize visible solar wavelengths from about 400 to 700 nanometers for photosynthesis. Within this region, red light at 700 nanometers accounts for almost all energy conversion processes of photosynthesis.[1] As a result, only a minor amount of solar energy is captured by photosynthesis processes.

$$H_2O + CO_2 + \text{Sun Light} \rightarrow \text{Organic Tissue} + O_2 \qquad \text{Equation 1}$$

Slash pine converts about 0.24-0.30% of available solar energy to chemical fuel potential energy in the form of various plant tissues. Corn converts about 0.44-0.69%. An exceptional plant is sugar cane which may convert as much as 0.82 to 1.18% of solar energy to plant tissue.[2]

Therefore fossil energy represents a very low conversion efficiency for storing ancient sun light as chemical fuel energy. Fresh biomass represents a slightly higher efficiency, but at best it is about 1% efficient. Conventional refinery practices for making fuels such as gasoline from fossil or fresh biomass feedstocks further reduces the amount of solar energy represented in the final product.

POWERING CARS WITH FOSSIL RESERVES

Burning petrocarbons in conventional automobiles or power plants converts only about 20 to 40% of the released heat to motive propulsion or electricity. Assuming that ancient photosynthesis captured light at about the efficiency of slash pine, driving a gasoline-fueled car provides an energy conversion efficiency of about 0.06% or less. Assuming a battery-charging efficiency of 85%, and a propulsion efficiency of 85%, the same fossil fuel in a conventional plant could recharge an electric vehicle for an overall conversion efficiency of about 0.09% or less. Equations 2 and 3 illustrate:

$$\begin{array}{llll} \text{(Photosynthesis)} & \text{(Refine/Dist.)} & \text{(Engine)} & = 0.05\% \\ (.30\%) & (.90) & (.20) & = 0.05\% \end{array} \qquad \text{Equation 2}$$

$$\begin{array}{llllll} \text{(Photosynthesis)} & \text{(Refine/Dist)} & \text{(Power Plant)} & \text{(Bat.)} & \text{(E-motor)} & = 0.08\% \\ (.30\%) & (.90) & (.40) & (.85) & (.85) & = 0.08\% \end{array} \qquad \text{Equation 3}$$

Operating gasoline-engine driven or electric cars from fossil fuel processes results in small differences in unacceptably low overall efficiencies. We now burn the fossil fuel equivalent of more than 100 million barrels of oil each day. We are wasting our fossil bank account at these low efficiencies.

SOLAR-ELECTRICITY and SOLAR-HYDROGEN

Electricity and hydrogen can be produced by using solar energy and water. Hydrogen can also be produced from wind energy or hydro-electricity and water. When hydrogen is produced from such forms of solar energy it is called *solar hydrogen.* Solar hydrogen is more efficient than fossil fuels.

Testing at Southern California Edison's test site has proven the ability of Stirling Dish Gensets to produce grid-quality electricity at about 30% efficiency.[3] In about 20 months of operation in a desert region of the U.S., a Stirling Dish Genset produces enough energy to manufacture a duplicate Genset. These solar breeders can produce enough energy to make 15 or more new Stirling Dish Gensets or Wind Generators before being recycled. Work is underway to improve the efficiency and manufacturability of Stirling Dish Genset components. It is possible that solar to electricity energy-conversion efficiencies of 35% to 40% will be achieved. Capital costs and electricity production costs for Stirling Dish Gensets are projected to be considerably below fossil-fired or nuclear power plants.[11]

In addition to already being about 30-times more efficient in energy-conversion efficiency, solar hydrogen or renewable electricity from Stirling Dish Gensets is far cleaner than fossil-fired power plants. Externality costs including health care, replacement of corroded transportation components, military protection of supplies from the Middle East, and various other subsidies from the U.S. government drive the true cost of a gallon of gasoline to more than $5.00 per gallon.[4, 5, 6, 7]

NEAR-TERM SOLUTION TO CITY SMOG

Most of the world's 400 million vehicles are in cities. Cars produce about 80% of airborne pollution in cities. Living standards are compromised by carbon monoxide, unburned hydrocarbons, peroxyacetylnitrate, and poisonous particles from Diesel engines and tire wear. The commonality of all of these components to smog is airborne carbon compounds.

Elimination of fuels that contain carbon and controlling the combustion temperature will prevent smog in cities. Cars that operate in cities can be served by existing electric power grids. One way to fuel cars with hydrogen is to add tanks and fuel controls for using hydrogen produced by electrolyzing water. Grid-delivered electricity or renewable electricity from on-site production will release hydrogen from water for use in cars that clean the air as they are driven. The equipment needed is a high-pressure electrolyzer and a system for metering the hydrogen into the existing engine. Cars using hydrogen produce no hydrocarbons, no carbon monoxide, no carbon dioxide, and no particulates.

Combustion engines using hydrogen can be "*minus emissions*" power plants. A system for precisely injecting and spark igniting fuel as it enters the combustion chamber to produce stratified-charge combustion has been developed. It is called Precision Spark Injection or PSI Technology. Burning hydrogen as a stratified-charge in internal combustion piston engines or turbines can reduce atmospheric concentrations of carbon monoxide, hydrocarbons, tire particles, and Diesel soot. Producing the predominant quantity of heat by hydrogen combustion allows airborne contaminants such as hydrocarbons to be forced to chemical destruction in the hydrogen flame. The result is cleaner air.

H_2 + AIR → H_2O @ 4000°F (1930°K) Equation 4

H_2O @ 4000°F + AIR *HxCy → CLEANER AIR + 0.5X(H2O) + Y(CO_2) Equation 5

H_2O @ 4000°F + AIR * CO → CLEANER AIR + CO_2 Equation 6

NET RESULT: H_2 + AIR/SMOG → H_2O @ 4000°F + CLEAN AIR Equation 7

Renewable hydrogen and carbon can be derived from biomass by anaerobic extraction techniques. Using hydrogen derived from biomass that would ordinarily be allowed to rot into the atmosphere extends minus-emissions benefits by preventing formation of greenhouse gases.

Equipping sizeable portions of the present vehicle population with PSI systems will be particularly beneficial to development of new fuel sources. Hydrogen from solar, wind, garbage, or farm waste conversion processes can be used in vehicles equipped with PSI Systems.

PSI Technology allows precise control of the temperature of combustion to minimize oxides of nitrogen production in conventional engines. This approach to pollution control allows existing piston engines in motor vehicles to be rapidly retrofitted with devices that overcome the problems of reduced power and ignition difficulties of past attempts to use hydrogen.[8] For a given storage volume, range is improved. Conversion to PSI operation does not require tear-down of the engine.

Vehicles can be changed-over to PSI operation in about the same time as tune-ups on standard fuel/ignition systems.[6] Conversion of existing low-compression engines with spark ignition is accomplished by replacement of spark plugs with devices called "*spark-injectors*" that provide direct hydrogen injection and spark ignition. Conversion of existing compression-ignition engines (Diesels) is accomplished by replacement of Diesel fuel injectors with *spark-injectors* that provide precision hydrogen injection and spark ignition.

PSI Technology reduces the "storage-volume" problem of hydrogen by providing greater range through a combination of the following interrelated factors:

1. Increasing the thermal efficiency of the engine.

2. Increasing the part-load efficiency of the vehicle.

3. Increasing the mechanical efficiency of converted engines.

Fuel metering and ignition are electronically controlled to produce the best use of hydrogen for the duty cycle of the moment.

OPERATION OF ENGINES CONVERTED TO PSI-HYDROGEN OPERATION

Figure 1 shows components of the PSI System. In engines converted to PSI operation, intake air is allowed to enter the combustion chamber without throttling. Increased breathing capacity of the engine results. More air is present in the combustion chambers to support combustion during all phases of operation. Unthrottled air entry greatly reduces "backwork" due to ordinary vacuum conditions in the combustion chamber. This eliminates the negative work required to move the piston against the pressure difference between crank-case pressure and the intake-manifold vacuum produced by air throttling.

After substantial compression, hydrogen is delivered through the PSI spark-injector plug directly into the combustion chamber and is ignited by a precisely-timed plasma discharge as it enters. Hydrogen burns in the compressed air with an extremely rapid flame speed. (See Table 1 for comparisons.) This allows delayed ignition and avoidance of compression-cycle back work and heat losses characteristic of long-lead ignition of gasoline and other hydrocarbon fuels. Delayed ignition of hydrogen by spark-injectors near top-dead-center conditions allows engines to run cooler with higher mechanical efficiency and greatly reduced heat losses during compression.

Power control is accomplished by metering the precise amount of fuel needed at the instant that the piston nears top-dead-center conditions. Delayed ignition provides smoother knock-free engine operation by reducing backwork due to combustion during compression. Delayed ignition also reduces backwork due to pressure rise from combustion before top dead center. Hydrogen combustion within excess air results in greatly reduced heat losses due to radiation and conduction. Specialized hydrogen injection patterns are provided for best air utilization in each combustion chamber type. PSI spark-injectors are designated for each engine combustion-chamber geometry and provide the appropriate thread, seal, and reach for replacement of spark plugs.

Fuel consumption is less than in ordinary homogeneous-charge engines because of improvements in heat management and mechanical efficiency. Further improvements in vehicle range result from stopping fuel injection during deceleration. During deceleration there is no fuel consumption until the engine speed reaches idle or a desired cruise speed.

PSI-Hydrogen conversion provides the following characteristics:

1. More air is in the combustion chamber than with homogeneous-charge operation. This enables greater fuel combustion and increased power development when desired.

2. Faster combustion rates provided by hydrogen within excess air enables ignition after substantial completion of the air-compression cycle. This greatly reduces or eliminates backwork, heat losses, detonation, back firing, and engine wear.

3. Initial combustion of hydrogen as it enters the combustion chamber provides locally fuel-rich conditions within an envelope of excess air. Combustion proceeds into excess air to provide reduced formation of nitrogen oxides per horsepower hour of operation.

4. By combustion of hydrogen within excess air there are no fuel-sourced hydrocarbons or carbon compounds in the exhaust of engines converted to PSI operation.

5. Cold start-up of PSI-Hydrogen engines provides the same clean-burning, low-emission characteristics as with warm-engine operation. During deceleration

there is no fuel consumption. During other transient operations each combustion chamber receives precise amounts of hydrogen at the most optimum injection and ignition timing.

POTENTIAL FOR IMPROVED THERMAL EFFICIENCY

Advanced engines that are designed to set new records for thermal efficiency are under development. Presently, the world thermal-efficiency record for a heat engine is held by a stratified-charge engine. This record has been set by a compression-ignition engine burning Diesel fuel. Over 50% thermal efficiency has been accomplished by compression-ignition of Diesel fuel at excess air-to-fuel ratios in combination with exhaust heat recovery by a turbo-generator.[9,10,16]

As shown in Table 1, hydrogen burns about 6- to 8-times faster than hydrocarbon fuels. Hydrogen is a small molecule that does not have to undergo endothermic evaporation and thermal cracking processes in order to enter into heat-releasing reactions. Peak combustion temperature is controlled by igniting entering hydrogen and metering the rate that hydrogen is injected into the combustion chamber. Keeping the combustion temperature below about 2000°K minimizes oxides of nitrogen.

Hydrogen combustion radiates much less heat than hydrocarbons. If hydrogen is burned as a stratified charge within excess layers of insulating air, losses due to conduction and radiation are greatly reduced compared to combustion of homogeneously-mixed hydrocarbons as in normal operation of today's cars.

New engines presently under development expected to achieve more than 55% thermal efficiency by burning hydrogen as a stratified charge in far excess air-to-fuel ratios. These engines will delay ignition to a point after which Diesel fuel would have to be ignited in order to complete combustion. Compared to the present thermal efficiency record on Diesel fuel, this faster burning advantage will result in a greater production of brake-mean-effective pressure for the same heat release. Combined with the advantage of reduced heat loss due to radiation it seems likely that new thermal efficiency records will be set by engines burning hydrogen instead of hydrocarbon fuels.

HIGH EFFICIENCY CARS WILL CLEAN THE AIR

Energy conversion starting at 30% efficiency with Stirling Dish Gensets can deliver electricity to home electrolyzers operating at 90% efficiency to store hydrogen at high pressure on-board a vehicle. PSI technology allows present automobile efficiencies of about 20% to be improved to 30 to 40%. Therefore overall solar to propulsion efficiencies of 8% are expected. This is about 130 times greater than the overall efficiency of burning fossil fuels in conventional engines and about 80 times more efficient than burning fossil fuels in conventional power plants to recharge electric vehicles. Starting with wind generated electricity is even more efficient.

In addition to being more efficient than battery-powered cars charged from fossil-fired power plants, the solar-hydrogen powered cars with piston engines will clean the air as they are driven. Tests have shown that combusting hydrogen as a stratified-charge in piston engines actually reduce atmospheric concentrations of carbon monoxide, unburned hydrocarbons, tire particles, and Diesel soot. Converting cars to operation on PSI-Hydrogen will allow motorists to clean the air as they drive.

The utility grid system can facilitate the use of cars presently in the World's cities to clean the air. Advanced engines using PSI Technology and solar hydrogen will improve overall propulsion efficiencies to about 15% to 20%.

POSSIBLE ROLE OF FUTURE PHOTOVOLTAICS

Present photovoltaics cannot compete with Solar Dish Gensets in energy conversion efficiency.[3] Just as green plants favor a narrow band of sun light for photosynthesis, semiconductors are able to convert a narrow band of light to electricity. Recent developments show promise for adding several semiconductors in ways that allow several narrow bands of light to produce electricity. Eventually overall efficiencies for converting sunlight to grid-quality electricity of about 30% seem reasonable.

It is possible that homes of the future could be built with photovoltaic roof tops to collect solar energy to run the home and charge electric batteries or operate electrolyzers for hydrogen-powered cars of the future. Overall propulsion efficiencies of about 16% to 25% may be possible.

RANGE

It is probable that cars of the future will be powered by hydrogen in advanced internal combustion engines or fuel cells. These options provide far greater range and life than battery-powered cars. The author has helped demonstrate a 1979 pick-up that has operated on hydrogen for more than 10 years. The piston engine of this vehicle has been modified to use hydrogen. Hydrogen can be stored at 400 atmospheres in a compact cylindrical canister measuring about 0.6 m by 1.2 m (2 feet x 4 feet) for providing a range of about 300 miles for this pick-up. The storage tank and internal combustion engine are expected to last 30 years compared to 3 to 5 years for batteries.

Work is underway to increase hydrogen storage capacity by higher pressures, utilization of hydrides, and liquid storage as cryogenic hydrogen.

IMPORTANCE OF ENERGY CONVERSION EFFICIENCY

We cannot sustain an acceptable economy for present and growing populations by continuing to deplete fossil resources, clean air, and potable water. Burning fossil reserves has resulted in atmospheric carbon dioxide levels that are about 30% greater than any time in the last 160 thousand years.[12, 13, 14, 15] Scientists are divided about the ultimate result in terms of global warming, but it is undisputed that the atmosphere acts as a giant heat engine. Virtually all weather on Earth is solar driven. Adding heat to the global atmospheric engine by increased presence of greenhouse gases that trap solar energy allows the atmosphere to do more work such as: evaporation of the oceans, generation of winds, and resulting occurrences of floods, hurricanes, and tornadoes.

In order to change from the economy of the Industrial Revolution which has been predicated on exploitation of air, fossil reserves, and glacial-age water deposits, we must adopt more-efficient, renewable energy conversion practices. This will enable U.S. energy conversion products to be exported to the 4/5ths of the world's population that demand the U.S. standard of living. Supplying more efficient energy conversion technology for U.S. and emerging nation's needs will improve the global environment and allow fossil reserves to be utilized for more valuable products than heat generation in power plants, transportation, and air-conditioning. It is far more valuable to use fossil petrocarbons for products such as polymers, paint, pigments, and the family of emerging carbon products such as graphite fibers, diamond plating, Fullerenes, and various forms of activated carbon.

CONCLUSION

There is no reason to wait for more cost-effective photovoltaics to be developed. Stirling Dish Gensets and Wind Generators have proven capabilities for supplying grid-quality electricity at efficiencies sufficient to warrant large-scale installations for base-load purposes. Renewable hydrogen can be produced from electricity supplied by Solar Dish Gensets and used in existing motor vehicles at overall efficiencies that far exceed presently operated gasoline-powered vehicles. Solar Dish Genset electricity supplied to an automobile with an electrolyzer and a PSI system for improving engine efficiency provides about 130 times greater overall efficiency than burning fossil fuels in conventional engines and about 80 times greater efficiency than burning fossil fuels in conventional electricity power plants to recharge electric vehicles. Vehicles using renewable hydrogen will provide environmental protection advantages and have greater range, convenience, and life than electric-battery driven cars.

References:

1. Lehninger, A.L. *Principles of Biochemistry*, pp. 647-655. ISBN: 0-87901-136-X, (1982).

2. Szego, G.C. and Kemp C.C., *Chemtech*, pp. 275-284. (May 1973).

3. Lopez, C.W., Stone, K. *Design and Performance of the Southern California Edison Stirling Dish*. (1992).

4. Heede, H.R., Morgan, R.E., Ridlet, S., *The Hidden Costs of Energy*, Center for Renewable Resources. (1985).

5. Battelle Memorial Institute, *An analysis of Federal Incentives Used to Stimulate Energy Production*, U.S. Department of Energy, (1978).

6. Heede, L., Lovins, E., *Hiding the True Costs of Energy Sources*, Wall Street Journal, (September 17, 1985).

7. Gordon, D., *Steering A New Course*, A Report By the Union of Concerned Scientists, (1991).

8. McAlister, R.E., *Method and Apparatus for Fuel Injection-Spark Ignition System for an Internal Combustion Engine*, U.S. Patent 4,066,046, (1978) and numerous pending patents.

9. Watson, H.C., *Heavy Vehicle Performance and Fuel Consumption* 'Tabesam' SAE-A's Validated Computer Model, SAE Australia, Special Publication, (March 1989).

10. Kuehn, Steven E, *Technology Advances Boost Reciprocating Engine Efficiency*, Power Engineering, pp. 30-33, (February 1993).

11. U.S. Department of Energy Report: *Solar Thermal Electric Technology Rationale*, (August 1990).

12. Stone, Peter, *Prediction of Near-Term Climate Evolution: What Can We Tell Decision Makers Now?* Proceedings of the First North American Conference on Preparing for Climate Changes, Government Institutes, Inc., Washington, D.C. (October 27-29, 1987).

13. NASA Goddard Institute for Space Studies, Hansen, James E., *The Greenhouse Effect: Impacts on Current Global Temperature and Regional Heat Waves, Presented to the United States House of Representatives* Committee on Energy and Commerce, (July 1988).

14. Rodhe, Henning, *A Comparison of the Contribution of Various Gases to the Greenhouse Effect*, Science, Vol. 248, (June 1990).

15. Houghton, R.A., Woodwell, G.M., *Global Climatic Change*, Scientific American, Vol. 260, No. 4 (April 1989).

16. New Sulzer Diesel Ltd., engine performance data.

17. Smock, R.W., *Generation In Focus*, Power Engineering, (November 1985).

18. U.S. DOE *HYDROGEN DEVELOPMENT PLAN FY 1993-1997.*

TABLE 1: FUEL COMBUSTION CHARACTERISTICS

FUEL	LOWER FLAME LIMIT	UPPER FLAME LIMIT	LOWER HEAT RELEASE (BTU/lb)	HIGHER HEAT RELEASE (BTU/lb)	AIR-FUEL RATIO *	FLAME SPEED (Ft/Sec.)
HYDROGEN	4% VOL.	74% VOL.	51,593	61,031	34.5 LBS/LB	30,200
CARBON MONOXIDE	12	74.2	4,347	4,347	2.85	
METHANE	5	15	21,518	23,890	17.21	4,025
ETHANE	3	12.5	20,432	22,100	16.14	4,040
PROPANE	2.1	9.4	19,944	21,670	15.65	4,050
BUTANE	1.8	8.4	19,679	21,316	15.44	4,060
BENZENE	1.4	7.1	17,466	18,188	13.26	4,150
METHANOL	6.7	36.5	7,658	9,758	6.46	3,900
ETHANOL	3.2	19	9,620	12,770	8.99	4,030
OCTANE			19,029	20,529	15.11	4,280
HEXANE	1.18	7.4				4,200

*AT ONE ATMOSPHERE, ROOM TEMPERATURE

FIGURE 1

PRECISION SPARK INJECTION (PSI) SYSTEM

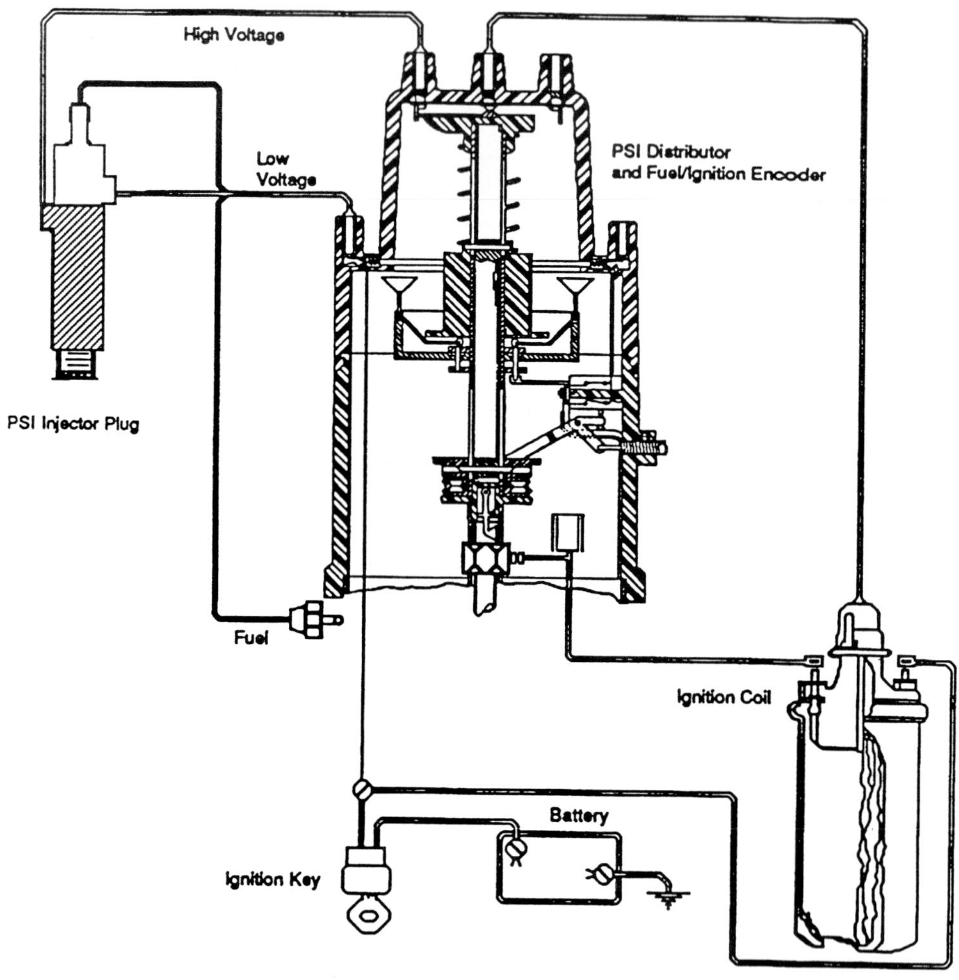

Trans Energy Corporation

Phoenix, Arizona

CHAPTER 8 -- ENERGY STORAGE SYSTEMS

Advanced Near-Term Battery Technology

Dwaine K. Coates
Senior Project Manager
Eagle-Picher Industries, Inc.
Joplin, Missouri, U.S.A.

Background:

Mr. Coates has been with Eagle-Picher for over eleven years, working in advanced battery research and development for aerospace and commercial applications. He holds a B.S. in Chemistry, and an M.S. in Chemistry from Purdue University.

ADVANCED NEAR-TERM BATTERY TECHNOLOGY

Dwaine Coates and Ron Repplinger
Advanced Systems Operation
Eagle-Picher Industries, Inc.
Joplin, Missouri USA

ABSTRACT

Several advanced, rechargeable (secondary) battery systems are currently under development, based on environmentally benign chemistry and non-toxic materials. Less toxic alternatives to lead-acid and nickel-cadmium batteries are being developed to minimize the environmental impact of battery manufacture, use and disposal/recycling. The Electronics Division of Eagle-Picher is recognized world wide as the leading producer of advanced battery systems for aerospace, defense and commercial applications. Eagle-Picher is actively developing non-toxic battery systems such as nickel-metal hydride, nickel-iron, nickel-zinc, nickel-hydrogen, silver-zinc, silver-metal hydride and sodium-sulfur for a wide variety of aerospace and commercial applications, including electric vehicles.

INTRODUCTION

Current trends in today's market stress the benefits of rechargeable battery technologies, in contrast to single-use (primary), throw-away batteries. Most of the rechargeable batteries in use are based on the sealed lead-acid (SLA) and nickel-cadmium (NiCd) battery chemistry. Lead and cadmium are considered toxic materials and are strictly regulated by the U.S. Environmental Protection Agency (EPA). The environmental aspects of toxic battery materials are multiple, including worker exposure to materials during manufacture, user exposure to battery materials and the impact of handling toxic battery materials during battery disposal and/or recycling. It is essential that less toxic battery materials be developed, to minimize the environmental impact of the manufacture, use and eventual disposal or recycling of these rechargeable battery systems. There is no way to eliminate the lead in a lead-acid battery, so alternative battery systems must be developed to replace the lead-acid battery. Nickel-based rechargeable batteries offer superior performance to lead batteries, and nickel is considerably less toxic than lead. The most commonly used nickel battery is the NiCd. Unfortunately, the cadmium contained in the NiCd is as undesirable as lead. Fortunately, the cadmium can be replaced by less toxic materials such as metal hydride, iron, zinc or hydrogen. Therefore, nickel-based batteries provide the opportunity to develop environmentally benign battery systems while improving performance, compared to lead-acid.

Nickel-based alkaline rechargeable batteries are extremely important in today's market. Total world-wide battery sales represent approximately $8 billion per year. The rechargeable battery market accounts for about 25% of this figure, or approximately $2 billion. Nickel-based batteries (primarily NiCd) make up over 80% of rechargeable battery sales, more than $1.6 billion per year. Nearly 50% of this market is consumed by Japan, with the U.S. share at about 30% and the remainder split between Europe and the rest of the world. Total rechargeable battery sales are increasing rapidly due to the recent surge in battery-operated consumer devices such as laptop computers, cellular telephones, cordless power tools, camcorders and home appliances. The U.S. Government and the military are also large consumers of rechargeable, secondary batteries.

There is an increase in interest in rechargeable batteries, as opposed to single-use, disposable primary batteries. This is partly due to increased environmental awareness, interest in conservation and recycling and changes in consumer attitudes towards the waste associated with disposable, single-use items. Environmental considerations play an important role in shaping battery technology development. For example, the nickel-metal hydride (NiMH) battery system has developed primarily because of its potential to replace the NiCd battery. Cadmium is a toxic material which has been heavily legislated against by the U.S. Environmental Protection Agency (EPA) and most state agencies as well. Similar restrictions have been imposed on lead, which is contained in lead-acid rechargeable batteries. Legislation has restricted and/or eliminated the use of lead in gasoline, plumbing and paint. Restrictions are already in place regarding disposal of the lead contained in lead-acid batteries. The cost of building lead and cadmium-based batteries will continue to increase due to the increased cost of compliance with environmental legislation and waste treatment and disposal. Eventually lead, and perhaps cadmium, may be completely eliminated as a battery material, if not by advancing technology, by legislation.

LEAD-ACID BATTERIES

Eagle-Picher manufactures a wide range of sealed lead-acid battery systems for consumer applications. Due to the current state of battery technology, lead-acid is the only battery that is economically feasible for many commercial applications. Sealed, maintenance-free lead-acid batteries are manufactured for both stand-by and cyclic applications including telecommunications, computers, fire and security systems, uninterruptable power supplies (UPS), emergency lighting, toys and medical equipment. Eagle-Picher batteries use an immobilized electrolyte design which allows the batteries to be used in any position. A lead-calcium grid is used to maintain constant charging characteristics throughout battery life. Lead-acid batteries continue to be the most economical battery system. It is important to collect and recycle these batteries to minimize the environmental impact of the lead materials used in battery manufacture.

RECHARGEABLE NICKEL-BASED BATTERY SYSTEMS

Eagle-Picher has more than forty years of experience in nickel battery systems. This experience is based in the design and manufacture of multiple nickel battery systems for military and aerospace applications. Eagle-Picher has produced tens of thousands of aerospace nickel-cadmium batteries in both sealed and vented configurations. These batteries are used in a variety of applications including earth-orbital satellites, and military and commercial aircraft. Eagle-Picher is currently building NiCd batteries for the B-1 and B-2 bombers, the AC-130 gunship and retrofit batteries for the B-52 bomber. Eagle-Picher also has experience with commercial NiCd batteries, including the design and construction of a commercial NiCd battery plant in the People's Republic of China. Nickel-zinc (NiZn) batteries have been manufactured in volume for various military applications over the years. Nickel-iron (NiFe) batteries are currently in production for the Chrysler TE electric minivan. This is the only advanced electric vehicle battery in production in the world. Nickel-hydrogen batteries have been in production for more than 20 years. Eagle-Picher currently has NiH_2 batteries flying aboard more than 50 communications and surveillance satellites. This amounts to over 90,000,000 cell-hours in orbital spacecraft operation without a single failure. Nickel-metal hydride batteries are currently being developed for aerospace and commercial applications, including electric vehicles. NiMH is considered the most promising replacement system for both lead-acid and NiCd batteries in many applications.

Nickel-Cadmium Batteries

NiCd's have been in production at Eagle-Picher for more than 40 years. Cells and batteries of many different designs and types have been manufactured. Sealed aerospace NiCd cells and batteries have been built for both low-earth-orbit (LEO) and geostationary-earth-orbit (GEO) communications and surveillance satellites for more than 30 years. These batteries are completely sealed and maintenance-free and have exceptional cycle life and performance characteristics. A large amount of testing has been completed, including a cycle life test which completed over 65,000 charge/discharge cycles under a LEO test regime without failure. Sealed aerospace NiCd technology, pioneered by Eagle-Picher, was used by many commercial battery companies in developing sealed commercial cylindrical cells such as the "C", "D" and "AA" size cells. Vented NiCd batteries are used in aircraft starting applications. The batteries must be capable of delivering in excess of 2000 amperes for short periods of time. The batteries must be rugged, durable and operate under extremely adverse conditions, such as in military aircraft in wartime. NiCd's typically deliver around 45 Whr/kg and good deep-cycle capability. Eagle-Picher also has considerable commercial NiCd experience. Eagle-Picher designed and assisted in the construction of a commercial NiCd battery manufacturing plant in the People's Republic of China. This was a large and complex program involving the design, development and technology transfer of commercial NiCd technology and manufacturing equipment and expertise.

Nickel-Zinc Batteries

NiZn batteries have been produced since the 1960's as an outgrowth of the silver-zinc (AgZn) and NiCd battery systems. NiZn is produced primarily for military and aerospace applications such as remotely piloted vehicles, spacecraft, missiles, missile silo applications, portable communications equipment, underwater propulsion and electric vehicles. Eagle-Picher set the international water speed record for an electric boat using NiZn batteries. Batteries have been produced in sizes ranging from 2 Ahr to 2000 Ahr. NiZn batteries exhibit excellent energy density, in the range of 80Whr/kg. This is primarily due to the high energy density of the zinc electrode, which delivers around 400 mAhr/g, or nearly four times the energy density of the nickel electrode. The other reason is the relatively high voltage developed by the nickel-zinc electrochemical couple, about 1.6 VDC, due to the high electromotive force (EMF) of the zinc/zinc-oxide reaction.

Zinc electrodes in the past were typically amalgamated with mercury to extend the cycle life. More recently, however, the trend has been to reduce or eliminate the mercury due to the environmental hazards associated with mercury. NiZn batteries exhibit excellent specific energy, peak power capability and operating efficiency. The drawback to the system is the relatively short cycle life compared to other nickel battery systems. NiZn batteries typically deliver around 500 deep discharge cycles due to failure of the zinc electrode. The system has come into use occasionally over the years, particularly for electric vehicle applications, but has never achieved full commercialization. If the cycle life of the system could be increased it would become viable for electric vehicle or other commercial uses. Development work is still being done at Eagle-Picher and other organizations, particularly in the area of improving the cycle life of the zinc electrode. Perhaps the strongest feature of the system is the lack of toxic materials in the battery. This factor, and the high specific energy available, may eventually rekindle interest in the nickel-zinc battery.

Nickel-Iron Batteries

NiFe electric vehicle batteries are currently being produced in a pilot scale operation for Chrysler Corporation for the TE electric minivan. This facility is currently the only electric vehicle battery manufacturing plant in the world. NiFe is being produced in a 6 volt, 200 ampere-hour battery module. This battery currently yields about 50 Whr/kg with a cycle life in excess of 1000 cycles at 100% depth-of-discharge. NiFe batteries have demonstrated an energy density of more than 120 Whr/l, and a specific peak power capability over 150 W/kg. Eagle-Picher has been producing nickel-iron EV batteries since 1978, for a variety of programs including Northrop/DOE, General Motors, Eaton/DOE Dual Shaft Electric Propulsion (DSEP), the Chrysler T-115 van and for EPRI/DOE programs. The NiFe battery is very rugged and dependable and has excellent power capability. The disadvantage of the battery is that current technology requires vented operation, which allows venting of gases during charging and requires the periodic addition of water to the battery to replenish the electrolyte. The NiFe system is one of the most environmentally benign nickel battery systems and has the additional advantage of being relatively low in cost. The nickel electrodes cost the same for all of the nickel-based systems and iron is the cheapest of the electrode materials. This system is viable for electric vehicle and other uses and will continue to be produced.

Nickel-Hydrogen Batteries

Nickel-hydrogen batteries are the system of choice for both LEO and GEO communications and surveillance satellites. NiH_2 batteries have almost completely replaced sealed aerospace NiCd batteries for these applications. The NiH_2 battery system has been in production at Eagle-Picher for more than fifteen years. The first NiH_2 batteries were launched in 1976 aboard the U. S. Navy NTS-2 satellite and a U. S. Air Force Flight Experiment satellite. There were several GEO satellites using EPI NiH_2 batteries launched throughout the 1980's including INTELSAT V, G-Star, AmericanSat, Spacenet and SatCom K1 and K2. Eagle-Picher now has NiH_2 batteries flying aboard more than fifty satellites, including several European spacecraft such as Olympus (British Aerospace), Eutelsat II (Aerospatiale), TVSat II (AEG), Telecom II and HispaSat (Matra). EPI NiH_2 batteries are also being used in demanding LEO applications such as the Hubble Space Telescope (HST). This flight database amounts to more than 90,000,000 operational cell-hours in space without a single failure.

NiH_2 batteries are also being developed for terrestrial military and commercial applications such as telecommunications, utility load leveling, remote location power, standby power applications and electric vehicles. The NiH_2 technology and its advantages have been firmly established in the aerospace industry. The only requirement to satisfy the commercial market is to make the batteries economical as compared to competing systems such as SLA and NiCd. Low cost NiH_2 cell designs and cell components have been developed to satisfy these requirements. These include new low cost catalytic hydrogen electrodes, low cost replacements for many of the standard aerospace cell components, novel manufacturing methods which can be automated for labor reduction and advanced cell and battery designs. Common pressure vessel (CPV), dependent pressure vessel (DPV), spiral-wound geometry and low pressure (LPV) NiH_2 battery designs, developed for critical space applications, are being adapted for low cost commercial applications. NiH_2 batteries are very low in toxicity and provide the best overall performance, cycle life, reliability and abuse tolerance of any battery system in existence. With the cost reductions being developed, and the performance and cycle life advantages of the system, NiH_2 is commercially viable for many applications.

Nickel-Metal Hydride Batteries

NiMH batteries are one of several battery systems currently being developed by Eagle-Picher Industries for aerospace and commercial applications. There is a high level of interest in the NiMH battery. The goal is to develop technology that will replace NiCd and lead-acid in many commercial applications. There is a growing market for higher energy density power sources than aerospace NiCd at a lower cost than NiH_2. These include small LEO satellites with scientific payloads, tactical military satellites, and satellite constellation programs such as Iridium and Brilliant Pebbles. NiMH batteries have nearly twice the gravimetric energy density of aerospace NiCd and twice the volumetric energy density of NiH_2. In addition, the projected cost for a NiMH cell is significantly lower than either aerospace NiCd or NiH_2. The nickel electrode used in NiMH cells is similar technology to those used in NiCd, NiFe, NiH_2 and NiZn. The hydride electrode can be either of two alloy systems classified as AB_2 or AB_5. The AB_2 alloy contains primarily iron-titanium or nickel-titanium with other metals added to improve performance. The AB_5 alloy is primarily lanthanum-nickel or cerium-nickel. AB_5 type mischmetal-nickel alloy is also used as battery electrode material and provides excellent battery performance. NiMH technology is being developed for electric vehicle (EV) and other commercial applications at Eagle-Picher.

The NiMH module will use an immobilized electrolyte design which will eliminate free electrolyte in the system. This will help prevent spillage in the event the battery container is damaged. The NiMH battery is capable of high rate charge and discharge and is a very low impedance system which will greatly improve vehicle performance. Normally, as the current is increased, batteries rapidly decrease in efficiency and begin to generate excessive amounts of heat due to IR drop and electrochemical processes in the battery. This trend is minimized in the NiMH battery due to the low internal impedance and the efficiency of the electrodes. The hydride electrode consists of a metallic alloy which has good conductivity and catalytic activity. AC impedance spectroscopy performed jointly with EPI and Texas Research Institute-Austin has shown to be useful in accelerated cycle life testing of NiMH and NiH_2 cells. Impedance spectroscopy work is also being conducted jointly with Florida Atlantic University in conjunction with both NiMH, NiH_2 and AgMH batteries.

SILVER-BASED BATTERIES

Silver-Zinc Batteries

Silver/silver-oxide batteries have been around for at least 200 years and were documented by Volta around 1800. Rechargeable silver batteries were produced as early as 1880. The silver-zinc system has developed into a specialty battery in modern times, primarily used for military and aerospace applications due to the relatively high cost of silver. However, silver batteries were used occasionally in weight and volume critical commercial applications such as hearing aids and cameras. Silver-zinc has one of the highest specific energies available in a rechargeable battery system. Batteries deliver more than 100 Whr/kg with very good power density. Cycle life is relatively short at 100 to 200 cycles at high DOD. Eagle-Picher has produced thousands of silver-based batteries in many different designs and configurations. Applications have ranged from manned spaceflight missions such as Mercury, Gemini and Apollo to SkyLab, the Lunar Rover and many assorted missile and weapons systems such as the Patriot missile and the Tomahawk cruise missile. Silver batteries are produced in primary and secondary versions and in remote and manually activated configurations. Eagle-Picher AgZn batteries have been used in many solar and electric race vehicles for GM Sunrayce, the Solar/Electric 300 and the World Solar Challenge in Australia.

Silver-Metal Hydride Batteries

Silver-metal hydride is a new battery system under development for aerospace and possible commercial applications. Silver has been combined in other electrochemical couples with varying degrees of success including silver-cadmium, silver-iron, and silver-hydrogen. Silver battery systems typically yield high energy densities, up to three times that of comparable nickel systems. There are applications in which the high energy density available with a silver battery offsets the relatively shorter cycle life or in which a long cycle life is not required. These are typically military and aerospace applications, but the silver-hydride system may be practical for premium commercial applications such as laptop computers or cellular telephones. The silver-metal hydride system delivers nearly the highest energy density currently available in a rechargeable battery.

HIGH TEMPERATURE BATTERIES

There are a number of high temperature battery systems currently under development in industry and the government laboratories. There are several variations of the lithium-metal sulfide (LiMS) system, including lithium-aluminum/iron-sulfide, which have been under development for several years. One of the more promising high temperature battery systems currently being proposed for commercial applications is the sodium-sulfur system, which is being considered for electric vehicles.

Sodium-Sulfur Batteries

Sodium-sulfur (NaS) batteries are currently under development and in limited production for aerospace applications such as communications satellites. EPI has been working with the NaS system since 1987. Considerable work has already been done, both at EPI and other organizations, towards developing the system for electric vehicles. Aerospace NaS cells have demonstrated over 150 Whr/kg. NaS also offers good cycle life and rate capability. Aerospace NaS cells have achieved more than 9000 charge/discharge cycles under a low-earth-orbit test regime. EPI has established an in-house manufacturing capability for the beta double-prime electrolyte. This ceramic electrolyte material is the key to NaS technology. A disadvantage of the NaS system is that the battery must be operated at high temperature (around 350°C). This means that the battery must be preheated to this temperature before it can be operated, which requires an input of energy from an external source. However, once the battery is operating, it generates enough heat to maintain operating temperature if adequate thermal insulation is provided. This insulating material required for efficient operation serves to add weight to the battery and therefore decreases the net specific energy (Whr/kg) of the system. Safety is an additional disadvantage of high temperature batteries such as NaS. Having a hot battery in a vehicle poses some problems in a collision or other catastrophic event which might breech the battery container. With high temperature batteries, not only electrical cycling must be considered, but also thermal cycling. Each time the battery is brought to operating temperature and then allowed to cool represents a freeze/thaw cycle. The NaS system does have many advantages such as high specific energy, low self-discharge, abuse tolerance and sealed, maintenance-free operation. The disadvantages are significant but can be overcome by engineering the system for the application. The real challenge, which is the same for any battery system, is to manufacture batteries at a low-enough cost to make them practical for EV's and other commercial applications.

ELECTRIC VEHICLES

There is currently a considerable interest in electric vehicles. EV's provide one of the best opportunities to improve the air quality and reduce airborne pollutants in urban areas. Electric vehicles are not new, but recent advances in technology have brought practical EV's much closer to a reality. This recent interest has been generated primarily by new legislation such as the Zero Emissions Vehicle (ZEV) regulations in California. The California Air Resources Board (CARB) has mandated that 2 percent of the new vehicles sold in the state of California by each manufacturer in 1998 must be zero emission. With current technology, only electric vehicles qualify as zero emission. The CARB regulations increase the percentage of ZEV's to 5 percent by 2001 and 10 percent by 2003. The regulations apply to any vehicle lighter than 1700 kg and to any manufacturer who sells more than 5000 vehicles per year in the state. Similar legislation is pending in 35 additional states. This trend is in part due to enhanced environmental awareness in general and in response to the problem of air pollution in particular. With over 100 cities in the U.S. currently in violation of federal clean air standards, the situation is near critical. And the problem is not unique to the U.S. Air quality is also a serious concern in many places such as Mexico City and Europe. The problems with automobile emissions are multiple. Besides the obvious immediate health concerns, auto exhaust emissions are also destructive to the environment through acid rain, the greenhouse effect and the release of toxins into the environment. Auto exhaust is a serious problem in European cities with narrow streets, tall buildings and inadequate ventilation. The emissions are causing serious damage to irreplaceable buildings, statues and other historic monuments.

Recognizing that the batteries are the limiting factor in electric vehicle performance, the big three U.S. auto makers have formed the U.S. Advanced Battery Consortium (USABC). Ford, Chrysler and General Motors, along with the Electric Power Research Institute (EPRI) and the Department of Energy (DOE) have made a substantial commitment to EV technology with the formation of the USABC. The consortium is also working in conjunction with electric utilities such as Southern California Edison, and with the national labs such as Argonne and Sandia National Laboratory. The importance of the electric vehicle issue in the United States is exemplified by the fact that three of the largest, most competitive corporations in the U.S. are co-operating on a single issue. The foundation of the consortium provides for the substantial commitment of funding for the development of practical EV batteries. The stated purpose of the consortium is to work with advanced battery developers to conduct R&D on advanced battery technologies to provide increased range and improved performance for electric vehicles through the year 2000. The USABC has broken down their technical approach into two basic categories, mid-term and long-term. Mid-term is defined as developing production capability of full vehicle battery packs in the 1995 to 1998 time period. Long-term is defined as developing battery technologies which will be directly competitive with ICE vehicles.

Eagle-Picher Industries (EPI) has been involved in EV development since 1967, with the conversion of a Renault 10 to electric operation. In 1971, an electric vehicle designed and built by EPI set 21 land speed records (top speed 158 mph) at the Bonneville Salt Flats in Utah. The same year, Eagle-Picher also designed and built the batteries which powered another unique electric vehicle, the Lunar Rover. The Rover was driven by the Apollo astronauts on the surface of the moon. This EV had a top speed of 6 mph and a 150 mile range. This vehicle (and it's batteries) remain on the moon. EPI was involved in many EV battery programs between 1974 and 1978, including lithium/metal sulfide and nickel-zinc with Argonne National Laboratory. An EPI vehicle set the world record for distance driven on a single charge (350 miles) using silver-zinc batteries in 1977. Another unique EV, designed and built by EPI, set the world water speed record (59 mph) for an electric powered boat in 1978 with nickel-zinc batteries. Nickel-iron (NiFe) battery development began in 1978, sponsored by DOE. NiFe batteries were supplied to many test programs such as the Eaton Corporation Dual Shaft Electric Propulsion Program (DSEP), the Northrop Site Operator Program, a General Motors electric van program and the Chrysler T-115 Van. This work has been continued through the present with the construction of a nickel-iron EV battery pilot plant in 1990. This plant is currently supplying NiFe batteries to Chrysler Corporation for the TE electric minivan. This is currently the only advanced EV battery manufacturing facility in the world.

CONCLUSIONS

Existing commercial batteries have not kept pace with other aspects of electronics, high technology and consumer products. Rechargeable battery systems provide an economically and environmentally sound alternative to single-use, disposable batteries. Advanced rechargeable battery systems, with reduced toxicity of materials and chemistry, must be developed. The development of improved battery technology represents an excellent opportunity to reduce the environmental impact of current lead and cadmium-based batteries. Advanced battery technology can also make a positive impact on environmental quality through applications such as electric utility load leveling and electric vehicles. Both applications reduce petroleum usage and the associated emissions. Electric vehicles eliminate noxious auto exhaust, which is a major environmental hazard in many urban areas. Nickel-based batteries, particularly nickel-metal hydride, represent the best near-term possibility for low-toxicity, rechargeable consumer batteries to replace lead-acid and nickel-cadmium batteries. Eventually lithium-polymer or other long-term battery systems will come into use and will offer improved performance and other advantages.

Thin-Film, Solid-State Polymer Electrolyte Batteries

Steven J. Visco
Staff Scientist
Lawrence Berkeley Laboratory
Berkeley, California, U.S.A.

Background:

With a Ph.D. in Chemistry from Brown University, Dr. Visco spent two years in a post-doctoral appointment at the University of California at Santa Barbara working on Lithium-conductive glasses. He has been a staff scientist at Lawrence Berkeley Laboratory since 1984, and is the Founder and Vice President of PolyPlus, Inc.

SOLID STATE SODIUM AND LITHIUM POLYMER BATTERIES FOR ELECTRIC VEHICLES

Steven J. Visco, Marca M. Doeff, and Lutgard C. De Jonghe

Materials Science Division
Lawrence Berkeley Laboratory
University of California
Berkeley, CA 94720 USA

ABSTRACT

The advantages of solid polymer technologies for electric vehicle (EV) batteries is presented. The particularly challenging goal of EV battery cost is addressed with regards to solid polymer electrolyte (SPE) batteries. The use of polyorganodisulfide positive electrodes have the dual advantages of low cost and low toxicity. Substitution of sodium metal negative electrodes for lithium electrodes is presented as an additional parameter to lower the cost of SPE batteries. The low cost of raw materials combined with the low cost of polymer processing should enable SPE batteries to meet the cost targets for electric vehicle batteries.

INTRODUCTION

The fascination with rechargeable lithium batteries is driven by their potential superior performance, by environmental necessity, and by the fundamental challenges these technologies present. The most immediate applications for rechargeable lithium batteries are for consumer electronics, and for electric vehicles. The general consumer market for rechargeable batteries is quite large and rapidly growing. The electric vehicle market is less defined, but has enormous potential in terms of size and benefits to mankind.

The exceptionally high energy density of rechargeable lithium batteries coupled with mounting concerns regarding the toxicity of cadmium gives rechargeable lithium batteries makes secondary lithium technologies an attractive alternative to nickel/cadmium technologies. The predicted growth rate for rechargeable batteries is 12 to 15% per year through the year 2000. Although the size of the electric vehicle industry can only be roughly predicted, the potential market is enormous. In the United States, California has passed legislation mandating that 2% of the cars sold in the state be zero-emission (electric vehicles) by 1998, increasing to 5% in 2001 and 10% by 2003. Several eastern states in the U.S. passed similar legislation. This translates to a new traction battery market of $375 million in 1998 and $1,873 million in 2003. Assuming other countries will soon follow suit, the economic driving force to develop suitable battery technologies for electric vehicles will be formidable.

Clearly, the high volumetric energy density (Wh/l) of lithium based battery systems makes them very attractive for consumer electronic applications. However, early enthusiasm for rechargeable lithium batteries based on lithium metal electrodes and liquid organic electrolytes must be tempered by the serious safety problems these cells present. In response to the issue of safety, several alternatives to lithium metal/liquid organic electrolyte batteries have been pursued. The most prominent of these are the lithium-ion (Li_xC) and the solid-state lithium polymer batteries. The high volumetric energy density and rather remarkable cycleability of the lithium-ion (or "rocking-chair") battery makes it an ideal candidate for consumer electronic applications. Notably, lithium-ion batteries have already been successfully commercialized, and are currently finding application in cellular phones.

It seems less clear that lithium-ion based batteries will present the definitive solution for electric vehicles. In substituting intercalated carbon (Li_xC) for the lithium metal electrode, there is a substantial weight penalty to pay, reflected in the somewhat mediocre specific energy of lithium-ion cells (relative to lithium cells). Although this does not present a major obstacle for consumer electronic devices, it is a critical parameter for electric vehicles since the specific energy directly affects the curb weight of the vehicle. In this sense, the lithium polymer batteries and the high temperature lithium-alloy batteries, although not yet in commercial production, may present more viable solutions for electric vehicles.

The United States Advance Battery Consortium (USABC) has established a set of performance goals for electric vehicle batteries (table 1)[1] . Although the USABC performance criteria present a fairly challenging goal for battery developers, the most daunting task will undoubtedly be to meet the cost requirements. The maximum cost for electric vehicle batteries is set at $150/kWh. This is an realistic upper limit to cost in that a 35 kWh battery for a passenger vehicle will cost $5250.0000. Clearly, if electric vehicles are to present a reasonable alternative to internal combustion engine (ICE) vehicles, the cost of the vehicle including the battery has to be competitive. This puts very serious constraints on the cost of raw materials and ultimate fabrication costs for EV batteries. Typically, high-end nickel/cadmium batteries for applications such as cellular phones cost approximately $1/Wh or $1000/kWh. Although this is appropriate in the consumer electronics market, a 35 kWh EV battery would then cost $35,000 which is clearly inappropriate. In fact, very few advanced battery systems will be able to meet the cost targets for EV batteries. Fortunately, the lithium/polymer and sodium/polymer-based battery systems represent low-cost, high performance technologies capable of fulfilling the demanding requirements for electric vehicles.

APPROACH

Research on advanced batteries at our laboratory has been focused primarily on two variants of solid polymer technology; lithium polymer and sodium polymer batteries. The use of sodium negative electrodes is motivated by the much lower cost and greater availability of this raw material. Since their inception by Armand[2], solid polymer lithium batteries (figure 1) have captured the interest of numerous researchers. Much of the early work on solid polymer electrolytes was carried in the temperature range of 60 to 100 ^{o}C, where polyethylene oxide has suitable ionic conductivity. Since then, a number of groups have reported solid polymer electrolytes with acceptable conductivity at ambient temperature[3], significantly broadening the scope of applications for solid polymer electrolytes.

One of the approaches in this laboratory has included the use of solid polymer electrolytes in combination with polymeric organodisulfides and/or polythiols (figure 2). These materials are termed solid redox polymerization electrodes (SRPE's) to reflect the dynamic process of polymerization/depolymerization that occurs as these electrodes are oxidized and reduced. Chemical synthesis of the polymers is easily accomplished by oxidation of di- or tri-mercaptans by an appropriate oxidant as shown below,

$$\text{nHS-R-SH} + \text{nH2O2} \Rightarrow (\text{SRS})_n + 2n\ H_2O, \qquad \text{Eq.1}$$

where R is an organic group such as ethyl, phenyl, etc. The synthesis of a number of such polymers has been described earlier[4],[5],[6]. The polymers are very inexpensive (*$3-4/lb.), non-toxic, and biodegradable. The low toxicity of the polyorganodisulfides is an important asset for electric vehicle batteries, particularly in view of the inevitable rupture of EV batteries in highway collisions. Among the advantages of the polymeric organodisulfides are the tremendous chemical flexibility in altering the physical, chemical, and electrochemical properties of the polymers through synthetic means. For SRPE's where R is a small organic group such as CH_2CH_2, the equivalent weight of the polymer can be as low as 46 grams/equivalent making these materials attractive for use in high energy density batteries (table 2).

A number of solid redox polymerization electrodes have been prepared and described in terms of solid-state lithium battery performance[4-6],[7] . The majority of these batteries have used polyethylene oxide as the solid polymer electrolyte, although, for ambient temperature testing, oxymethylene-linked polyethylene oxide[3] (amorphous PEO) was found to exhibit acceptable ionic conductivity as well as good mechanical properties. In principle, SRPE's can function as positive electrodes for a number of metal negative electrodes. In fact, early tests on SRPE's were done using liquid sodium electrodes and β"-alumina electrolytes[5]. Sodium negative electrodes offer advantages over lithium including greater abundance, ease of processing, and much lower cost. The theoretical energy densities shown in table 3 for the sodium/SRPE cells are somewhat lower than for the lithium cells, but are still in excess of the requirement for electric vehicle and/or consumer electronic devices. The performance of Li/PEO/SRPE cells has been reported in detail elsewhere[4-7]; practical specific energy and power can exceed 200 Wh/kg and 1kW/kg, respectively. These batteries meet or exceed most of the goals targeted by the USABC in table 1. In addition to high levels of performance, lithium solid polymer technologies can meet the cost goals set forth by the USABC. The cost of raw materials for generic lithium polymer cells is shown in table 4 and figure 3. Clearly, the use of expensive perfluorinated electrolyte salts such as lithium trifluoromethane sulfonate would make it very difficult to meet target costs. However, perchlorate salts are well within the cost envelope. Figure 4 shows estimated costs to fabricate a thin-film solid polymer lithium battery using techniques such as coextrusion, and continuous roll-to-roll lamination. The cost to manufacture this cell is approximately $70/kWh. Assuming that the finished batteries would then be sold to the consumer at approximately twice this price, the retail price of the battery would be about $140/kWh. This is below the USABC goal, and would represent a total cost of $4800 retail for a 35 kWh battery.

Further reduction in costs for solid polymer EV batteries can be realized through the use of sodium metal negative electrodes. The behavior of Na/PEO/SRPE cells has been investigated to a lesser extent[8], but the results are encouraging[9]. Recent work at our laboratory has broadened the work on Na/PEO batteries to include the use of metal oxide cathodes. Although the choices for positive electrodes in Na/PEO batteries are more limited than for the lithium analogs, our laboratory has had great success with a number of candidate materials. To date, the best performance has been achieved with the sodium cobalt bronzes, Na_xCoO_2. Solid polymer batteries of the type Na/PEO/Na_xCoO_2 have demonstrated high energy density, and power density, and stable behavior over hundreds of cycles[10]. Although the cobalt bronze cathode exhibits excellent behavior in the solid polymer battery, we are currently investigating alternative lower cost materials for the positive electrode in solid sodium batteries. Still, the very low cost of sodium may offset the expense of the cobalt electrode to yield an affordable technology. Figure 5 shows the cost savings in moving from lithium-based solid polymer batteries to sodium-based batteries. The cost to produce solid sodium polymer batteries could be as low as $44/kWh. The major concern with this approach is the low melting point of sodium metal, 98 ^oC. Given that solid polymer batteries are typically operated in the range of 60 to 90 ^oC, thermal management of solid sodium batteries would be more critical. This problem can be overcome by the use of sodium-lead alloy (Na_4Pb) which melts above 200 ^oC, at the expense of some lowering of the specific energy of the battery.

CONCLUSIONS

Solid polymer technologies offer realistic solutions to the problem of affordable, high performance EV batteries. Potential positive electrodes include the polyorganodisulfides and various metal oxides. The polyorganodisulfides have the advantages of low cost and low toxicity. The choice of negative electrodes is limited to lithium or sodium. Sodium metal has a tremendous cost advantage over lithium, however, the low melting point of sodium may be a problem. The ability to use continuous roll-to-roll processing to fabricate solid polymer batteries, as is done in the manufacture of plastic laminates, should allow this technology to meet the USABC's ambitious cost goals.

ACKNOWLEDGMENTS

This research was supported by the Assistant Secretary for Conservation and Renewable Energy, Deputy Assistant Secretary for Utility Technologies, Office of Energy Management, Advanced Utility Concepts Division of the U.S. Department of Energy under Contract No. DE-AC03-76SF00098.

REFERENCES

[1] E.J. Cairns and F.R. McLarnon, "Status of Batteries for Energy Storage Applications," *183rd Meeting of the Electrochemical Society, May 16-21, 1993, Honolulu, Hawaii*, Ext. Ab. No. 21.
[2] M. B. Armand, J. M. Chabagno and M. J. Duclot, "Fast Ion Transport in Solid," ed. P. Vashista, J. N. Mundy and G. K. Shenoy, Elsevier, North Holland, New York, 131 (1979).
[3] C. Booth, C. V. Nicholas, and D. J. Wilson, "Polymer Electrolyte Reviews," ed. J. R. MacCallum and C. A. Vincent, Elsevier, New York, 229 (1989).
[4] S. J. Visco, M. Liu, M. B. Armand, and L. C. De Jonghe, *Mol. Cryst. Liq. Cryst.*, **190**, 185 (1990).
[5] M. Liu, S. J. Visco, and L. C. De Jonghe, *J. Electrochem. Soc.*, **138**, 1896 (1991).
[6] M. Liu, S. J. Visco, and L. C. De Jonghe, *J. Electrochem. Soc.,* **138**, 1891 (1991).
[7] M. M. Doeff, M. Lerner, S. J. Visco, M. Ue, , and L. C. De Jonghe, *J. Electrochem. Soc.*, **139**, 2077 (1992).
[8] S.J. Visco, M. Liu, M.M. Doeff, Y.P. Ma, C. Lampert, and L.C. De Jonghe, *Solid State Ionics*, **60**, 175 (1993).
[9] Solid polymer/polyorganodisulfide technology has been recently transferred to a privately held company in the Bay Area, and as such this work is no longer pursued by the Lawrence Berkeley Laboratory.
[10] Y.P. Ma, M.M. Doeff, S.J. Visco, and L.C. De Jonghe, "Rechargeable Na/PEO/Na_xCoO_2 and $Na_{15}Pb_4$/PEO/Na_xCoO_2 Solid State Cells," *183rd Meeting of the Electrochemical Society, May 16-21, 1993, Honolulu, Hawaii*, Ext. Ab. No. 1834.

TABLE 1.

USABC Performance Goals	
Specific energy (Wh/kg)	>80
Energy density (Wh/l)	>135
Specific power (W/kg)	>150
Energy efficiency (%)	>60
Cycle life	>600
Calendar life (years)	>5
Cost ($/kWh)	<150

TABLE 2. ENERGY DENSITY OF VARIOUS Li/SRPE COUPLES

ID	SRPE	Description	E.W. equiv/g	mAh/g	mAh/cc	Vnom vs. Li	Wh/kg	Wh/l
Li	Lithium		7	3,829	2,045			
X0	$(SCH_2CH_2 0CH_2CH_2S)_n$	clear, viscous liquid	76	353	564	2	**646**	**885**
X1	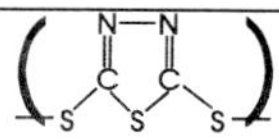	pale yellow powder	74	362	652	3	**993**	**1,483**
X5	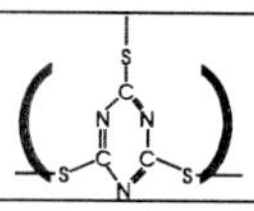	pale, yellow powder	58	462	739	3	**1,237**	**1,629**
X8	$(SCH_2CH_2S)_n$	white powder	46	583	932	2	**1,012**	**1,281**

TABLE 3. ENERGY DENSITY OF VARIOUS Na/SRPE COUPLES

ID	E.W. equiv/g	mAh/g	mAh/cc	Vnom vs.Na	Wh/kg	Wh/l
SODIUM	23	1,165	1,142			
X0	76	353	564	1.7	**460**	**642**
X1	74	362	652	2.7	**746**	**1,121**
X5	58	462	739	2.7	**894**	**1,212**
X8	46	583	932	1.7	**660**	**873**

TABLE 4. RAW MATERIALS COSTS FOR SOLID POLYMER BATTERIES

Raw Material	Cost
Lithium	$35/lb.
Sodium	$0.20
Polyethylene Oxide	$2/lb. ($8/lb. for high purity)
Lithium Perchlorate	$10/lb.
Fluorinated Salts	$120/lb.
Positive Electrode (Metal Oxide or SRPE)	$4/lb.
PET	$5/lb.

Figure 1. Schematic of a solid polymer battery.

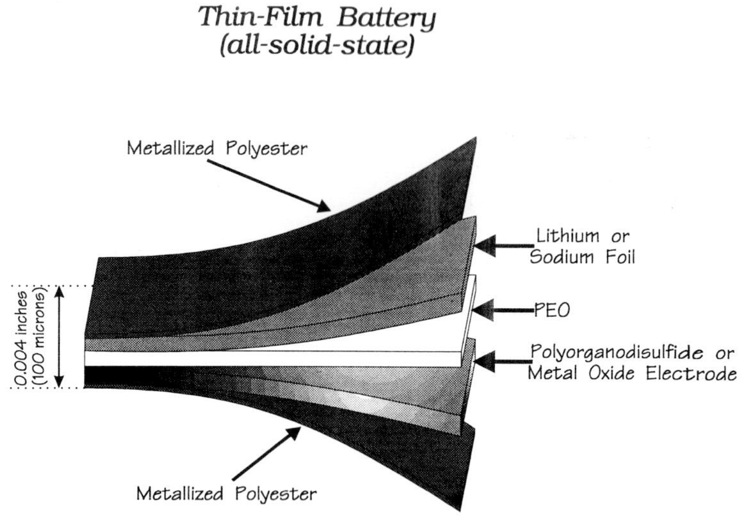

Figure 2. Polythiol and polyorganodisulfide electrodes

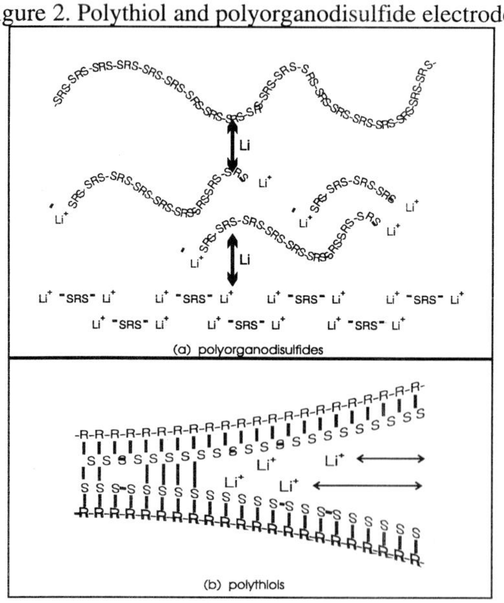

Figure 3. Cost of Raw Materials for Solid Polymer Batteries

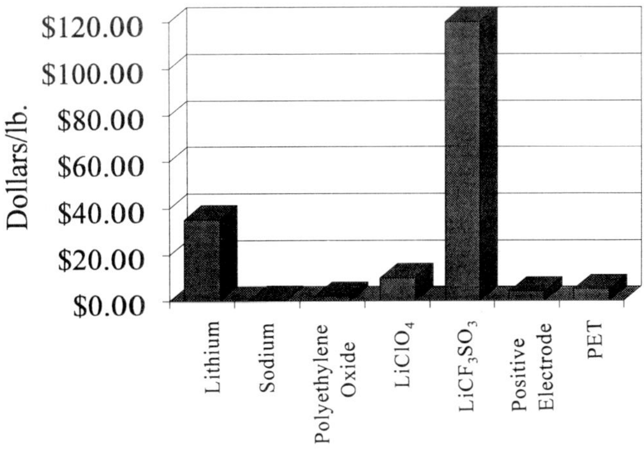

Figure 4. Cost estimates to manufacture lithium polymer batteries using coextrusion techniques

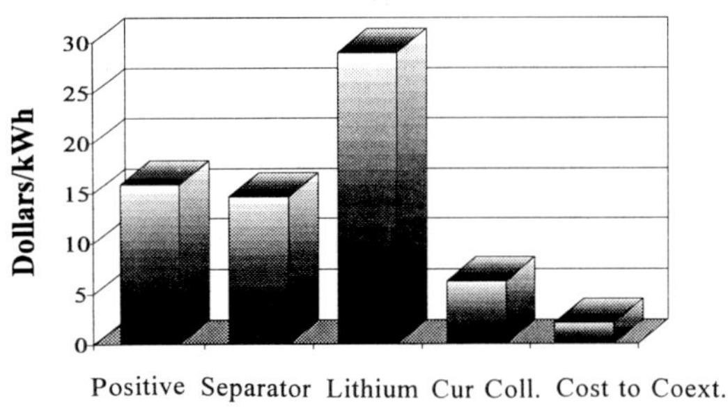

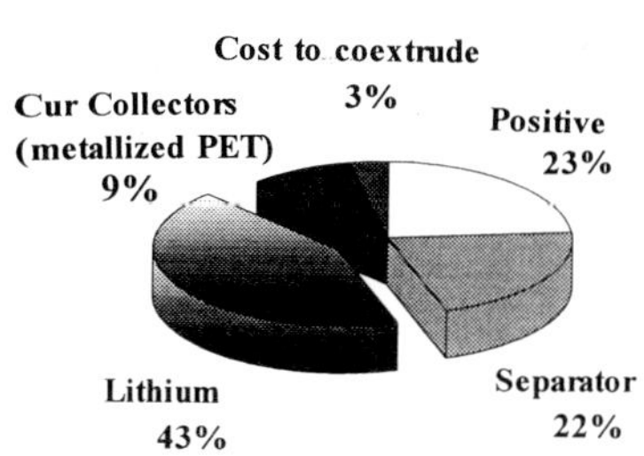

Total cost of Solid Polymer Laminate: $68/kWh
Cost of 35 kWh battery for EV: $2400

Figure 5. Cost estimates to manufacture sodium polymer battery as above.

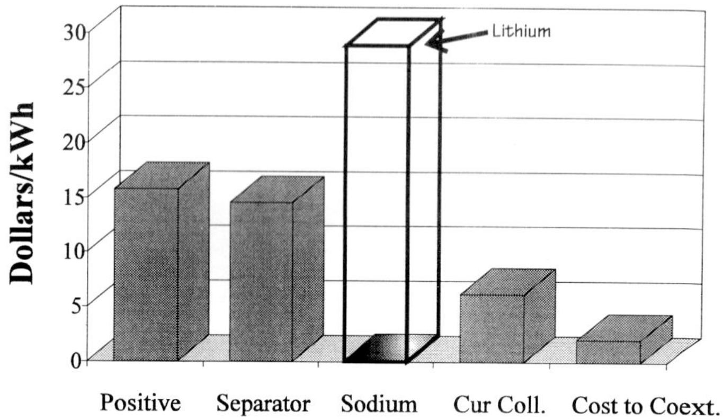

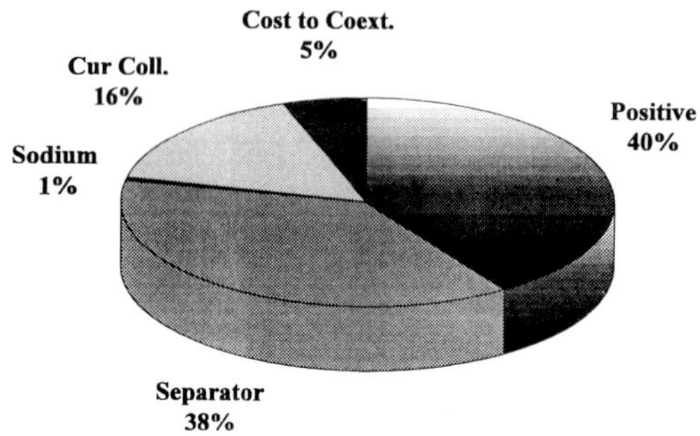

Cost of Solid Sodium Polymer Laminate: $40/kWh

Cost of 35 kWh Battery: $1400

Electromechanical Battery Development at Lawrence Livermore

Richard F. Post
Senior Scientist
Lawrence Livermore National Laboratory
Livermore, California, U.S.A.

Background:

Dr. Post is a senior scientist at the Lawrence Livermore National Laboratory and Professor (emeritus) of the University of California, Davis/Livermore. A physicist by training, his fields of expertise are magnetic fusion research and inertial energy storage systems.

ELECTROMECHANICAL BATTERY RESEARCH AND DEVELOPMENT AT THE LAWRENCE LIVERMORE NATIONAL LABORATORY

Richard F. Post
Lawrence Livermore National Laboratory
P.O. Box 808, L-644, Livermore, CA 94550 USA

ABSTRACT

New materials and new design concepts are being incorporated in a new approach to an old idea - flywheel energy storage - to create an important alternative to the electrochemical storage battery for use in electric vehicles or for stationary applications, such as computer back-up power or utility load-leveling.

INTRODUCTION

Efficient and cost-effective means for storing electrical energy is becoming an increasing need in our electricity-oriented society. For the transportation sector the need for energy storage, now acute, is that for better batteries to power electric vehicles. As yet no electrochemical battery exists that provides a wholly satisfactory answer to this need. For the electric utilities the need for means to store electrical energy is growing and is taking a new direction — distributed storage systems. That is, energy storage at substations, at solar or wind-power sites, or for load-leveling at the site of major consumers of their electricity. One of the important savings of such an approach would be the reduction in transmission losses that would result from having a local source of load-leveling power or "spinning reserve". In California these diverse uses are becoming increasingly important as we move toward the use of alternate energy sources, and as the time approaches when State-mandated electric vehicle quotas must be met.

One possible answer to the entire spectrum of needs just outlined is the "electromechanical battery". The E-M battery has the potential to solve both of the above energy storage problems in a manner superior to the electrochemical battery in all important attributes, namely, energy storage density, power density, energy recovery efficiency, cycle lifetime, and amortized capital cost.

By an electromechanical battery (EMB) we mean an energy storage module consisting of a high-speed rotor, fabricated from fiber composite, and having an integrally mounted generator/motor. The rotor operates at high speed, in vacuo, inside of a hermetically sealed enclosure. In our proposed embodiment of the EMB the rotor is supported by a "magnetic bearing", that is, a bearing that uses magnetic forces to support the rotor against gravity. Considerations of achieving long service life, of the problems of heat removal in a vacuum, and of the consequent need to minimize frictional losses, mitigate against the use of conventional bearings, making the magnetic bearing a virtual necessity.

The storage of energy in flywheels is, of course, a very old idea. In fact, the concept of using fiber composites to achieve higher energy densities in flywheel energy storage was studied in the U.S. in government-sponsored programs in the 1970's. The general problems of designing such flywheels [1] were well understood at that time and substantial progress toward implementing the new ideas had been made before the programs were canceled in 1982.

Recent years have seen the rebirth of interest in the subject, based on new materials, new design concepts, and new needs. In the remainder of this article we will briefly describe the particular features of the new breed of EMB we are developing at the Laboratory, discuss some aspects of its use in electric vehicles and in bulk storage, and conclude with a summary of the results to date in our funded program to demonstrate a 5 kW-h EMB module in a practical setting (backup power for a subset of the computers in the Livermore Computer Center). Some of the material in this paper has been presented in previous papers [2,3].

DESIGN FEATURES

The salient design features of our version of the EMB are illustrated in Fig. 1, a cutaway drawing that represents schematically a sectioned view of a small module. For use in an all-electric or a hybrid-electric auto the module would be cubical in shape, having outside dimensions of order 25 cm and storing approximately 1 kW-h. The rotor of the module is made up of nested cylindrical shells [1] fabricated of high-strength graphite-fiber/epoxy composite. These shells are coupled to each other by "separators". As noted earlier, the rotor is housed in a sealed, evacuated, chamber and is supported on magnetic bearings.

The innermost shell of the rotor has mounted on its inner surface a special assembly (the "Halbach Array" — described later) of high-field permanent magnet material (NdFeB). Inside of this magnet array is a re-entrant cylinder of insulating material that serves as a vacuum barrier through which the field of the rotating array is coupled to three-phase stator windings. In this way the windings are made directly accessible for cooling, either by forced air or by a liquid coolant. The entire generator/motor assembly is "ironless", a feature that leads to high efficiency and low standby losses (no hysteresis effects). Because of the high rotation speed of the rotor (of order 200,000 RPM for the 1 kW-h module) it also leads to a very high specific powers (5 to 10 kW/kg; up to two orders of magnitude greater than those typical of electrochemical cells).

Also shown in Fig. 1, albeit only schematically, are the magnetic bearing assemblies, with their permanent magnet exciters. These bearings, which will not be discussed here in detail, also benefit from the use of NdFeB magnet material. Only a relatively few grams of this material will be required to provide the magnetomotive force needed to support the rotor. However, the magnetic bearings represent one of the most challenging of the design problems for compact EMBs. As a consequence we have made a special effort to develop simplified versions of these bearings, one embodiment of which will be employed in the EMB modules now under development.

The chamber within which the rotor/bearing assembly contained is a sealed, evacuated, metal housing reinforced with fiber composite. This reinforcement is both to sustain vacuum-induced loads, and also to provide a first level of defense against the consequences of rotor failure. Safe containment of the debris and of the energy released upon rotor failure is an essential requirement for any vehicular use of EMBs. To address this issue fully will require an extensive test program. However, we believe that experience gained in the aircraft industry, where similar problems are encountered in the failure of turbo-jet engines, can be applied to help solve the problem in vehicular EMBs. As described in a NASA report [4], structures containing only a few kilograms of polymeric fiber provide containment in the event of blade failure in large jet turbines. We therefore visualize the EMB modules in a vehicle, in addition to having their own containing features, to be surrounded by a light-weight "egg crate" composite structure that both provides mechanical support and additional containment in the case of rotor failure. Of substantial help is the fact that the failure of rotating graphite fiber composite structures under hoop tension has been found in earlier tests [5] to result in an amorphous mass of low-strength material, rather than in shrapnel, as is the case of the failure of steel rotors.

Figure 1 shows the spherical housing supported on springs (symbolically representing gimballing action). If gimbals are used they greatly reduce the restraining forces that must be provided by the magnetic bearings, at the expense of increasing the volume of the module. Whether or not gimballing is employed we believe that it will be necessary to include "backup" bearings in the overall design to accommodate to shock loads (as in impacts or in encountering "potholes" in the road). Not shown in the figure is the solid-state power electronics needed to convert and control the power flowing out of and into the EMB module. It appears that the preferred vehicular drive systems will be ac-based ones. For example, a recent drive train (in the GM Impact) employs variable-frequency ac induction motors. While electrochemical battery systems must use combinations of rectifiers and inverters to accomplish their power transfer, in the case of the EMB with its multi-phase ac output a more attractive possibility arises: the use of a so-called "matrix converter" [6] to directly convert multi-phase, variable frequency ac to another (variable) frequency. The matrix converter, which is bidirectional, uses pulse-width modulation techniques to chop and resynthesize ac waveforms. Matrix converters can be made to be more efficient than rectifier/inverter-based systems. As a result of the high efficiency of both the generator/motor and the matrix converter it is predicted that the recovery efficiency of EMBs based on our design concepts will be very high — of the order of 95 percent. This figure is to be contrasted with the typical 60 to 70 percent efficiency of electrochemical cells. As a result, both electric and hybrid-electric vehicles, when driven in typical urban cycles, are predicted to have substantially increased range (for the same amount of stored energy expended or fuel consumed) if EMBs are used in place of electrochemical cells.

The module just described is one that is sized for vehicular applications. For other applications we would expect to scale up the module size somewhat. For example, for the computer-backup application and for power-conditioning applications 5 KW-h modules would be desirable, while for load-leveling and bulk-storage applications 25 kW-h units (used in a "battery-bank" mode) appear to be attractive from a manufacturing viewpoint. For the rotors of the latter units the fiber of choice would likely not be graphite fiber (presently too expensive per kW-h stored), but instead the workhorse fiber, "E-glass", would be used, under temperature controlled conditions to minimize stress-corrosion effects.

SPECIAL DESIGN ISSUES

A. The Rotor

Our choice of a multi-shell configuration for the rotor of the EMB is based on both practical and economic considerations. The first consideration is a materials-related one, having to do with the mechanical properties of fiber composites, as follows: To achieve maximum strength in hoop tension, the fiber composite of which the cylindrical shells the rotor are composed is "uni-directional" in nature. That is, the shells are to be fabricated (by automated filament winding) with the filaments oriented azimuthally. As a consequence, although maximal strength against centrifugal hoop tension is achieved, it is at the sacrifice of strength in the radial direction, where the tensile strength is only that of the epoxy matrix. However, owing to the radial gradient of centrifugal forces internal radial stresses will arise within the body of the ring. As was found in early experiments with thick-ring rotors [7] these stresses can cause rotor failure (by delamination) at hoop stress levels far below the tensile strength limit of the composite. Theory shows that these stresses vary roughly as the square of the radial thickness of the shells. However, by limiting the thickness of the rings to approximately 10% of their radius the internal tensile stresses can be kept well within allowable limits. We have incorporated this feature in our designs. Another feature of our particular multi-shell rotor design is that of mass loading the inner rings. That is, the use of moderate amounts of a high-density material on the inner surface of these rings. The effects of mass loading are threefold: first, loading increases the amount of energy stored in the rotor per unit volume, with only a modest decrease in the specific energy. Second, loading diminishes the amount by which the gap between adjacent rings grows under centrifugal stress, thus simplifying the job of designing the separators. Third, loading can reduce the radial tensile stress within the shells by introducing a compressional term, even including transforming those stresses to pure compression. Thus, the addition of even modest amounts of loading can allow the use of thicker rings without fear of failure by delamination. Note that mass loading of the innermost shell of the rotor occurs naturally because of the presence of the permanent magnets of the Halbach Array.

Using the stress-strain equations for an anisotropic material [8] we have calculated the azimuthal and radial stresses in thin cylindrical shells without loading and thicker shells with mass loading. Figures 2A and 2B show two such cases. In each case the stress levels are normalized to the hoop stress level that would result in an infinitesimally thin shell having a radius equal to the inner radius of the shell. Case A is an unloaded shell with a radius ratio of 1.1, Case B is a loaded shell with a radius ratio of 1.2 (i. e., twice as thick radially) and with a mass loading sufficient to give a normalized compressional radial stress at the inner radius equal to .05 (i. e., 5% of the normalized hoop stress at that radius). As can be seen the effect of the mass loading is to reduce the radial tensile level in the thicker shell to about the same value as that of the unloaded, thinner, shell.

Additional issues associated with the multi-shell rotor design are those of resonances and rotational instabilities of the rotor body itself. One such class of unstable resonance response arises from the transverse oscillations of the shells of the rotor with respect to each other. These oscillations were analyzed theoretically early on and subjected to experimental test, with close agreement between the experiments and theory. [9] It was shown that instability can be avoided if the lowest mode of transverse oscillation (determined by the effective spring constants of the separators) is constrained to lie above the highest operating speed of the rotor.

The remaining class of rotational instabilities are so-called "whirl" instabilities, arising from dissipative losses (and consequent out-of-phase torques) occurring within the rotating system. The general prescription for eliminating these in a rotor supported by magnetic bearings is to incorporate compliant and/or dissipative elements in the bearing supports. [10]

Compliance also plays another important role in our designs. To avoid the necessity of precision balancing, which would be difficult (if not impossible) in a fiber-composite rotor of our configuration, operation is to be "super-critical". That is, by design the operating speed of the rotor lies far above the first critical speed determined by the mass of the rotor and the compliance of the suspension/bearing system. The validity of this concept for a multi-ring flywheel system was demonstrated in tests performed in the 1970's. [11]

B. Vacuum-Related Issues

It is necessary to operate the rotor in vacuo in order to avoid excessive aerodynamic losses. In our design the rotor is housed in an evacuated chamber that is sealed off at the time of manufacture. The degree of vacuum required for an acceptably long self-discharge time can be estimated approximately from kinetic theory. For our 1 kW-h rotor the calculated rundown time constant (from aerodynamic losses) at a residual gas pressure of 3×10^{-5} Torr is 60 days.

We have performed preliminary tests, in a sealed-off chamber, that indicate that we should be able to maintain the required vacuum without the necessity for continuous pumping. The tests were performed at various temperatures from ambient to 100° C, using a specially treated and pre-processed epoxy-graphite fiber composite structure of the same general size as the rotor. Figure 4 shows a pressure vs time curve for the system following seal-off, when maintained at a temperature of 50° C (to simulate the highest ambient temperature likely to be encountered in vehicular use). As can be seen, the pressure level remains acceptably low even when the chamber is no longer being pumped. This pressure was maintained for a period of months, before the test was terminated.

C. The Generator/Motor

A central feature of our version of the EMB is the generator/motor design. As mentioned previously, our design is an "ironless" one, utilizing a special array of permanent magnet bars to create a rotating dipole field. This array is one that was pioneered by Klaus Halbach of the Lawrence Berkeley Laboratory. [12] Figure 3 shows the dipole version of Halbach's array, together with the calculated lines of force of the field that it produces. Note the uniformity of the interior field, and its near-cancellation outside the array. The efficiency of the Halbach array is such that using NdFeB material for the magnets, dipole fields of order 0.5 T can readily be obtained.

Both the predicted power output and the efficiency of the generator motor in our example 1 kW-h module are remarkably high. There are no hysteresis losses in the ironless system, and since the inductances of the multi-phase windings are consequently very low, and rotation speeds are high, unusually high peak power outputs are possible, with stator copper losses that can readily be handled by conventional means (air or liquid cooling). Example parameters for such a module are as follows:

Rotation speed	200,000 RPM
Magnetic field of Halbach Array	0.5 Tesla
Length and width of windings, (turns)	.08 m, 04 m (10)
Output voltage (3-phase)	240 V rms

In a hybrid-electric vehicle application the calculations show that two of the above units, when at full speed, should together be able to deliver 75 kW (100 hp) for several seconds, at a generator/motor efficiency of .993, associated with conductor losses of 260 W. These losses lead to a calculated initial rate of rise of the temperature of the windings (ignoring cooling) of about 8°C/sec.

In our designs litzendraht wire is to be used in the windings. The purpose: to reduce residual eddy current losses that would occur even in the absence of a load. In the design above conductors composed of 6 paralleled strands of 175/46 gauge litz wire were used, leading to a calculated eddy current loss (at peak speed) of about 2.0 W per phase. This level of loss would result in an initial self-discharge time constant of about 160 hours for a 1 kW-h module.

D. Magnetic Bearings

Since the inception of our present program we have paid special attention to what we consider to be the key technical problem of the EMB — the magnetic suspension/bearing system. To that end we have initiated a combined theoretical/experimental program with the objective of developing new classes of magnetic bearings that would be simpler and less expensive than the presently available commercial systems. At this time it is premature to discuss the results of our program, but some comments on the general problem are in order.

As is well known, the first problem that must be faced in designing any magnetic suspension/bearing system is dealing with Earnshaw's Theorem. [13] This theorem asserts the impossibility of stably levitating an object by means of magnetic forces exerted between elements characterized by constant currents and/or constant magnetization. The theorem implies that if an object is levitated there will be at least one axis in space along which it is unstable. The conventional answers to this dilemma either involve mechanical means of stabilization (as in some commercial turbo-vacuum pumps) or the use of sensors and servo circuits to stabilize the bearing system. The latter approach is the one that was taken in a previous Laboratory study by Comfort et al. [14]

Examining the assumptions implicit in the derivation of Earnshaw's Theorem, circumstances can be listed under which it does not apply. For example, it does not apply to diamagnetic bodies, such as superconductors. Indeed magnetic bearings using high-temperature superconductors have been experimentally tested. [15] The theorem also does not exclude the use of time-varying forces to stabilize, such as may occur in systems subject to periodic forces. [16]

Our approaches to simplifying magnetic bearings utilize various combinations of "loopholes" in Earnshaw's Theorem to accomplish stability. We are presently testing these ideas and the results of these tests will determine our selection. Since our presently funded project is intended for a stationary application, its magnetic bearing system can be simpler than ones needed for vehicular applications. However, we expect to learn much that will be helpful toward solving the bearing problem for vehicular applications in executing this task.

APPLICATIONS TO ELECTRIC AND HYBRID-ELECTRIC VEHICLES

In the United States urban driving accounts for the lion's share of the petroleum fuel required for the automobile — and urban use of the automobile makes a major contribution to urban air pollution. Legislation in California and other states requires the progressive introduction of "zero pollution" autos (read "electric") to address both of these problems.

Owing to infrastructure requirements and other issues, the hybrid-electric vehicle is increasingly viewed as an alternative to a direct transition to all-electric autos. Whichever the choice, however, there is a clear need for better means for energy storage. Not usually understood, however, is the importance of the parameter "energy recovery efficiency" in this context (Fig. 5). The figure shows the calculated "Energy Conservation Factor" (ECF) achieved with and without regenerative braking for a particular case. The case: the figurative replacement of the internal-combustion engine of an example auto — the Porsche 912 — by an electric drive system with the indicated energy recovery efficiency, when "driven" (calculated) over an urban driving cycle.

As defined here, the ECF is the ratio of the primary energies required to drive the same vehicle over the given driving cycle in the two different modes — I-C engine or electric drive. Thus an ECF of 5.0 corresponds to the fact that it would take 5 barrels of oil delivered to a refinery, refined to gasoline and delivered by truck to a gas station, to drive the I-C version of the car over the same distance that the energy equivalent of 1 barrel of oil used as energy input to an electrical utility would drive the electrified version of that same car (i. e. one with the same weight and same drag coefficients).

Returning to the graph, three features are notable: first, in urban (stop-and-go) driving, the increase in efficiency from regenerative braking is strongly dependent on energy recovery efficiency. Second, for the same amount of energy storage (in an all-electric auto) in urban driving nearly twice the range would be achieved using EMBs as compared to that achieved with electrochemical batteries. Third, the use of EMBs in electric vehicles could result in a major reduction in the primary energy requirements of the transportation sector. The message is clear: In addition to energy storage capacity, it is equally important to consider energy recovery efficiency in the effort to extend the range of electric vehicles driven in urban settings beyond that available with present-day batteries.

STATUS OF LIVERMORE EMB PROGRAM

The funded program at the Laboratory to develop an EMB module for use in the Livermore Computer Center began in December 1992. Since initiation of the effort considerable progress has been made toward the goal of installing such a unit by the end of Calender Year 1993. Work to date has included the following:

- Design of of the rotor and its containment; procurement of materials;
- Theoretical analyses of the stability of the bearing system;
- Finite-element and theoretical analyses of rotor stress distributions;
- Vacuum tests of sealed chamber containing a simulated rotor;
- sub-system tests of mechanical "backup" bearings;
- Sub-system tests of support and damping elements of LLNL passive magnetic bearing designs;
- Design and "breadboard" construction of solid-state drive electronics;

In the work to date the emphasis has been on gaining experience and testing designs that will be of general value in future work where other applications, such as vehicular uses or large-scale energy storage systems, are contemplated. We consider that the most important step en route to proving the viability of our EMB concepts is the one we are now taking — designing, building, and delivering such a unit for use in a real-world application.

SUMMARY AND CONCLUSIONS

A program is underway at the Laboratory to design, build, and demonstrate a new breed of electromechanical battery. This program addresses critical design issues for EMBs, with special emphasis on magnetic bearings. We are taking a systems approach to the problem, since all elements of the EMB — rotor, motor/generator, and magnetic bearings — are intimately coupled. Our projections are that EMBs designed around our concepts have the potential to meet the needs of electric and hybrid-electric vehicles in all of their technical requirements. We also conjecture that the first entry of EMBs into widespread vehicular use is likely to be in the hybrid-electric vehicle, with all-electric vehicles to follow, as the cost of EMBs is driven down through large-scale production.

ACKNOWLEDGEMENTS

The contributions of D. E. Baldwin, D. A. Bender, S. J. Deteresa, G. R. Dreifuerst, T. K. Fowler, D. R. Hathaway, and B. T. Merritt to diverse aspects of the work reported here are gratefully acknowledged. This work was performed under the auspices of the U.S. Department of Energy by Lawrence Livermore National Laboratory under contract No. W-7405-ENG-48.

REFERENCES

[1] R. F. Post and S. F. Post, "Flywheels", Scientific American, **229**, (Dec. 1973), 17.
[2] R. F. Post, T. K. Fowler, and S.F. Post, "A High-Efficiency Electromechanical Battery," Proceedings of the IEEE (in press).
[3] R. F. Post, D. E. Baldwin, D. A. Bender, and T. K. Fowler, "Electromechanical Battery Research and Development at the Lawrence Livermore National Laboratory."
[4] C. L. Stotler, "Development of Advanced Lightweight Systems Containment — Final Report," NASA CR-165212 (May 1981).
[5] H. Dodson (private communication).
[6] L. Huber, D. Borojevic, "Conf. Record of the 1989 IEEE Industry Appliations Society Meeting," Part I (1989), 871 (IEEE Cat. No. 89CH2792-0).
[7] A. Svensson, Jr., A. E. Weatherbee, Jr., "Ninth Progress Report: Feasibility of Flywheel Energy Storage," Pratt and Whitney Report PWA-3391 (May 15 1968).
[8] R. M. Christensen, Mechanics of Composite Materials, John Wiley (1979).
[9] T. K. Fowler and R. F. Post, "Mechanical Stability of Multiple-Shell Energy Storage Rotors," presented at Western Applied Mechanics Conf., Stanford Research Institute, Sept 17-19, 1973.
[10] E. J. Gunter, Jr., Dynamic Stability of Rotor-Bearing Systems, NASA SP-113 (1966).
[11] R. F. Post and S.F. Post, "Spin Tests of a Multi-Ring Flywheel," (unpublished report).

[12] K. Halbach, "Specialty Magnets," Lawrence Berkeley Laboratory Report LBL-21945 (July 1986).
[13] S. Earnshaw, "On the Nature of the Molecular Forces which Regulate the Constitution of the Luminiferous Ether," Trans. of The Cambridge Philosophical Society, Vol VII, Part I (1839) 97.
[14] W. J. Comfort, S.E. Bumpas, T. A. Edmunds, et al., "Feasibility Assessment of Electromechanical Batteries for Electric Vehicles," 25th ISATA Conf. Proceedings, paper 920279 (June 1992) 157.
[15] B. R. Weinberger, L. Lynds, J. R. Hull, and U. Balachandran, "Low Friction in High Temperature Superconductor Bearings," App. Phys. Lett. **59**, August 1991, 1132.
[16] L. Brillouin, Wave Propagation in Periodic Structures, Dover (1953).

FIGURES

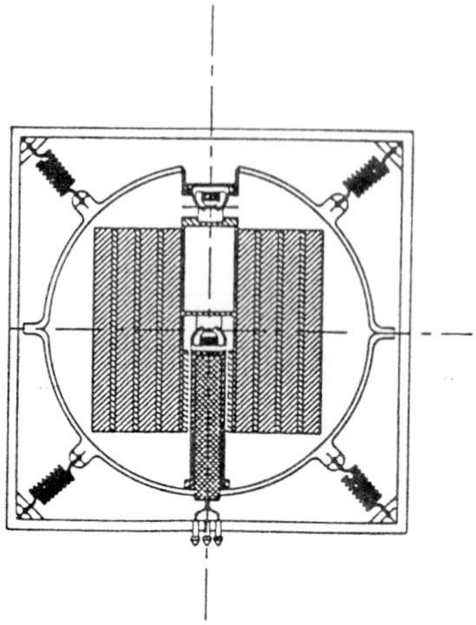

Figure 1. Schematic cutaway drawing of modular EMB

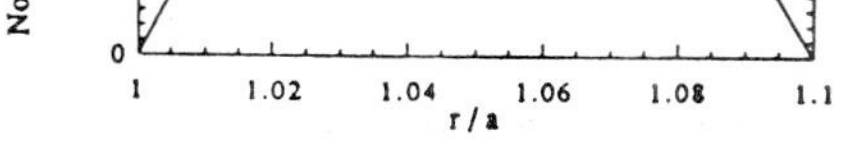

Figure 2A. Normalized radial stress in unloaded composite cylinder, b/a - 1.1 - (η is square root of ratio of hoop to radial elastic modulus.)

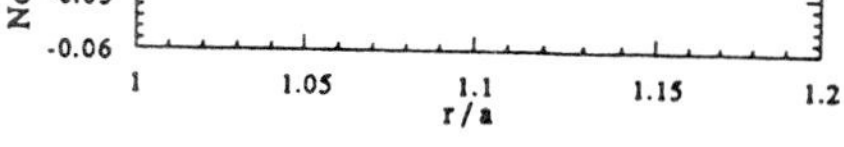

Figure 2B. Normalized radial stress in composite cylinder with mass loading on inner surface equal to 5% of normalized radial stress, b/a = 1.2 - (η is square root of ratio of hoop of radial elastic modulus.)

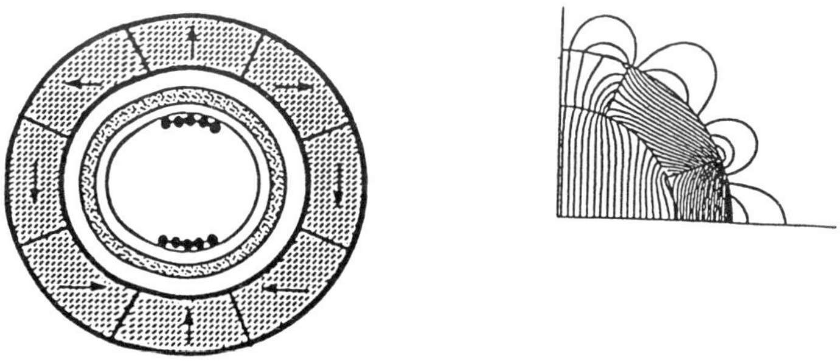

Figure 3. Left: End view of Halbach array showing vacuum barrier and some of windings. Right: Quadrant of calculated field lines.

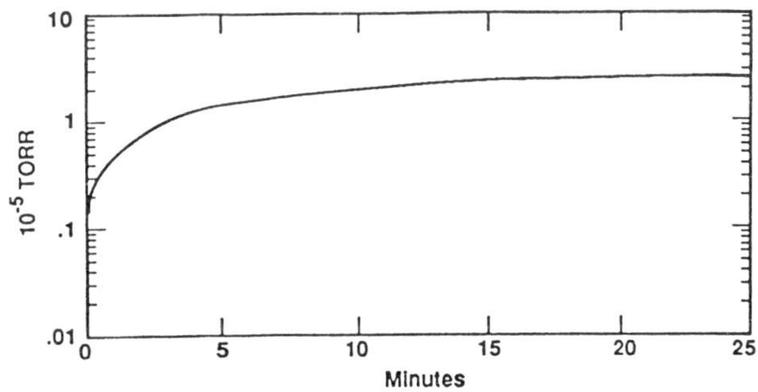

Figure 4. Pressure vs time curve for sealed-off vacuum system containing specially treated epoxy-graphite simulated rotor structure.

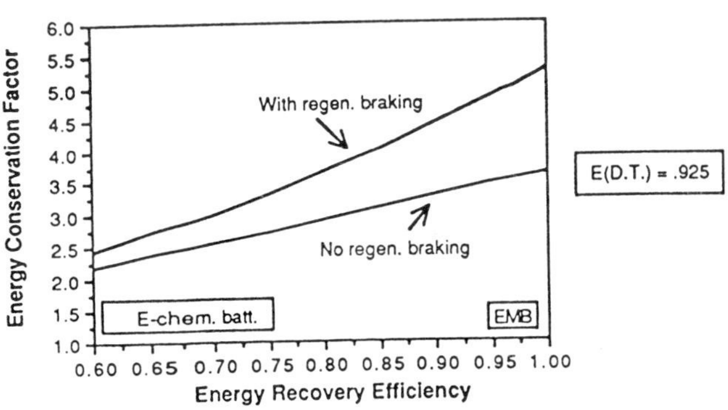

Figure 5. "Energy Conservation Factor" calculated over Federal 7-Mode urban cycle showing importance of energy recovery efficiency. Power-time averaged drive-motor efficiency of 0.925 assumed.

Ovonic Nickel-Metal Hydride Batteries: Here and Now for Electric Vehicles

Dennis A. Corrigan
Director of Advanced Battery Development
Ovonic Battery Company
Troy, Michigan, U.S.A.

Background:

Dr. Corrigan joined Ovonic Battery Company in 1992 after twelve years of research and development on electric vehicle batteries at GM Research Labs. He is currently Program Manager of the USABC development contract.

Abstract:

Ovonic nickel-metal hydride batteries are the enabling technology for modern electric vehicles as demonstrated by energy density, power, and durability performance.

No paper. Oral presentation only.

CHAPTER 9 -- FUSION ENERGY

What is Fusion?

Stephen O. Dean
President
Fusion Power Associates
Gaithersburg, Maryland, U.S.A.

Background:

Dr. Dean has been President of Fusion Power Associates since 1979. Previously he was Director of Magnetic Confinement, Fusion Division, U.S. Department of Energy. He is a Fellow of the American Nuclear Society.

WHAT IS FUSION?

Dr. Stephen O. Dean
Fusion Power Associates
2 Professional Drive, Suite 249
Gaithersburg, MD 20879

I'm Dr. Stephen Dean, President of Fusion Power Associates in Gaithersburg, Maryland. I'm chairing the opening session here this morning, and I will be introducing the speakers as we go. I'm also going to be the first speaker. I will be giving a general introduction to fusion to set the stage for the speakers that follow.

Fusion is the energy source of the sun and the stars. There are a number of reasons why people pursue fusion. The primary fuel, deuterium, comes from water and is therefore universally available and essentially inexhaustible. There's very low radiation hazard from fusion and very little proliferation potential. It has a negligible impact on the atmosphere from emissions. In general, fusion has high safety and low ecological impact, and everything we know suggests that it will have costs comparable to other energy sources.

Fusion is the opposite process of the well known fission process. Fission is the splitting of a large atom into two smaller atoms. If you think of this in the reverse order; taking two smaller atoms, putting them together into a larger atom, this is the process of fusion. Fusion works at the very low end of the periodic table, using hydrogen -- combining hydrogen to form the next heaviest element in the periodic table, namely helium. Again, this is the opposite end of the periodic table compared to fission, where one uses the very heaviest of elements (uranium) and splits it into lighter elements.

There are several isotopes or forms of hydrogen that are important to fusion. Hydrogen occurs in three isotopes. The simple hydrogen atom has a proton in the nucleus and an electron in orbit. There are two heavier forms of hydrogen called deuterium and tritium. They are differentiated by the fact that, although they still have only one proton in the nucleus, they have one neutron in the case of deuterium and two neutrons in the case of tritium. These isotopes are important to fusion because it is very difficult to create fusion using simple hydrogen, but its quite easy to do it with deuterium and tritium.

When deuterium and tritium are fused to form helium, there are two isotopes of helium that are important. One is the normal isotope of helium, which has two protons and two neutrons. The other is called helium-three which has two protons and one neutron.

The easiest fusion reaction involves the fusion of deuterium and tritium, and when the fusion reaction occurs, the product one gets is a normal helium and a neutron. The nuetron carries three quarters of the energy that's released in the reaction. This is a nuclear reaction. Mass disappears in this reaction and is released in the form of kinetic energy. If you were to weigh the initial deuterium and tritium nuclei and compare that to the weight of the products after the fusion reaction, you would find that mass has disappeared. This mass has been converted into energy in accordance with the Einstein formula, $E=MC^2$.

You have to have many such reactions in order to produce a useful amount of energy. The medium in which these reactions occur is a gas containing a large number of these nuclei. This gas is called a plasma. The energetic neutrons that are produced in the fusion reaction, leave the reaction region and are slowed down and captured in a region which is called a blanket or a moderator, which is typically lithium. The reason that lithium is used is that deuterium occurs naturally, but tritium does not and therefore the tritium must be manufactured or made in the process. The easiest way to do that is to capture the neutron in lithium. Tritium is formed in that process. If a neutron combines with lithium in the blanket, you get tritium formed and you get additional helium.

The overall system then consists of a fusion power source which generates heat, basically by the slowing down of that neutron in the blanket, thereby heating the blanket; circulating a coolant through that blanket, which produces steam through a heat exchanger, to run a steam turbine that then produces electricity. This would be the conventional way to do it. There are a variety of more sophisticated power conversion systems that one looks at in the program, but this is the most straightforward technique that people have in mind.

There are other fusion reactions, other than deuterium and tritium, which are a little bit more difficult to do, or produce less energy. For example, deuterium will fuse with itself. In this case it makes the $Helium_3$ nucleus, and again an energetic neutron. $Helium_3$ itself is an interesting nucleus. It will fuse with deuterium and in this case it forms a normal helium nucleus and an energetic proton. This is a very interesting reaction because both the proton and helium nucleus are charged, and hence one can imagine a power conversion scheme in which the charged particles acting against a magnetic field produce electricity directly, thereby bypassing the thermal steam cycle and enhancing the efficiency of the plant.

There are a large number of ideas of how to configure a chamber so as to make fusion, and we do not have time to go into them all. The most popular and successful is a magnetic confinement concept called a Tokamak. We are going to limit ourselves today because of time, to discussing the Tokamak concept.

Magnetic confinement takes advantage of the fact that, just as a normal magnet exerts forces on iron filings, a magnetic field also exerts a force on charged particles. The deuterium and tritium, being charged particles, can be acted upon magnetic forces. Therefore, by taking certain shapes of magnetic structures, and creating what we call "magnetic bottles", we can create confinement vessels in which to contain this plasma. Because of the fact that the plasma particles are charged, they have a repulsive force between them. That force must be overcome in order for fusion to happen; that translates into giving them a lot of energy, which translates into having to raise this plasma to a very high temperature, comparable to the temperature that exists on the sun. By taking magnetic fields and making different shapes you create different concepts, and the Tokamak is one such magnetic field configuration or concept.

If you were to just put your plasma in a vessel without any magnetic field, the particles would simply move around, hit the vessel walls, and in hitting the vessel walls they would cool, and you would not be able to maintain the very high temperature that you need for fusion. But if you put on a magnetic field, the particles tend to stick to the magnetic field lines. By this mechanism you can inhibit their motion towards the walls and thereby insulate them, heat-wise, and maintain a higher temperature. This is the concept of magnetic confinement. If there is no magnetic field, the plasma would completely fill the vessel. But if you put a magnetic field on, you can now shape the plasma.

There are a variety of ways in which one can raise the temperature of a plasma up towards the temperatures that exist in the sun. One is to produce a very large electric current through the plasma -- millions of amperes. Another is to inject energetic particles into the vessels using particle accellerators. This is what we call neutral beam injection. Another way is to increase the strength of magnetic field. This creates a compression force that imparts energy and heat to the plasma. A very elegant way is to use radio frequency waves which resonantly couple with the charged particles in the magnetic field, thereby imparting energy and heat to the plasma. All of these techniques have proven successful.

The most successful geometry today is a donut-shaped vessel in which the magnetic field lines spiral around inside the vessel. The plasma following these lines tends to then stay away from the walls of this vessel for a relatively long time. There are a variety of different shapes of field structures inside of these donut-shaped vessels that go by different names, but as I mentioned, the Tokamak is the most successful.

In the Tokamak, one starts with a donut-shaped vessel. You put deuterium-tritium gas in the vessel and run an electric current through it. You also put magnetic coils around the outside, and by doing this you now have two types of magnetic fields in the chamber. One is produced by the magnets around the vessel, and the other is produced by the magnetic field that's created by the electric current in the plasma. The combination of these two fields creates a sheared magnetic structure, which has been found to be very effective in confining the plasma. This is the Tokamak configuration.

Safety and Environmental Aspects of Fusion Power

Kathryn A. McCarthy
Engineering Specialist
EG&G Idaho, Inc.
Idaho Falls, Idaho, U.S.A.

Background:

In addition to her work at EG&G Idaho, Dr. McCarthy received significant experience abroad as a guest scientist at Karlsruhe, Germany, and as a participant in the U.S. Department of Energy's US/USSR Young Scientists Program. Dr. McCarthy received her M.S. and Ph.D. in Nuclear Engineering from UCLA.

SAFETY AND ENVIRONMENTAL ASPECTS OF FUSION POWER

Kathryn A. McCarthy
Idaho National Engineering Laboratory, EG&G Idaho, Inc.
P.O. Box 1625
Idaho Falls, ID 83415-3880 USA

ABSTRACT

Fusion power has the potential to be a safe and environmentally friendly energy source. Materials and design can limit hazards from accidental release of radioactive material and minimize waste disposal problems. In addition, no emissions are produced to degrade visibility, increase greenhouse gases, cause acid rain, or reduce the ozone layer. Because of the flexibility in materials choice, recycling and near-surface burial are potential options for radioactive waste management.

INTRODUCTION

We are exposed to radiation, both from natural and manmade sources, every day of our lives. Figure 1 shows the level of radiation from various sources. The level shown for nuclear power is the approximate radiation dose that would result from a low probability, severe accident in a fusion power plant. As the graph shows, the dose from this type of accident is quite small in comparison with radiation sources we are exposed to every day.

The most commonly discussed potential fusion fuel cycles produce energy according to the following equations:

$$D+T \rightarrow {}^{4}He(3.5MeV)+N(14.1MeV) \quad (1)$$

$$D+D \rightarrow {}^{3}He(0.82MeV)+N(2.45MeV) \quad (2)$$

$$D+D \rightarrow T(1.01MeV)+H(3.02MeV) \quad (3)$$

$$D+{}^{3}He \rightarrow {}^{4}He(3.6MeV)+H(14.7MeV) \quad (4)$$

In these equations, D is deuterium, a heavy isotope of water, T is tritium, a radioactive isotope of water, N is neutron, He is helium, H is hydrogen, and one MeV is one million electron volts (an average X-ray is 90,000 electron volts). First generation fusion reactors will most likely utilize the D-T fuel cycle because this reaction takes place at lower temperatures than do the other reactions, thus it is easier to cause it to take place. The important thing to notice in these equations is that with the exception of tritium, which has a short half-life (the amount of time in which the radioactivity decays to half it's original value), there are no radioactive materials generated in the fuel itself.

Fusion technological differences allow more flexibility to incorporate safety in design than does fission. In fission, neutrons fission uranium creating fission products, transuranics (elements with a higher atomic number than that of uranium), and radiological inventories in structural material that present a significant long term radiological hazard. In D-T and D-D fusion, the neutrons generated in the plasma create radiological inventories that are tied up in structural materials whose long term hazard depends strongly on material choice. Thus, material choice has a much larger impact on the induced radioactivity and safety and environmental attractiveness of fusion than it does in fission.

FUSION REACTOR SAFETY

The goal of fusion is to design reactors with inherent/passive safety. A facility that is "inherently/passively safe" is designed such that public protection is assured under all credible accident circumstances. In this type of a facility, safety systems do not depend on human/computer control. Inherently or passively safe designs eliminate high-consequence accidents. If safety is designed into a facility, we can avoid "add-on" safety systems that produce complicated systems such as that shown in Figure 2.

Inherent safety features include energy sources in the design that could potentially release radiation in an accident. This can include limiting chemical energy and stored electromagnetic energy in magnets, assuring that natural circulation will keep materials at temperatures cool enough to avoid release of radioactivity in case of a loss of flow accident, and choosing structural materials that do not become highly activated to name just a few [1]. The improved safety and environmental performance should translate into economic benefits via faster and more assured licensing and reduced need for safety grade (N-stamp) components.

FUSION REACTOR WASTE

The radioactive inventory in a fusion reactor results from the production of tritium, and the activation of structural materials by the neutrons produced in the D-D and D-T fuel cycles. Fusion does produce radioactive waste, however, through material choice and design, radioactivity can be minimized. Radioactive waste is produced during operation and decommissioning of the reactor. Certain reactor components must be changed out several times during the reactor lifetime due to radiation damage, and at the reactor's end-of-life, something must be done with structural materials activated by neutrons.

Waste Disposal

The primary regulation for the disposal of low-level radioactive waste in the U.S. is 10CFR61 [2]. It was written primarily for the disposal of wastes produced by fission reactors and nuclear medicine. Waste disposal is divided into two main categories, near-surface burial and deep geological disposal. There are three categories of near-surface burial. Table 1 shows the categories defined by 10CFR61. Current waste disposal guidelines do not allow for any "intermediate" level waste. All waste that does not qualify for near-surface burial must currently be treated as high-level waste, and disposed of in deep geological repositories. There are currently no deep geological repositories operating in the U.S., and political pressures may continue to delay the opening of such a facility in the near future. Thus a goal in the development of fusion power is to choose materials such that all waste qualifies for near-surface burial.

Recycling of Fusion Materials

Many candidate fusion materials offer great potential for recycle because of their low residual long-term radioactivity. By recycling, space in waste repositories is conserved. Additionally, radioactive waste disposal is becoming increasingly expensive. Recycling conserves both material and energy resources. Another motivation for recycling is the uncertainty in waste disposal regulations. It is quite likely that regulations concerning near-surface disposal will become more strict, making it extremely difficult to bury material as low-level waste.

There are several studies currently in the U.S. to look at the feasibility of recycling radioactive materials. Figure 3 shows material use from production to disposal/recycle. Several factors must be considered in such a study. Some of these factors are shown in Table 2. Particularly important to consider is what kind of additional waste the recycling produces. If chemical processes are necessary to separate materials, is the chemical waste hazardous? How must it be disposed of? The overall requirement in recycling is that it must be economical.

Recycling of radioactive materials is receiving increased attention. Several U.S. universities, national laboratories, and companies in private industry are investigating recycling radioactive materials. A recent study in Europe by the steel industry and a government laboratory looked at the feasibility of recycling steel [3]. This analysis included all waste streams, and radiation exposure to the public and workers at all steps. The economics of recycling low activation steels were shown to be favorable and the technology feasible.

A current study at the Idaho National Engineering Laboratory is focusing on the recycle of vanadium alloys. Vanadium alloys are a prime candidate for the structural material in commercial fusion plants because of their desirable material safety properties. The first phase of the study looked at the impact of recycling on accident safety, and at waste disposal and recycling options [4]. This study showed that the accident safety was not adversely affected by recycling because radioactivity did not increase with each additional cycle in the reactor. The reason for this is that accident safety tends to be dominated by short-term radioactivity, and while the material is being stored before recycle, this radioactivity decays away. Waste management issues are dominated by long-lived radioactivity, and this tends to come from material impurities. The numbers of cycles that the vanadium alloy can be reused in the reactor is influenced by the achievable impurity levels in the alloy.

Recycling studies done to date are encouraging. Further studies are necessary to better understand the technology, impact on the public/workers/environment, and economics. Fusion materials show great potential for recycle.

CONCLUSIONS

Fusion power can take advantage of the flexibility available in materials selection, thus providing a safe and environmentally friendly energy source. By incorporating safety into the reactor design, this energy source can be passively safe, and not rely on human or computer actions for mitigating accidents. Additionally, through correct material choice, waste disposal problems can be minimized. In particular, fusion materials show great potential for recycle, thus conserving repository space, resources, and energy.

ACKNOWLEDGEMENTS

This work supported by the U.S. Department of Energy, Director of Energy Research, Office of Fusion Energy, under DOE Contract No. DE-AC07-76ID01570.

REFERENCES

1. S. J. Piet, "Passive Safety Features of Some Recent Fusion Designs," Proceedings of the 12th Symposium on Fusion Engineering, Monterey, CA, October 12, 1987, pp. 557-560.

2. 10 CFR, Part 61, Federal Register, 47(48), December 27, 1982.

3. C. R. Gomer, D. Dulieu, K. W. Tupholme, and G. J. Butterworth, "The Feasibility of Recycling Activated Steel First Wall/Blanket Materials from Fusion Reactors," Fusion Engineering and Design 11, 1990, pp. 423-440.

4. G. J. Butterworth, K. A. McCarthy, G. R. Smolik, and C. B. A. Forty, "Safety and Environmental Aspects of Vanadium Alloys," to be presented at the Sixth International Conference on Fusion Reactor Materials, Stresa, Italy, September 27 - October 1, 1993.

FIGURES

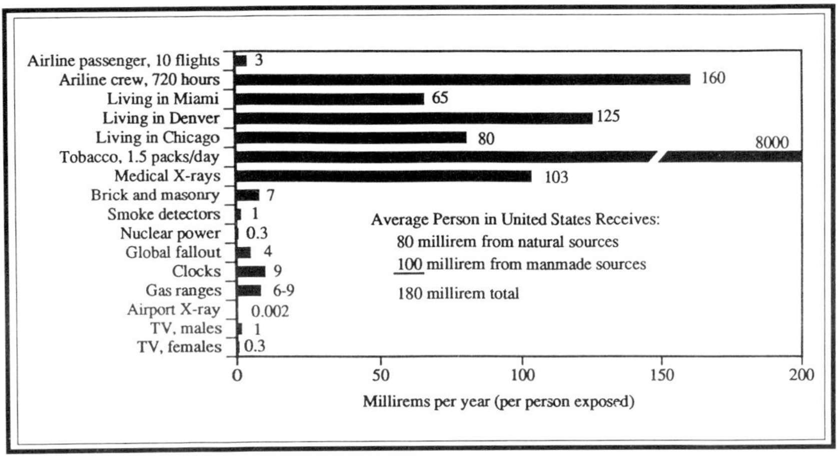

FIGURE 1: DOSES FROM EVERYDAY RADIATION

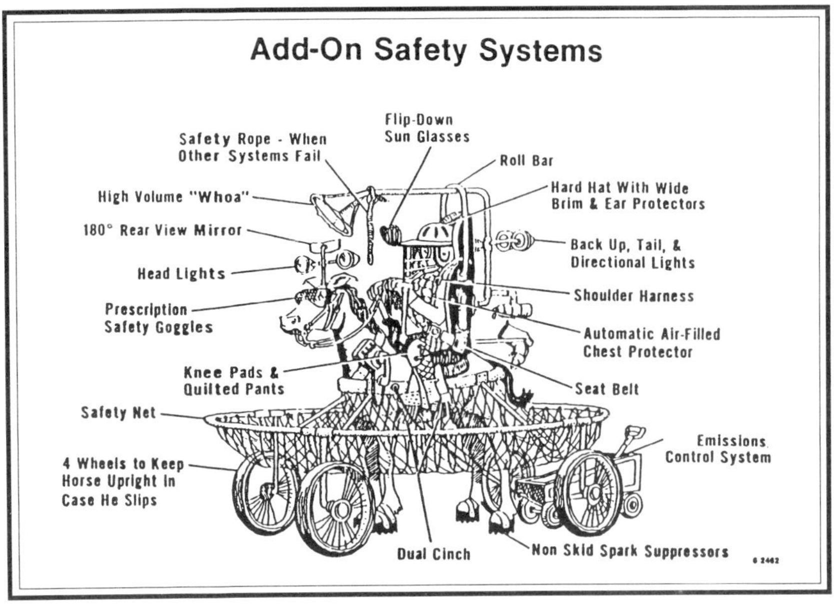

FIGURE 2: ADD-ON SAFETY SYSTEMS

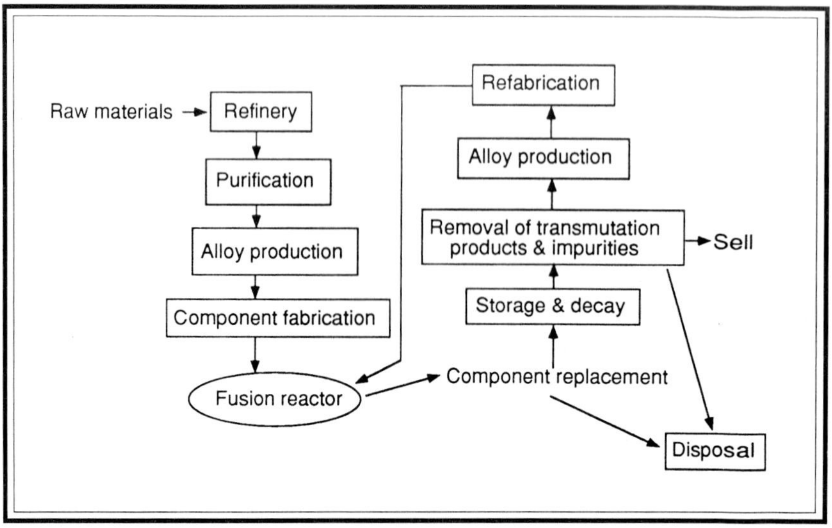

FIGURE 3: MATERIAL USE FROM PRODUCTION TO DISPOSAL/RECYCLE

TABLES

TABLE 1: DESCRIPTION OF WASTE DISPOSAL AS GIVEN BY 10CFR61

Waste Class	Definition	Disposal
Class A Segregated Waste	Decays to acceptable level during site occupancy	Segregated, minimum requirements
Class B Stable Waste	Stabilized and decays to levels that do not pose a danger to public health and safety in 100 years	Covered to reduce surface radiation to a few percent of natural background
Class C Intruder Waste	Does not decay to safe level in 100 years. Decays to acceptably safe level in 500 years	Five meters below surface with natural or engineered barrier
Waste that does not meet Class C definition	Does not qualify for near-surface disposal. Proposed disposal methods are considered on a case-by-case basis.	Geologic

10CFR61 defines "acceptably safe levels" to mean that inadvertent entry into the waste will result in a whole-body dose of less than 5 mSv/year (500 mrem/year)

TABLE 2: FACTORS THAT MUST BE CONSIDERED IN DETERMINING TRADEOFFS BETWEEN WASTE RECYCLING AND WASTE DISPOSAL

Waste Recycling	Waste Disposal
Cost of facilities	Cost of facilities
remote handling	
hands-on	Disposal costs
Worker/public exposure	Cost of replacement materials
storage facilities	
recycling facilities	Worker/public exposure
Waste stream	Hazards of mining/purifying new materials
chemicals used in processing	
can radioisotopes removed be used	Impact on resources
Effect of recycling on accident hazards	
in the reactor	
in recycling facilities	

The International Thermonuclear Reactor Project

Thomas R. James
Division Director
Office of Fusion Energy
Department of Energy
Germantown, Maryland, U.S.A.

Background:

Division Director of the ITER and Technology Division in the Office of Fusion Energy at the Department of Energy, Mr. James is responsible for ITER and near-term technology development to support current research programs.

Abstract:

The International Thermonuclear Experimental Reactor Project is a 6-year collaborative effort involving the U.S., Europe, Japan, and the Russion Federation to produce a detailed engineering design for the next-step fusion device.

No paper. Oral presentation only.

Recent Results and Near-Term Expectations in Tokamak Fusion Research in the U.S., Europe, and Japan

Dale M. Meade
Deputy Director
Plasma Physics Laboratory
Princeton University
Princeton, New Jersey, U.S.A.

Background:

Prior to assuming his position of Deputy Director of the Princeton Plasma Physics Laboratory, Dr. Meade headed the TFTR Project and the Experimental Division at PPPL.

Recent Results and Near-Term Expectations in Tokamak Fusion Research in the U.S., Europe, and Japan

Dr. Dale Meade
Deputy Director, Plasma Physics Laboratory
Princeton University
U.S.A.

It is a pleasure to be here this morning to talk about the status of experimental results and our near term expectations for fusion power research.

The development of fusion is often thought about in terms of three different activities: scientific feasibility, engineering feasibility, and economic feasibility. This morning, I would like to concentrate on where we are regarding scientific feasibility of fusion. You heard from the previous speaker about where we are on the engineering feasibility of fusion. At the end, I would like to talk about the economic feasibility of fusion.

A schematic of the Tokamak Fusion Test Reactor (TFTR) is shown in Fig. 1. The plasma is located in the central region. The large coils are the main magnetic field coils. This vacuum vessel is a very high vacuum system with a base pressure of a few 10^{-9} torr. A pinch of hydrogen and deuterium is injected as fuel for the plasma; later this year we will also inject some tritium as part of the fuel. On the side of Fig. 1 are the neutral beam injectors that are used to heat the plasma. They inject neutral atoms at energies corresponding to 120 kV at a total power of 33 MW. Most of what can be seen on the outside of TFTR has to do with auxiliary heating, cooling systems and diagnostic systems.

Figure 2 shows TFTR in the actual experimental arrangement. Neutral beam injectors, vacuum pumping systems and diagnostic systems are on the side of the tokamak. For perspective there is a person standing on the top of the tokamak.

The inside of the TFTR vacuum vessel (Fig. 3) shows the toroidal nature of the vacuum vessel. The central column has tiles that are similar to the tiles on the NASA space shuttle. TFTR tiles are made of carbon fiber composite materials and they define the edge of the plasma. The plasma is held inside the vacuum chamber in the shape of a donut by magnetic fields. A plasma current flows around the chamber in a loop that helps provide the magnetic fields. The plasma current of about 2 million amperes in TFTR is formed by inductive or transformer action. The plasma current is maintained in space to a positional accuracy of about 2 mm.

On the far side of the vacuum chamber are rectangular or square objects that are the front ends of radio frequency launchers that inject waves into the plasma at frequencies corresponding roughly to that of the ion particles gyrating around the magnetic field.

Table I shows general parameters for other large tokamaks. The first column shows the plasma volume of JT-60 in Japan at 80 m^3, JET at 150 m^3, TFTR at 35-37 m^3, and ITER at 1000 m^3; this shows the size and progression of these experimental facilities.

The magnetic field in TFTR is 5.2 Tesla (1 Tesla is 10,000 gauss). The earth's magnetic field is about a half a gauss. Therefore, TFTR's field is about 100,000 times the earth's magnetic field.

The fusion researchers in the U.S. work collaboratively with their colleagues in Japan. The large Japanese tokamak, JT-60U, shown in Fig. 4 is about the same scale as the TFTR device (Fig. 2).

Figure 5 shows JET, the Joint European Torus Tokamak, located just outside Oxford, England. The large components are part of the iron core of a large transformer. All of these large tokamaks are addressing issues at the forefront of scientific feasibility magnetic fusion. In present-day tokamaks, the magnetic fields are made using copper coils; as such, the length of the time which we have a magnetic field to confine the plasma is relatively short, about a few seconds.

Table II shows some of the requirements that must be satisfied to achieve scientific feasibility. First and foremost, the greatest challenge over twenty years has been to produce reactor temperatures. Temperatures greater than 100 million degrees centigrade are required. One has to also produce high density and the energy confinement simultaneously with high temperature, and that is really what we have been striving to do during the last twenty years. In these particular experiments the plasma is heated by neutral beams as it is held by the magnetic field. The power that heats the plasma is slowly transported to the plasma edge. In the case of TFTR, the power comes out at the carbon fiber composite tiles at the edge of the plasma. For experiments that are a few seconds long, inertial cooling can be used; but, for the future, one of the challenges is going to be removing the plasma heat on a continuous basis.

The last item in Table II is the most interesting and important of all of the scientific issues for a fusion reactor; it has to do with plasma self-heating. In the D-T fusion reaction the neutron goes into the blanket and generates heat there. The helium ion (alpha particle) is electrically charged and stays in the magnetic field. If there are enough helium ions at 3.5 MeV (million electron volts) of energy, they will heat the plasma. Self-heating is what we are working to see for the first time on TFTR. If the plasma is completely self-heated, it is called an ignited plasma, and ITER is the project that will be the first to demonstrate ignition.

The road map for magnetic fusion is shown in Fig. 6. The Lawson Chart for fusion was derived by looking at the energy balance of a plasma. The energy needed to heat the plasma is compared to the fusion power produced.

The parameter that is of interest to us and our experimental results is the parameter when n is the fuel density and τ is the plasma energy confinement time. In a magnetic fusion device, fuel densities are about 10^{14} particles/cm^3. That is about 1/100,000 the density of atmospheric air and corresponds to a very tenuous plasma.

The temperatures required for fusion are about 100 million degrees centigrade. In our experiments, we have measured temperatures in electron volts. So we generally talk about temperatures of about 10 to 20 keV. The energy confinement time is the time that energy resides within the plasma and then comes out to the wall and the plasma has to be reheated. For magnetic fusion, energy confinement times are typically one second.

On this graph, the $n\tau$ values go vertically and the central ion temperature goes horizontally. To get to the fully ignited plasma, one has to reach the upper right-hand corner of Fig. 6. In the mid-1970's, we were in the lower left-hand corner region and the ignition goal was very far away. The construction of TFTR, JET and JT-60 was initiated in the mid-1970's. These machines have achieved results close to those required for an ignited fusion reactor. These results are generally achieved in deuterium, the low octane fuel. If D-T is put in TFTR and JET, we expect to generate about 10 MW of fusion power which is near energy break even within the plasma. That has been our goal. There is one medium-sized step left to go to get into the region of self-heated plasmas, and that step will be achieved by ITER.

The present generation of experiments have achieved the reactor conditions that we set out to achieve. The most important goal was to produce temperatures greater than 10 keV or 100 million degrees centigrade. You can see that we have done that with a large margin. Not just on one machine, but several machines are doing that.

Another way to look at this is to look at the fusion power produced. Most of our experiments are done using deuterium fuel which has a low fusion cross section and produces low power. It is a low octane fuel. In Fig. 7, the peak deuterium fusion power in kW is along the vertical axis and the injected plasma heating power in MW along the horizontal axis. TFTR produces about 65 kW, JT-60 is at a similar level, and JET is about 30% lower in the 45 kW range.

We have estimated how much deuterium-tritium fusion power would be produced when a reactor fuel mix of 50/50 deuterium-tritium is put in TFTR this fall. That is the scale over on the right hand side. The best TFTR points achieved in deuterium would correspond to something a little over 10 million watts of fusion power.

It is worth comparing how fusion is doing relative to other achievements in science. Everybody is aware of the tremendous progress in the increase of density of computer memory chips as shown in Fig. 8. We had kilobits in 1970 and in the range of about 16 million bits in 1993. Fusion power has had extremely rapid growth from very low powers (hundredths of a watt) in 1975 and 1.7 million watts in 1991. These achievements are the results of investments that were made during the mid-1970s. We are now coming to the completion of those experiments and the critical issue is how to proceed to the next step of fusion development.

When D-T is put in TFTR to make 10 MW of fusion power, one of the things we will be looking for is the self-heating of the plasma by the alpha particle, the helium nucleus that is injected into the plasma by the fusion process.

Figure 9 shows the electron temperature of the plasma which is typically around 100 million degrees centigrade (about 10 keV) for a deuterium plasma. We expect in a 50/50 D-T plasma, that self-heating will take place and produce about a 10% increase in the electron temperature which we can easily detect experimentally. This self-heating is one of the main experimental objectives of the D-T experiments which will start this fall, and carry on through September of 1994.

As we study this self-heating, we will also look for other waves or instabilities that are produced in the plasma by the alpha particles. We have done simulation experiments by using neutral beam systems to inject fast particles and have shown that these waves exist. Figure 10 shows the waves on the bottom trace and their deleterious effect on the neutron production and hence on the confined fast ions which simulate the alphas (upper trace).

With regard to scientific feasibility (Table 1), achieving reactor temperatures has been accomplished. Reactor densities and confinement have also been achieved. Establishing particle control over a longer period of time and handling the power over a longer period of time have not been done yet. The self-heating is the most critical item yet to be done.

Some of the items that are relevant to economic feasibility are shown in Table III. We would like the reactor to be compact, to have a high power density and to be operating continuously, or at least for very long pulses. I mentioned earlier we would like to have durable materials and low activation materials. Let me focus on the first two items because they are related to plasma physics aspects. The first step is to achieve the high temperatures and the confinement and densities that are required for a fusion reactor. Next we want to optimize the magnetic field structures and make them more efficient.

Tokamaks have large currents (several millions of amperes) within the plasma that provide part of the magnetic field. Technique is needed to efficiently drive that current. Fortunately, the plasma generates part of that current itself and, if you optimize the profiles of the plasma, the plasma will generate a large fraction of its total current. Also, particles injected from the beams or radio frequency waves can be used to drive these currents.

We are now in the design phase for the Tokamak Physics Experiment (TPX). This device would replace the Tokamak Fusion Test Reactor, the TFTR at

Princeton. The mission of the TPX is to develop a scientific basis for a compact and continuously operating Tokamak Fusion Reactor. The objective is to get smaller unit size for the ultimate reactor, and also to have a reasonably high power density in the fusion core.

The TPX device is being designed to have a fully superconducting coil system. Since the schedule would bring this device into operation around the year 2000, it would be the first tokamak with a fully superconducting coil system. It is designed to take advantage of advanced techniques to drive the plasma current. TPX would have a construction cost of about $530 million.

TPX is shown schematically in Fig. 11. TPX is almost exactly the same size as TFTR. It will be located in the same place in the experimental hall and makes use of all of infrastructure and the power systems from the TFTR facility. TPX will have plasma currents that would be similar to TFTR.
Figure 12 shows a cutaway view of TPX. The plasma is in the center, and will have a current of about 2 million amperes. The plasma edges are peeled off at the X points at the top and bottom of the plasma. Think of this plasma as the layers of an onion. We are magnetically peeling off the outer layers, the dirty areas, to pull impurities out of the plasma and pull the power into the divertor chambers.

TPX will do mainly deuterium experiments which will produce deuterium fusion neutrons. So, we will have to deal with the activation of the device, which will require remote handling inside the vacuum vessel. It is interesting to note that TPX is being designed with a low activation titanium alloy vacuum vessel so it is more easily maintained.

The U.S. Fusion Program is planned to provide a path to the best Demonstration Fusion Plant (DEMO) (Fig. 13). ITER will be the main link between the present devices and the DEMO and will do long pulse ignition and nuclear technology development. In parallel we are looking toward devices, such as TPX, that are smaller and have continuous operation as their goal.

In summary, if you look back over the last twenty years, progress in magnetic fusion has been outstanding. We have gone from 1/100 W of fusion power in 1975 to 1.7 MW in 1991, and we are expecting to increase this to 10 MW in 1994. If we are successful in our experiments on TFTR later this year, this could be the springboard moving into the next phase of fusion energy development. World wide we are coming to the completion of the phase that was started in the mid-1970s to look at scientific feasibility. These experiments have achieved their objectives. As has been mentioned earlier, technical issues have been identified and a world-wide plan has been developed to demonstrate the engineering feasibility of fusion, and to develop the design basis for a practical demonstration fusion reactor.

The bottom line is that by investing now in fusion research, the U.S. will move toward a technology that can pay enormous dividends, measured in the quantity of energy available and in the quality of the environment in which our children will live.

This work was supported by U.S. Department of Energy Contract No. DE-AC02-76- CHO-3073.

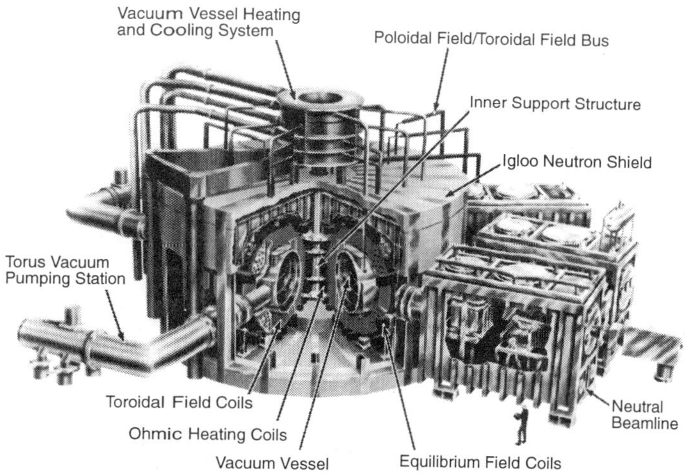

Figure 1 Schematic of the Tokamak Fusion Test Reactor (TFTR).

Figure 2 TFTR Experimental Arrangement.

Figure 3 Interior of the TFTR Vacuum Vessel.

Figure 4 Large Japanese Tokamak (JT-60U).

Figure 5 Joint European Torus (JET).

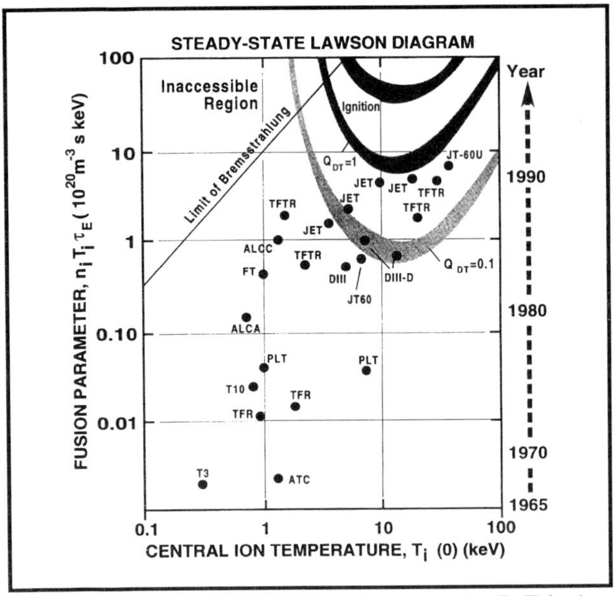

Figure 6 Lawson diagram showing requirements for D-T fusion reactors.

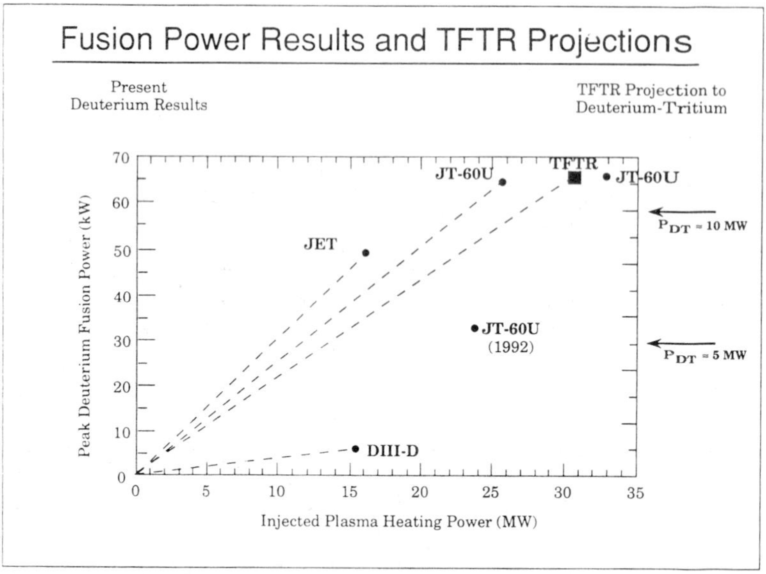

Figure 7 Fusion power produced in deuterium plasmas and expected in D-T plasmas.

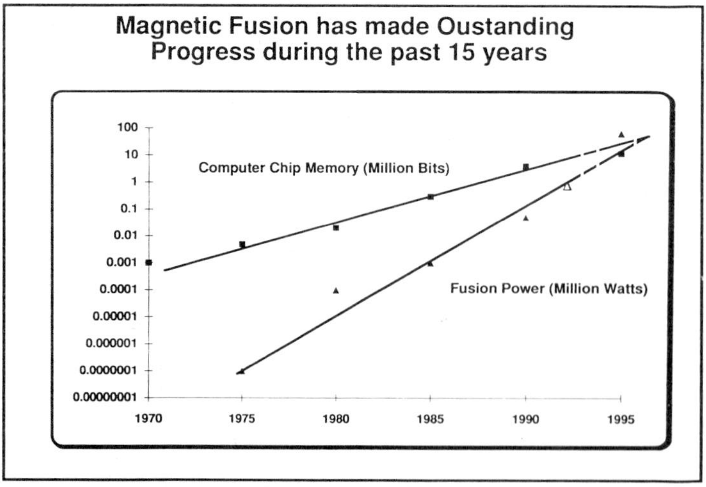

Figure 8 Comparison of progress in computer memory chip density and fusion power.

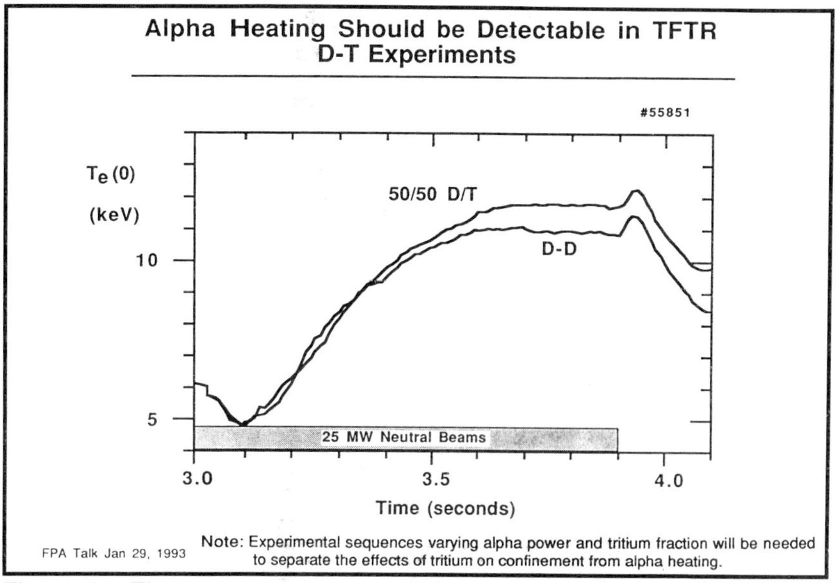

Figure 9 Electron temperature in TFTR deuterium plasmas (D-D) and that expected in D-T plasmas.

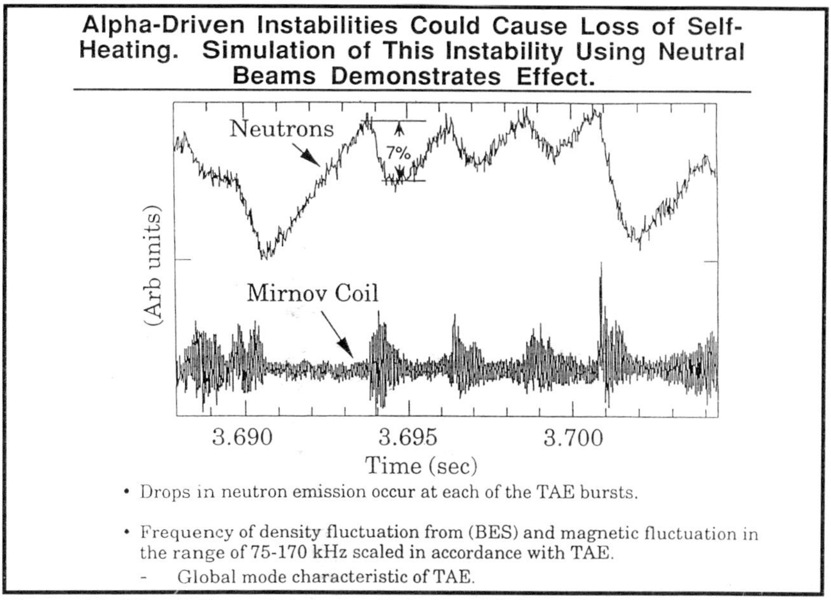

Figure 10 Simulation of alpha particle instabilities using neutral beam ions.

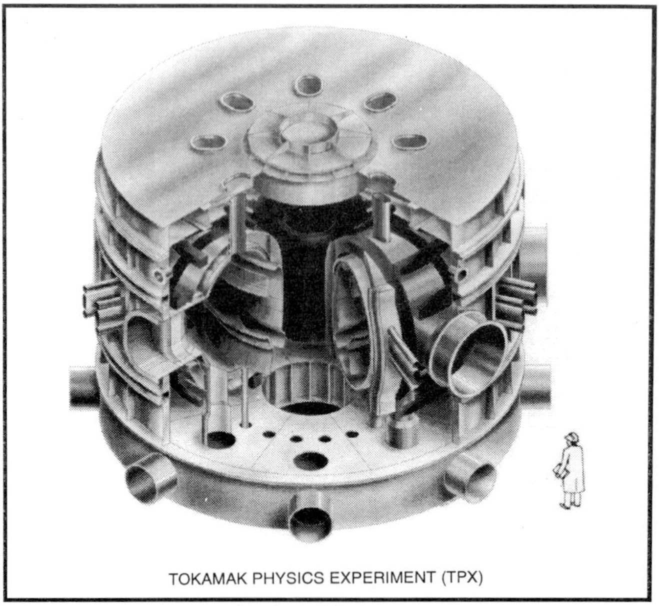

Figure 11 Schematic of the Tokamak Physics Experiment (TPX).

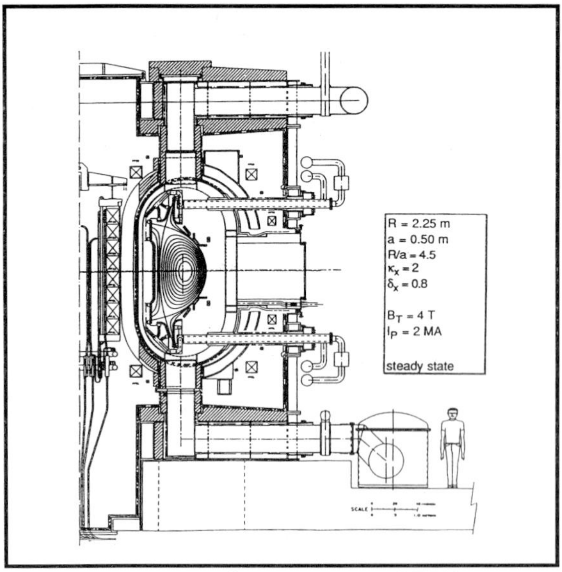

Figure 12 Cutaway view of TPX.

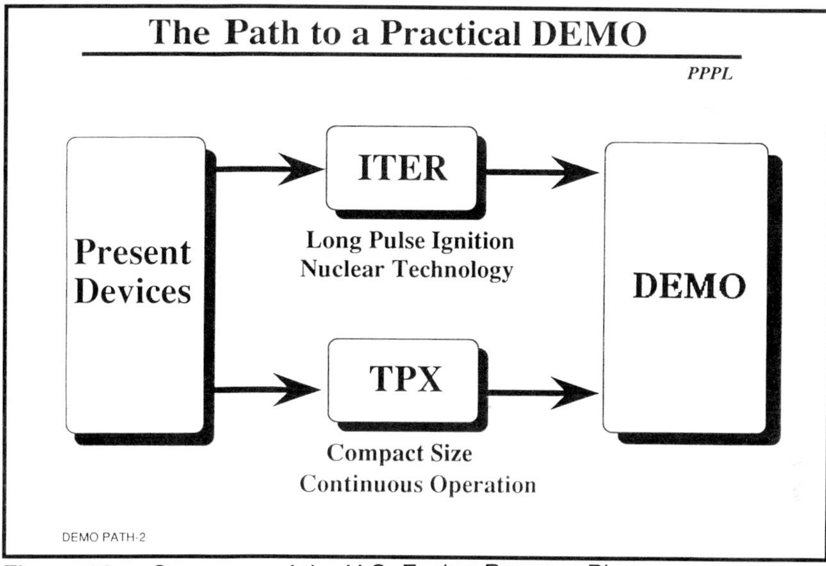

Figure 13 Summary of the U.S. Fusion Program Plan.

Large Tokamak Parameters

PPPL

	Major Radius (m)	Volume (m^3)	Magnetic Field (T)	Fusion Power(D) (kW)	Operating Budget ($M)
JT-60U Japan	3.4	80	4.4	65	140
JET Europe	2.96	150	3.4	45	120
TFTR USA	2.5	35	5.2	65	80
DIII-D USA	1.67	37	2.2	6	40
TPX USA	*2.25*	*23*	*4*	*50*	*100*
ITER	*7.75*	*1000*	*6*		*400*

Table I

FUSION DEVELOPMENT STEPS

SCIENTIFIC FEASIBILITY

REACTOR TEMPERATURES

REACTOR DENSITIES AND CONFINEMENT

PARTICLE CONTROL

PLASMA POWER HANDLING

SELF HEATING

FUSION DEVELOPMENT-1, DMM

Table II

FUSION DEVELOPMENT STEPS

ECONOMIC FEASIBILITY

COMPACT REACTOR

CONTINUOUS OPERATION

DURABLE MATERIALS

LOW ACTIVATION MATERIALS

FUSION DEVELOPMENT-3, DMM

Table III

Prospects for Commercial Fusion Power

Stephen O. Dean
President, Fusion Power Associates
Gaithersburg, Maryland, U.S.A.

Background:

Dr. Dean has been President of Fusion Power Associates since 1979. Previously he was Director of Magnetic Confinement, Fusion Division, U.S. Department of Energy. He is a Fellow if the American Nuclear Society.

PROSPECTS FOR COMMERCIAL FUSION POWER

Dr. Stephen O. Dean
Fusion Power Associates
2 Professional Drive, Suite 249
Gaithersburg, MD 20879

I'd like to summarize a little bit, and since we have a little extra time I will make a few remarks about inertial confinement and some other aspects of fusion. As you've heard, fusion has been going on for some time. It started back in the early fifties. In fact,I'd like to acknowledge the presence out in the audience of Dick Post who spoke to you yesterday about a totally different topic. He is one of the founders of fusion research in the United States and one of the pioneers of fusion in the world. Fusion has been going on for a long time. Because it has been going on for so long, a lot of people wonder if it will ever pay off. Sometimes people joke that it's always twenty years in the future, and a few years ago the Department of Energy created a policy advisory committee that put fusion thirty-five years into the future. So, it seems like we're actually losing ground in terms of when is it that the public is going to derive a benefit from fusion.

Nevertheless, in spite of that image, over the years there's been a steady progress in terms of producing the achievements and the conditions necessary to demonstrate fusion. There is an international plan to take the next big leap up to the thousand Megawatt, very long pulse or steady state level in an engineering test reactor called "ITER". It's not designed to make electricity or to be a commercial plant, but in every other way it is a full scale power reactor. So shortly after the year 2000, there should be in operation a Tokamak (ITER) that makes a large amount of fusion energy. In the process of operating that machine, industry and electric utilities will gain experience with operating and maintaining fusion equipment.

There are a number of issues associated with whether or not, and when, fusion will become commercial. One of these has to do with the cost of the development. These machines are quite expensive. Government budgets are very tight, and as a result of that, in order to make progress towards a machine like ITER, the United States has banded together with other countries with the idea that if we didn't get other countries to band together on this project, probably we could not get our country to fund it on our own. Therefore, the cost of this development is inhibiting a more rapid payoff towards commercialization.

Other aspects have to do, not just with the cost of development, but the capital cost of the reactors themselves. The ITER reactor may cost 6 - 7 billion dollars. This is for a test reactor. One hopes that from that learning experience the commercial reactors might be somewhat less costly. But, nevertheless, from the point of view of an electric utility, these are sizeable capital investments. It is going to be somewhat difficult for fusion to enter the marketplace in competition for capital to put power on the grid against technologies that might be able to produce power at smaller sizes, or with less capital-intensive type plants. So, that's another inhibition to the commercialization of fusion which we are all working on.

Another aspect of fusion is the safety and environmental aspects of fusion. Fusion is a nuclear power source. It is a much less hazardous, much safer technology than fission technology. Many people believe that fission is a safe and acceptable technology, but other people believe that it is not. In fact fission technology today is dead in the water because of the anti-nuclear sentiments of many people and because of the very complicated regulatory structure that has been created that allows nuclear power plants to be stopped during their licensing process, when in some cases they are half-built.

Fusion should have an easier time with licensing and public acceptance, but nevertheless, because it's nuclear and because there is radioactivity involved in fusion, fusion will have to address and work through the licensing and regulatory procedures and the public relations process that will allow the public to accept it and to allow the utilities to be willing to brave the licensing process in order to purchase such plants. These are all substantial difficulties. Nevertheless, the scientific community is moving ahead to resolve these issues one at a time, and in many cases in parallel, trying to get ready for the more substantive, political, financial and safety issues of becoming a commercial power source.

The Department of Energy has a plan that's been written into the Energy Policy Act of 1992, in which, through a series of steps, one gets to the operation of a first demonstration power plant around the year 2025.

These major steps consist of the deuterium-tritium operation in our largest Tokamak, the Tokamak Fusion Test Reactor; the experiments in the Tokamak Physics Experiment, which are aimed at solving some of the problems having to do with the size and cost of Tokamak reactors, to make them smaller and less costly; the demonstration of the engineering and technology that will be done in the ITER engineering test reactor, which is aimed at operating between 2005 and 2010.

There are a number of other technology and engineering facilities that are believed to be needed in order to test the materials to be used in fusion, in order to ensure safety and environmental advantages. These consist of materials, test facilities, and facilities for developing blankets. All these facilities provide data which then goes into the design and construction of a demonstration plant and all of this is supported by a core of physics and technology program that runs along in parallel. So, this is the U.S. game plan for the development of the Tokamak as a commercial power source.
As I indicated at the beginning, the Tokamak is the primary, but not the only concept for producing fusion. There is another major effort in the United States to develop a technology called, "Inertial Confinement Fusion". In fact, there are a large number of different concepts for producing fusion, all of which have advocates, and most of which have very little funding compared to the Tokamak.

The next largest program in fusion is this inertial confinement, which uses lasers and ion beams to produce fusion. This program is funded in the United States by the defense portion of the Department of Energy, because the techniques here mimic on a very small scale the techniques which were developed for use in nuclear explosions of hydrogen bombs. Since we have a little bit of time here I'll just go over for you some of the concepts here.

In inertial confinement fusion one takes a what's called a pellet, which is a small hollow sphere. One puts the deuterium and tritium inside this sphere, which might be the size of a pea or a small beebee. Then one impinges energy on the surface of this sphere, perhaps from a laser beam or a beam of ions. When one heats the surface of the sphere, the surface blows off and from the rocket "action-reaction" principle, there is an inward force that compresses and heats the fusion fuel inside. When it gets to a very small but very dense spot in the middle, it ignites and fusion energy is released in a very rapid burst. This is often called a micro-explosion because, even though it's a fusion explosion, it is on such a small scale that it in fact can be contained in a vessel. If one then sets off these miniature explosions maybe five or ten times a second, one can produce continuous or semi-continuous release of fusion energy in a somewhat similar way that one produces pulsed power in the cylinders of your automobile.

Primarily, this work is done with lasers. The world's largest laser, the so called Nova laser is at the Lawrence Livermore National Laboratory. It has been in operation for almost a decade. You take a small amount of energy and you run it through what's called an amplifier chain, and then you focus all of the laser beams down around a single pellet.

There is another fairly large laser, the so called Omega Laser, at the University of Rochester. It is in the process of being upgraded to a larger size.

One of the problems is that the beam of energy that you put on these pellets must be extremely uniform because if it isn't, then the compression isn't uniform. If the compression isn't uniform you can't achieve the conditions that you need in order to ignite the pellet.

The Omega Upgrade Laser at Rochester has the potential of getting very close to being able to ignite pellets. Also there is a proposal for a new experiment which has been named the Nova Upgrade which would also be able to ignite pellets out around the year 2000.

So, you have here a parallel approach being funded by the Department of Energy to the Tokamak. It is being carried out as a smaller level of effort, both here and internationally. There is a sizeable program in Japan and in Russia and to a lesser degree in Europe, primarily in France. These programs are quite vigorous, nevertheless, and they hope to also be able to make commercial energy on more or less the same time scale as the Tokamak, although there's not the same degree of international collaboration, and this is in part due to problems of classification. Much of this work has been developed in weapons laboratories and coming out of work that was done for weapons development. Some of it is still classified and this has inhibited free exchange of information and international agreements from being carried out in this field.

So, that gives you at least a feel that Tokamaks are not the only game in town. Nevertheless, it is the primary thrust of the world effort and it is the technology which is expected to lead into the first generation of commercial fusion reactors.

CHAPTER 10 -- NATURAL GAS VEHICLES

Meeting the California Challenge: Natural Gas Vehicle Performance and Emissions Evaluations

Kenneth W. Tenure
Senior Consultant
Bevilacqua Knight, Inc.
Oakland, California, U.S.A.

Background:

Mr. Tenure brings to the clean fuels arena an economics and industrial organization background from the University of Maryland and The American University. Prior to taking his present position, Mr. Tenure was Program Development Manager for the Electric Vehicle Association of the Americas, and Business Development Manager with Toyota Motor Sales, U.S.A. in California.

Abstract

A presentation on NGV field and laboratory evaluation activities and the relevance of reactivity factors in meeting emission standards.

No paper. Oral presentation only.

Natural Gas Vehicles -- The Technological Frontiers

Christopher S. Weaver
President
Engine, Fuel, and Emissions Engineering, Inc.
Sacramento, California, U.S.A.

Background:

Mr. Weaver has more than ten years of experience in mobile-source emissions and alternative fuels. He received a B.S. from U.C. Berkeley, and an M.S. from Massachusetts Institute of Technology.

NATURAL GAS VEHICLES
The Technological Frontiers

Christopher S. Weaver, P.E., President
Engine, Fuel, and Emissions Engineering, Inc.
3050 Fite Circle, Suite 212
Sacramento, CA 95827 USA
ph. (916) 368-4775 fax (916) 362-2579

ABSTRACT

This paper describes some of the significant recent technological developments with natural gas vehicles (NGVs), as well as some of the promising technological frontiers now apparent. Significant recent developments in NGV technology include the development and commercial introduction of low-emitting, lean-burn heavy-duty natural gas engines, NGV catalyst systems, and advanced electronic fuel injection systems for natural gas, similar to those used in modern emission-controlled gasoline vehicles. These systems have made possible the achievement of emission levels well below the California ULEV emissions standards. New dual-fuel diesel technology has also produced substantial emission reductions and efficiency improvements in heavy-duty engines. Areas offering significant future promise include the demonstration of durable low emissions in light-duty vehicles, the application of feedback electronic controls to heavy-duty lean-burn and dual-fuel engines, the development of "zero emission" natural gas hybrid/electric vehicles, and the application of low-emission dual-fuel retrofit systems to existing diesel engines in trucks, locomotives, marine vessels, and other applications.

INTRODUCTION

The last three years have seen rapid progress toward the commercialization and large-scale use of natural gas vehicles. As the cost, supply, and environmental problems with competing alternative fuels have become clearer, natural gas has become widely recognized as the leading "clean" alternative fuel in the U.S. This success has been due, in part, to an active program of research and development supported by the natural gas industry. Recent technological developments have included produced new, low-emitting lean-burn natural gas engines for heavy-duty vehicles, improved catalysts for NGVs, lightweight cylinders for compressed natural gas, improved fuel metering systems, and other advances. Newly-developed electronic fuel injection systems for natural gas constitute an advance in emissions and fuel-efficiency comparable to the replacement of carburetors by fuel injection in gasoline engines. Industry support has also played a crucial role in one of the most significant developments of the last few years: the beginning of commercial production and sale of new, dedicated natural gas vehicles by the original vehicle manufacturers. Vehicle manufacturers now offering new, OEM natural gas vehicles include Chrysler Corporation, General Motors, Blue Bird Bus Company, and Bus Company of America.

The last few years have also seen substantial advances for natural gas vehicles in the regulatory and legislative arena. In November, 1992, the U.S. Environmental Protection Agency (EPA) proposed emissions standards and supporting regulations for NGVs. At the same time, the Clean Fuel Fleet requirements of the Clean Air Act Amendments of 1990 and the fleet alternative fuel requirements of the National Energy Policy Act of 1992 are encouraging (and in some cases mandating) fleet operators to convert to alternative fuels. A number of states have adopted similar requirements - especially for public fleets. The natural gas industry has responded with a substantial increase in infrastructure development. Some 700 compressed natural gas refueling stations are now in operation nationwide, and more are being built at a rate of four per week. In the future, continuing emphasis on improving urban air quality and increased emphasis on reducing emissions that contribute to global climate change will both tend to favor substitution of natural gas for gasoline and diesel as vehicle fuels.

This paper reviews some of the most significant recent developments, present technological frontiers for natural gas vehicles -- areas in which added effort offers significant potential for further advance. These areas include:

- demonstrating the durability of ULEV emission levels in commercial NGVs;
- incorporating NGVs into vehicle emission models and air quality planning;
- feedback control of air-fuel ratio, and increased efficiency for heavy-duty lean-burn engines;
- "Zero emission" natural gas/electric hybrid vehicles; and
- Low-emission dual-fuel diesel retrofit systems using electronic fuel control.

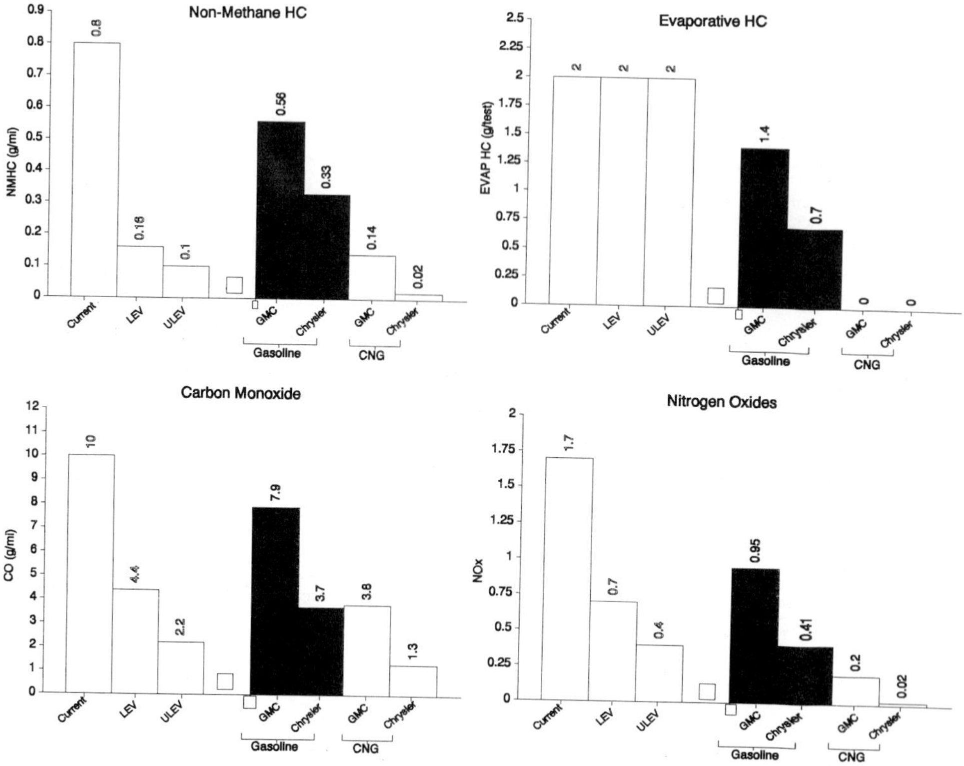

Source: U.S. EPA and CARB emissions certification data for MY 1992. Gasoline vehicle data are from federally-certified trucks of the same models and engine displacements as the NGVs. Current Federal emission standards only regulate total-hydrocarbons (THC).

Figure 1: Emissions certification data for MY 1992 California-certified production medium-duty natural gas trucks compared to federal certification data for similar gasoline vehicles and to current and future California emissions standards.

ULTRA-LOW EMISSIONS

In 1989, the California Air Resources Board (CARB) adopted very stringent, technology-forcing vehicle emission standards that were intended to accommodate (and, in the opinion of many observers, to force) the introduction of light- and medium- duty vehicles using alternative "clean" fuels such as natural gas. These rules established a set of categories for low emission vehicles, including "transitional low emission vehicles" (TLEVs), "low emission vehicles" (LEVs), "ultra-low emission vehicles" (ULEVs), and "zero-emission vehicles" (ZEVs). Sales of these vehicles are to be phased in over the next decade. The CARB rules are also unique in that they incorporate provision for "reactivity adjustment factors" (RAFs) to reflect the differences in ozone-forming potential of the pollutants emitted by different fuels.

Congress, in adopting the Clean Air Act Amendments of 1990, mandated that EPA set emissions standards for light-duty vehicles covered by the Clean Fuel Vehicle (CFV) fleet program identical to the California LEV standards. Additional credit under the CFV program is to be extended to vehicles meeting ULEV and ZEV standards. Congress also clarified the rights of other states to adopt and enforce the more stringent California vehicle emissions standards instead of the Federal regulations. Because of these provisions, the ability of auto manufacturers to meet the new California emissions standards is of great concern in many areas outside of California.

Emissions data from a number of natural gas vehicles (NGVs) using modern electronic emission control systems clearly show the ability to meet California ULEV standards. One such vehicle, the Chrysler B250 natural gas van, has been certified in California with what are believed to be the lowest 50,000 mile NOx and NMHC emissions ever reported. This vehicle could easily have met the California ULEV standards. The benefit of NGVs in reducing "real world" pollutant emissions from vehicles in use is expected to be even greater. In addition, NGVs emit far below the standards set for a number of pollutants, including carbon monoxide and evaporative hydrocarbons, and they are far cleaner with respect to a number of unregulated pollutants, such as air toxics.

NGV Emissions Performance - Until very recently, nearly all of the available natural gas fuel systems relied on mechanical air-fuel metering and mixing systems, analogous to the carburetors formerly used on gasoline engines. Because of their mechanical nature, mixers are sensitive to changes in adjustment, environmental conditions such as temperature, and gas composition, and thus have difficulty maintaining the precise air-fuel ratio control required to meet current emissions standards, especially for NOx. The failure of such mechanical systems to maintain good air-fuel ratio control in some retrofit vehicles led to high NOx emissions, and to the widespread but erroneous view that NOx emissions from NGVs are inherently high. NGV technology is now undergoing a metamorphosis similar to the change from carburetors to electronic fuel injection in gasoline vehicles. Currently, all of the leading prospective air-fuel metering systems for natural gas vehicles are based on fuel injection with full electronic control.

In 1992, both Chrysler and GMC truck began offering California-certified factory-original natural gas vehicles for sale - the first such vehicles produced in the U.S. Both vehicles are equipped with electronically-controlled fuel injection systems employing high-speed solenoid valve (HSV) injectors. These systems are very similar in design and operation to the electronic fuel injection systems that are now standard on gasoline vehicles. Despite the similarities in design, however, the difference in the emissions performance between gasoline and natural gas vehicles is very large, with NGVs emitting far less non-methane organic gas (NMOG), CO and CO_2 than gasoline vehicles. Figure 1 compares the emissions certification results for the Chrysler and GMC vehicles with those for the same model and year vehicles using gasoline, and with the applicable California emissions standards. As Figure 1 clearly shows, the emissions performance of the NGVs was far superior to that of the gasoline vehicles.

The Chrysler natural gas van is possibly the cleanest internal-combustion vehicle ever certified, and could easily have met California ULEV emissions standards. In 1993, the Chrysler B350 van using the same engine became the first and - as of this writing - only vehicle ever certified to the California LEV standards. The GMC natural gas truck also demonstrated good emissions performance compared to its gasoline counterpart. Although the performance of the gasoline trucks shown in Figure 1 is not quite representative of the "state of the art" in gasoline emissions control, the potential for significantly lower pollutant emissions from natural gas vehicles is clear.

Reasons for the favorable emissions performance of NGVs - In addition to the low emissions for certification, real-world emissions from NGVs should be significantly lower than from gasoline vehicles, even though both types of vehicles met the same emissions standards. In real-world driving, vehicles often run at higher speeds and acceleration rates than is reflected in the test (the Federal Test Procedure, or FTP) that is the basis of the emissions standards. Gasoline vehicles have high pollutant emissions during acceleration. The reason for these high pollutant emissions is that gasoline vehicles use more fuel (power enrichment) during acceleration. Natural gas engines do not require or benefit from power enrichment, so their emissions during high acceleration are not expected to differ significantly from their emissions in other modes, and emissions during real-world driving are expected to be much lower than gasoline vehicles.

Cold-start emissions from NGVs are also much lower than from gasoline vehicles. Unlike gasoline vehicles, cold-start emissions from NGVs are unaffected by temperature. Because liquid gasoline does not evaporate well at low temperatures, gasoline engine control systems are programmed to provide a rich mixture during starting and warmup, in order to ensure that enough gasoline vapor is formed to produce a combustible mixture. HC and CO emissions thus increase dramatically at low temperatures. Testing done by EPA showed that the average gasoline vehicle CO emissions were seven times higher at 75° F and 24 times higher at 20° F than the emissions from a natural gas vehicle. The NGV CO emissions were also much lower than the average for late-model multi-port fuel injected vehicles, representative of future gasoline technology.

Table 1: Breakdown of average NMOG and ozone reactivity from natural gas vehicles compared to gasoline vehicles (source: CARB, 1992).

	NMOG Mass		Carter Reactivity	
	mg/mi	percent	mg/mi	percent
Aldehydes	4.88	7.4%	32.67	29.4%
Aromatics	0.56	0.8%	1.86	1.7%
Olefins	7.90	11.9%	59.31	53.4%
Paraffins	52.83	79.8%	17.14	15.4%
Totals	66.17	100.0%	110.98	100.0%
Gasoline Vehicle Data				
From Table 8	104.4		357.7	

NMOG reactivity - Not only do natural gas vehicles have lower emissions of non-methane organic gas than gasoline vehicles, the NMOG that they do emit is significantly less reactive, gram-for-gram, in contributing to ozone formation. NMOG emissions from NGVs are composed primarily of ethane, which has very low reactivity. Aromatic hydrocarbons (a very important contributor to the reactivity of gasoline vehicle NMOG) are virtually absent - most of the reactivity in NGV NMOG emissions is due to the small amounts of ethylene and formaldehyde formed during combustion.

These observations are confirmed by CARB testing. Table 1 gives a breakdown of emissions and ozone reactivity for three gasoline vehicles converted to use natural gas, compared to the average of nine advanced-technology gasoline vehicles. Both sets of data were measured by CARB. As Table 1 shows, the retrofit converted natural gas vehicles, on average produced roughly 60% as much NMOG mass, and only 31% as much ozone-forming reactivity as the advanced gasoline vehicles. This means that the average NGV exhaust emissions will produce about 31% as much smog as the best advanced-technology gasoline vehicles.

Toxic air contaminants - The potential for exposure to toxic and carcinogenic compounds such as benzene, formaldehyde, 1,3 butadiene, and acetaldehyde is a significant source of public concern. Studies by EPA have shown that a large fraction of the total public exposure and cancer risk due to ambient air toxics are due to emissions from motor vehicles, and the Clean Air Act Amendments of 1990 included a number of provisions aimed at reducing air toxic emissions. Natural gas and gasoline vehicle exhaust emissions differ significantly in their air toxics content. Table 2 compares the toxic emissions from nine late-model gasoline vehicles tested by CARB with those from four retrofit NGVs. The gasoline vehicles were selected by CARB as being representative of future TLEVs, and were used to establish the reactivity baseline. Three of the natural gas vehicles were also tested by CARB, and the fourth by General Motors. As the table shows, on average, the gasoline vehicles emitted twice as much total mass of the four identified air toxics as did the natural gas vehicles. Since natural gas contains few hydrocarbons with more than three carbon atoms, emissions of benzene and 1,3-butadiene are very low.

Table 2: Emissions of toxic air contaminants from gasoline (CARB, 1991) and natural gas vehicles (CARB, 1992).

	Benzene	1,3 Butadiene	Form-aldehyde	Acet-aldehyde	Total
Gasoline Vehicles					
DeVille	7.24	0.49	1.26	0.71	9.69
LeSabre	6.84	0.54	1.17	0.99	9.54
Tempo	6.53	0.40	1.02	0.74	8.68
300E	5.29	0.36	1.11	1.10	7.86
Celica	4.96	0.32	0.47	0.71	6.45
Cougar	14.02	0.21	0.74	0.64	15.61
Cavalier	5.34	0.00	3.90	1.51	10.75
Shadow	7.36	0.64	1.81	1.39	11.20
Delta 88	7.97	0.53	1.15	0.76	10.41
Average	**7.28**	**0.39**	**1.40**	**0.95**	**10.02**
Natural Gas Vehicles					
Impco Taurus	0.04	0.01	1.51	0.27	1.83
S&S Taurus	0.05	0.01	2.64	0.27	2.96
Astrovan	0.54	0.02	7.86	1.05	9.47
GM Corsica	0.19	0.00	1.90	0.50	2.59
Average	**0.20**	**0.01**	**3.48**	**0.52**	**4.21**

<u>**Evaporative, running, and refueling emissions**</u> - Evaporative emissions, running losses, and refueling vapor emissions from gasoline vehicles are significant sources of VOC emissions. These emissions result from the fact that the liquid fuel storage system on these vehicles is designed to operate at atmospheric pressure, and must be vented to the atmosphere. Gasoline vapors escape as part of the venting process. Figure 2 shows the MOBILE5 projections of evaporative, running, and refueling losses combined for current-model vehicles,

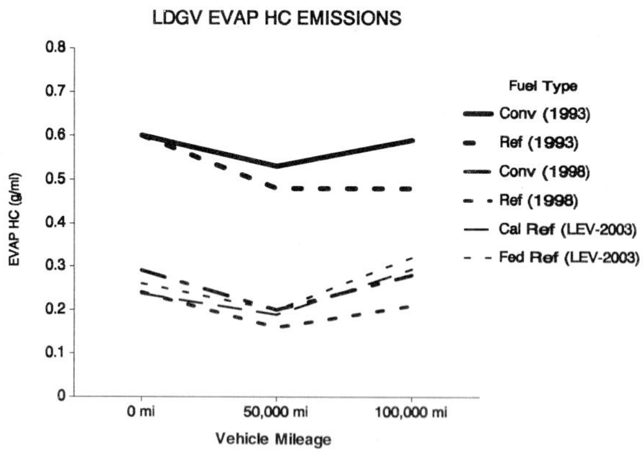

Figure 2: Evaporative emissions (including running and refueling losses) from light-duty gasoline vehicles (EPA MOBILE5 model).

1998 Tier 1 vehicles, and gasoline vehicles meeting California LEV emissions standards. These estimates were calculated assuming the use of Stage II refueling vapor control, and incorporate the effects of current (1993) and advanced future evaporative emission controls. Nonetheless, as Figure 2 shows, the combined emissions of gasoline vapor from the fuel system significantly exceed the estimated tailpipe emissions of VOC from each of these classes. Natural gas fuel systems are not vented to atmosphere under normal conditions, and therefore produce no evaporative emissions or running losses. While some present natural gas fueling systems do produce refueling emissions from venting of the high-pressure fuel hose, the NMHC content of the emissions is small. Future refueling systems will most likely be designed to eliminate these emissions by venting to the compressor suction rather than to atmosphere.

Emissions durability - A major contributor to air quality problems is the failure of many gasoline vehicles to meet the emissions standards they were designed to, especially as the vehicles age. Emissions tests conducted on aging vehicles in consumer use by EPA and CARB show a significant minority of vehicles with very high emissions, generally due to some malfunction affecting the emissions control system. This minority of malfunctioning vehicles raises the average emissions from the entire vehicle cohort substantially. Common malfunctions include damage to or deactivation of the catalytic converter, failure or deterioration of the oxygen sensor, and failure of the sensors used to control cold-start enrichment.

NGVs are expected to show much lower levels of emissions deterioration than gasoline vehicles. In the first place, their emissions will be significantly lower to start with, so that malfunctions in the emissions control system will result in a smaller emissions increase. Equally important, the emissions control system itself will be simpler, and thus less susceptible to failure. In particular, NGVs will not be subject to the types of failures that result from malfunctions linked to cold-start enrichment, power enrichment, or the evaporative control system. Coolant temperature sensor failures - a common cause of high CO and HC emissions in gasoline vehicles, will have little or no effect on NGVs. Similarly, complex air injection control algorithms and hardware - necessary on gasoline vehicles to avoid overheating the catalyst during rich operation - will not be required for NGVs. Failure of these systems is a common cause both of high HC and CO emissions and of catalyst damage due to overheating. Natural gas does not tend to form deposits on fuel injectors, and is less likely than gasoline to contain contaminants that could poison the catalyst and/or oxygen sensor. The palladium-based catalysts preferred for use with NGVs may also be less sensitive to damage from overheating than the platinum-rhodium catalysts used on gasoline vehicles.

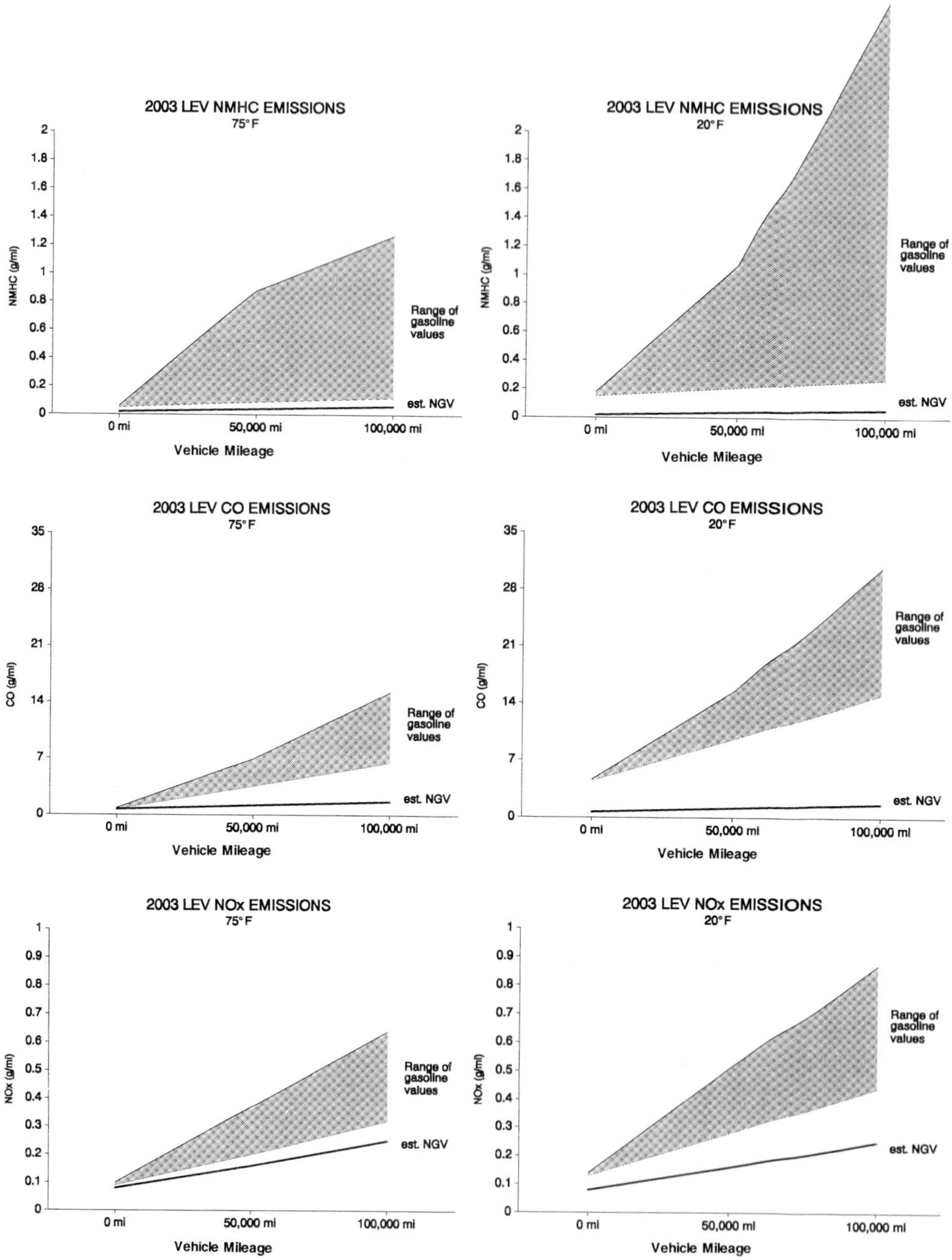

Figure 3: Range of MOBILE5 deterioration estimates for California LEV fleet.

For all of these reasons, NGVs are expected to demonstrate much lower rates of in-use deterioration than gasoline vehicles. The resulting emissions benefits are difficult to quantify, however, due both to the absence of adequate long-term surveillance data on factory-built NGVs, and to the potential for improvements in the deterioration performance of gasoline vehicles. Recent measures such as requiring on-board monitoring and diagnosis of emissions-related malfunctions, and the future use of improved inspection/maintenance programs and remote sensing techniques to detect high-emitting vehicles may significantly reduce the emissions deterioration by gasoline vehicles. The extent of the resulting improvements is controversial at present. Figure 3 shows the range of emissions deterioration estimates produced by MOBILE5 for the year-2000 California LEV fleet (assuming a mix of LEVs, ULEVs, and ZEVs, with LEVs and ULEVs being 100% gasoline fueled). MOBILE5 estimates are shown for two temperatures, 75° and 20° F, to show the impacts of low temperature on gasoline vehicle emissions. The lower end of the range of gasoline values reflects an extremely optimistic view of deterioration from the vehicles, while the upper end assumes deterioration rates similar to those for current late-model vehicles. While no one will know which is more correct until a sufficient sampling of aged LEVs can be tested, the judgement of the author is that the reality is likely to lie somewhere between optimistic and pessimistic views shown in the figure. Also shown are our best estimates of the likely emission levels for new, factory-produced NGVs. As discussed in the sections above, these are anticipated to have lower NMHC emissions than gasoline LEVs, lower CO emissions and substantially lower CO deterioration, and NOx emissions and deterioration somewhat less than those of gasoline LEVs. In addition, NGV emissions are insensitive to temperature, while emissions from gasoline vehicles show a strong temperature effect.

EMISSION MODELS AND AIR QUALITY PLANNING

In the real world, pollutant emissions from motor vehicles are affected by many different factors, including the emission control technology used, maintenance condition, driving pattern and speed, temperature, and fuel volatility, to name just a few. In order to prepare consistent estimates of motor vehicle pollutant emissions, the U.S. EPA has developed MOBILE5, the latest version of its MOBILE series of motor vehicle emission factor models. This computer program estimates the average emission factors (grams of pollutant emissions per mile travelled) for different classes of motor vehicles as functions of average speed, age structure of the fleet, temperature, inspection/maintenance program type, and other variables. Except for California (which has its own model, EMFAC7F), state air quality agencies are *required* to use emission factors produced by MOBILE5 in developing their state implementation plans (SIPs) for attaining air quality standards under the Clean Air Act.

A major shortcoming of the present MOBILE5 model is that it deals only with conventional vehicles burning gasoline or diesel fuel. The emissions properties of natural gas and other clean vehicle fuels are not reflected in the model. The omission of NGVs from MOBILE5 constitutes a significant obstacle to the NGV industry's efforts to capitalize on the air quality benefits of NGVs to establish and expand NGV programs. There are three reasons for this:

1. Air quality planners do not have a convenient and readily accessible way to estimate the emissions benefits of natural gas vehicles. Pressed for time and with limited budgets, they are likely to pass them by for more "accessible" emission control strategies.

2. State planning agencies and the EPA staff who review their plans have limited knowledge of NGVs. In the absence of a consistent and accepted methodology, views of their emissions benefits may differ, leading to controversy and risk of EPA disapproving the SIP. Given limited knowledge, there will also be a tendency toward "conservative" estimates of the benefits.

3. In the absence of a consistent and accepted methodology for estimating NGV benefits, vested interests may succeed in obscuring the issues - e.g. by equating NGVs with reformulated gasoline vehicles, which have much higher emissions.

To overcome these problems, there needs to be a consistent and accepted emissions factor model - similar to MOBILE5 - for NGVs. In order to support the development of such a model, it will be necessary to collect substantial data on the emissions performance of natural gas vehicles in use, under a variety of ambient conditions and driving patterns. Plans for the collection of such data are now being developed by EF&EE and the Gas Research Institute.

HEAVY-DUTY LEAN-BURN ENGINES

Heavy-duty vehicles are defined in EPA regulations as vehicles having a manufacturer's rated gross vehicle weight (GVW) in excess of 8,500 lb. These vehicles can be divided into a number of subclasses. Vehicles between 8,500 and about 14,000 lb GVW are mostly large pickups, vans, and specialty vehicles built on pickup and van chassis. These are referred to as "light-heavy" duty vehicles by EPA, and as "medium-duty" vehicles by CARB, which regulates them separately from other heavy-duty vehicles. These vehicles use a mix of gasoline and diesel engines. Vehicles with GVW between 14,000 and 33,000 pounds are mostly two-axle, single-unit trucks. Most of these use diesel engines, although some gasoline engines are also found in this category. Vehicles over 33,000 lb GVW are either three-axle single-unit trucks or tractor-trailers, and are almost exclusively diesel-powered.

Because of their overall design, diesel engines are well-adapted to lean operation using natural gas, in either a dual-fuel or spark-ignition mode. Many heavy-duty diesel engines in New Zealand, Italy, and other countries have been modified for lean natural gas operation by replacing the fuel injection system with a high-energy ignition, and adding an air-gas mixer to the intake. In many of the diesel engine conversions performed to date, emissions were of secondary or no concern, so that the mixture and spark timing were optimized for best efficiency, without regard to NOx. This led to a widespread (but erroneous) impression that heavy-duty natural gas engines have inherently high NOx emissions.

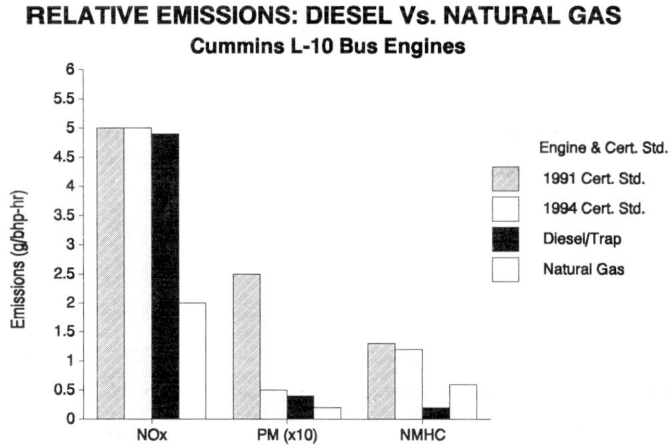

Figure 4: Emissions from Cummins natural gas bus engine compared to diesel engine and emissions standards.

In the last few years, a number of heavy-duty engine manufacturers have developed diesel-derived lean-burn natural gas engines for use in emissions-critical applications such as urban transit buses and delivery trucks. These engines incorporate low-NOx technology used in stationary natural gas engines, and typically an oxidation catalyst as well. They are capable of achieving very low levels of NOx, particulate, and other emissions (less than 2.0 g/BHP-hr NOx and 0.03 g/BHP-hr particulate) with high efficiency, high power output, and (it is anticipated) long life.

Two such engines - the Cummins L10 engine for transit buses, and the Hercules 5.6l engine for schoolbuses and medium trucks - have recently been certified in California, and several others are expected to be certified shortly. Figure 4 compares the emissions from the Cummins natural gas engine with the emissions standards and with emissions from the 1993 diesel version of the same engine with a particulate trap.

Precise, stable, responsive control of the air-fuel ratio is essential to achieving low emissions and good driveability in lean-burn engines. If the air-fuel mixture is too lean, the engine will misfire; if too rich, NOx emissions will increase and knock may occur. The air-fuel ratio control must be responsive enough to follow transients in engine performance, and stable over the full life of the vehicle, at all temperatures and all altitudes. This requirement implies some sort of feedback control. One promising approach is to monitor the cylinder pressure as a function of time during combustion. This monitoring will easily detect weak or misfiring combustion, or excessively energetic combustion which may lead to knock. This approach is already used by one Japanese auto manufacturer on its lean-burn gasoline engines. R&D on this and other potential feedback systems for lean-burn engines could pay substantial dividends in improved performance, emissions, and efficiency.

NATURAL GAS/ELECTRIC HYBRIDS

A hybrid vehicle is one that combines an electric drive system with an on-board electric generation system to supply at least part of the electricity. Most hybrid-electric vehicle designs also include some capacity for energy storage, in the form of batteries or ultracapacitors.

Views of the relative roles of battery storage and on-board power appropriate for a hybrid vehicle vary. Two polar cases can be defined, which we will refer to as type A and type C, to match the nomenclature used by the California Air Resources Board. A type A hybrid is essentially a battery-electric vehicle, with an on-board battery charger to extend its driving range. The expectation is that this vehicle would operate most of the time on battery power, with the batteries recharged from the electric utility grid. The on-board engine would be used only for when the battery charge was near exhaustion - primarily on extra-urban trips. In this view, hybrid vehicle is essentially a way to make battery-electrics acceptable to the public by extending their driving range.

A type C hybrid, on the other hand, would not rely on the batteries for primary energy storage to any great extent. Instead, the engine would run continuously, at a more-or-less constant load, and the batteries (or possibly an ultracapacitor bank, flywheel motor-generator, or other energy storage medium) would function as a accumulator - evening out the load on the prime mover, and accepting the energy recovered by regenerative braking. This would allow the prime mover to be sized for the *average* power demand over the driving cycle (with some reserve), rather than the *peak* power demand, as it must be in a mechanical drive system without energy storage. Since the prime mover would operate at 50-100% of its designed power output most of the time, its average efficiency over the driving cycle would be much greater than that of present engines (for which the average load over the driving cycle is typically less than 10% of maximum power rating). The higher efficiency and the decoupling of engine and wheels should result in much lower emissions than in a conventional vehicle. Decoupling the engine from the wheels (and thus from the need to respond to rapid changes in speed and power output) would also make possible the use of new, lower-emitting and more efficient prime movers such as Stirling engines, ultra-lean burn IC engines, gas turbines, and fuel cells.

A type C hybrid would not require very much energy storage - only enough to supply the kinetic energy required for a few accelerations, and possibly for climbing short, steep hills. To accelerate a 1500 kg car from zero to 80 MPH requires only about 0.3 KWH, while the extra energy required to climb a 500 meter hill is about 2.4 KWH. The total electric storage capacity would therefore need to be about 2-3 KWH - much less than for a type A hybrid. The energy storage system would need to be capable of very high rates of charging and discharging, however, which would require a different kind of battery (or other energy storage mode) from those typically used in dedicated battery- electric vehicles.

The maximum power output required from the prime mover on a type C hybrid would be determined by the maximum continuous cruising speed and continuous grade-climbing ability needed to meet customer requirements. Figure 5 shows the relationship between continuous cruising speed, road grade, and maximum power requirement for a 1500 kg car, assuming a combined motor/generator efficiency of 85% and Cd of 0.3. As this figure shows, to climb a 6% grade at a sustained speed of 50 MPH would require a maximum prime mover power output of about 33 KW, which would also suffice for cruising at 84 MPH on level ground. Prime mover power of 20 KW would allow sustained cruising at 70 MPH on level ground, and would allow the vehicle to ascend a prolonged 4% grade at 40 MPH. (Note that substantially higher speeds would be possible on grades less than about 1000 meters in height, as the prime mover power could be supplemented by the on-board electric storage until that was exhausted. Based on these data, 20 KW is probably a reasonable maximum power rating for a 1500 kg car. It is desirable to minimize the prime mover power rating, as the *average* power demand under most driving modes would be under 10 KW, and efficiency for nearly all engine types declines markedly at loads below about 50% of maximum. In addition, the lower the prime mover power rating, the lower the weight, bulk, and cost of the prime mover will be.

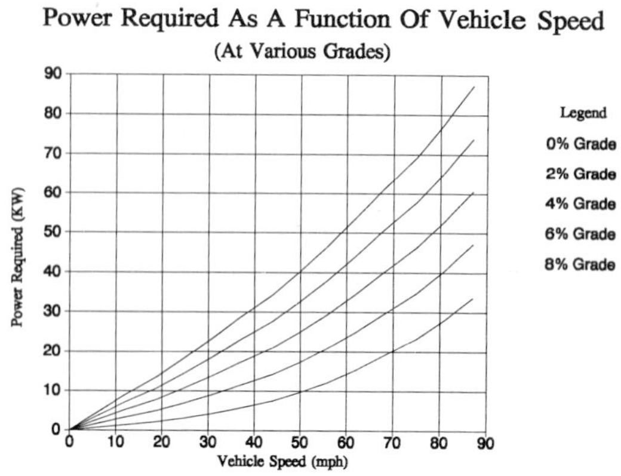

Figure 5: Sustained cruising speed for a hybrid electric vehicle as a function of prime mover power and road grade.

A hybrid-electric natural gas vehicle with a suitably low-emitting prime mover has the potential to qualify as a "zero emission vehicle" or ZEV. CARB regulations require manufacturers that 2% of California vehicle sales in 1998 be ZEVS. The required percentage goes up to 10% in 2003. Presently, CARB regulations require that a vehicle, to qualify as a ZEV, have emissions that are mathematically zero. The only vehicle technology that could foreseeably meet the proposed requirement of zero emissions from the vehicle would be battery-electric propulsion. But even battery-electric vehicles are not truly <u>zero</u> emissions, since the powerplant emissions associated with generating electricity for battery charging should be taken into account. These emissions can be very significant. For example, a coal-fired powerplant can produce around 2.6 grams of NOx per KWH. A recent study for NESCAUM[1] estimated that powerplant emissions for battery charging in the year 2000 would average 3.9 g/KWH of SOx, and 0.4 g/KWH of NOx in the Northeast. By 2010, SOx is projected to decrease to 1.5 g/KWH, but NOx will increase to 0.6 g/KWH. Note that these are emissions per KWH produced by the power plant. Additional losses will be incurred in electric transmission, in converting from AC line power to DC, in voltage drops in the batteries during charging and discharging, and

---

[1] M.W. Tennis, "Impact of Battery-Powered Electric Vehicles on Air Quality in the Northeast States", prepared for NESCAUM, Boston, MA, July 1992.

in electrical losses in the motors and power conditioning system. These losses amount to 30% to 50% of the total electricity supplied[2], so that powerplant emissions per BHP-hr of mechanical work actually delivered to the drivetrain will be correspondingly greater. It is these latter emissions that should properly be compared with those from other vehicle propulsion systems.

Besides contributing to powerplant emissions, battery-electric vehicles have many other well-known disadvantages, including high first cost, increased weight, limited range, the need for periodic replacement and disposal of the battery system, and the questionable crash safety of many of the battery types under development. These problems are especially severe for heavy-duty vehicles, for which the added weight of the batteries would directly affect payload capacity.

Table 1 shows the calculation of relative emissions from battery-electric, Stirling, gas-turbine, and fuel-cell technologies. The first four columns in the table show the emissions from each technology per kilowatt-hour of electric power output. These are compared with the NESCAUM projections for average emissions from electric generation in the Northeast U.S. The fifth column shows the estimated loss in going from the electricity source to the motor input on the vehicle. In the case of the on-board generation technologies, these losses would be negligible. For the battery-electric alternative, we have assumed that cumulative losses due to AC-DC conversion and electrical losses in the battery would amount to 40% of the power input, which is toward the low end of the range of values reported in the literature. These differences are reflected in columns 6 through 9, which show the emissions produced by the energy conversion system per BHP-hr of work output by the electric propulsion system.

The weight and cost of the engine/generator or fuel cell and fuel system would be much less than those of the batteries required to achieve similar range. These approaches would thus retain the major advantages of the electric drive system (regenerative braking, decoupling of the energy conversion system from the wheels), while providing longer range, lower costs, better performance, quicker and more convenient refueling, and higher energy efficiency than a battery-electric. Because of these characteristics, hybrid vehicles using these technologies are more likely to be

---

[2]see, for instance, G.D. Whitehead and A.S. Keller, "Performance Testing of the Vehma G Van Electric Vehicle", SAE Paper No. 910242. Between 42% and 49% of total AC electric power supplied was lost between the AC input to the charger and the DC input to the batteries.

Table 3: Comparison of emissions from energy conversion in hybrid and battery-electric HDVs.

	Emissions (g/kWh elec.)				Battery Loss	Emissions (g/BHP-hr mech)			
	NOx	CO	NMHC	SOx		NOx	CO	NMHC	SOx
Hybrid Vehicles - On-Board Generation									
Natural Gas Fuel Cell[1]									
Phosphoric Acid	0.066	na	na	0.00	0%	0.049	na	na	0.00
Solid Oxide	0.156	na	na	0.00	0%	0.116	na	na	0.00
Natural Gas Stirling Engines[2]									
MTI Stirling	0.097	0.425	0.0005	0.00	0%	0.073	0.317	0.0004	0.00
DDC-STM	0.201	2.856	0.038	0.00	0%	0.150	2.131	0.028	0.00
Natural Gas Gas Turbine									
24 KW	0.046	0.03	0.025	0.00	0%	0.034	0.022	0.019	0.00
Battery-Electric Vehicles - Fixed Power Plants									
Future Northeast Power Plants (NESCAUM, 1992)									
Avg. 1995	1.9	0.15	0.15	5.1	40%	2.387	0.187	0.187	6.32
Avg. 2000	0.4	0.15	0.013	3.9	40%	0.497	0.187	0.016	4.85
Avg. 2015	0.6	0.15	0.013	1.5	40%	0.746	0.187	0.016	1.87

[1] J.J. Hirschenhofer "Commercialization of Fuel Cell Technology", Mechanical Engineering, September, 1992.
[2] C.S. Weaver and L.M. Chan technical memorandum under GRI contract, October, 1992.
[3] R. MacKay "Hybrid Vehicle Gas Turbines", SAE Paper No. 930044.
[4] M.W. Tennis "Impact of Battery-Powered Electric Vehicles on Air Quality in the Northeast States", report to the NESCAUM, Boston, MA, 1992.

accepted by the public and commercial users than battery-electric vehicles.

To sum up, the most viable approach to a natural gas-electric hybrid vehicle is probably a type C hybrid, with only minimal on-board energy storage, and a highly efficient, inherently low-emitting prime mover of approximately 20 KW maximum output. Potential near-term prime movers include small Stirling engines, and possibly other high-efficiency, low-emission engine types. In the longer run, this architecture could easily be adapted to a natural gas fuel cell when these become available at reasonable cost. Such vehicles could be attractive compared to conventional NGVs due to their greater fuel efficiency, which would reduce the weight/volume/range penalties associated with natural gas. If, as we expect, CARB changes its rules in the future to allow such vehicles to qualify as ZEVs this would open a huge potential market. Of course, greenhouse warming and other emissions would be greatly reduced as well. Because the number of mechanical moving parts would also be greatly reduced, such vehicles would probably be durable and have very low maintenance expenses, further enhancing their attractiveness.

OFF-ROAD VEHICLES

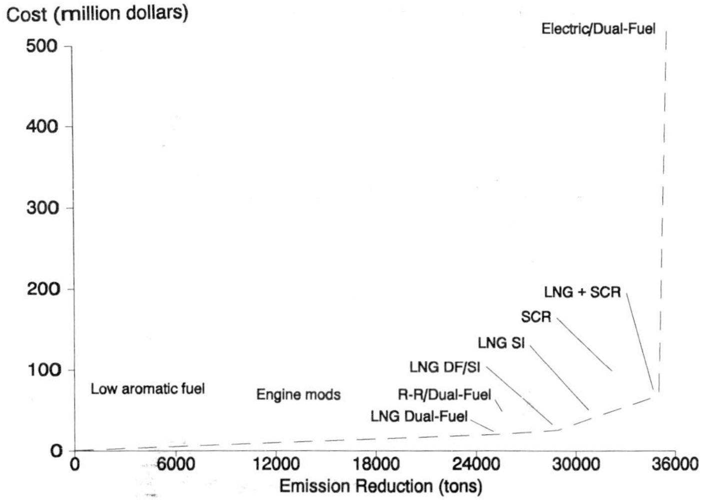

Figure 6: Cost vs. emissions reduction for potential line-haul, local, and switcher locomotive control measures.

In addition to the use of natural gas in cars and trucks, there has been increasing interest in natural gas use in off-road vehicles in the last few years. Promising applications for natural gas use in off-road vehicles include railway locomotives, tugboats and towboats, farm machinery, and some large construction and mining machines. All of these vehicles now use diesel engines, and all consume large quantities of fuel. Presently, the major reason that users of these vehicles are investigating natural gas is the potential savings in fuel cost, due to the low price of natural gas compared to diesel fuel. This creates a substantial economic incentive to switch to natural gas operation. Of increasing importance in the future will be the impact of new emission standards and control requirements on off-road vehicles. The Clean Air Act Amendments of 1990 gave EPA authority, for the first time, to regulate off-road vehicle emissions, and regulations to do so are under development. The California Air Resources Board has also adopted emission limits for off-road diesel equipment, and is developing regulations for marine and railroad emissions.

Figure 6 summarizes the results of a recent study of locomotive emission control options carried out by the author's firm. This figure compares the costs of a number of control measures, including converting existing diesel locomotives to natural gas operation either in dual-fuel mode (as further discussed below) or as spark-ignition engines. Converting existing diesels to dual-fuel operation could reduce emissions by 70% (more if advanced dual-fuel technologies allow a reduction in idle and light-load emissions), at a cost less than $1,000 per ton of NOx reduced. Use of low emission spark-ignition LNG engines in local and switch locomotives would increase the emission reduction to 80%, at an incremental cost of $1,052 per ton. Use of SI engines in line-haul units as well, or the addition of selective catalytic reduction to the dual-fuel engines, would produce even larger emission reductions, but at significantly higher incremental cost. Combining electrification of line-haul locomotives with dual-fuel+SCR in local and switch locomotives would produce the highest level of control efficiency - 99% - but the high price of electrification forces this combination into the realm of very poor cost-effectiveness.

LOW-EMISSION DUAL-FUEL ENGINES

A heavy-duty engine can be designed to operate on natural gas in one of three ways. Most large stationary engines at present are designed for spark-ignition (Otto cycle) operation, with a lean air-fuel ratio. This engine technology is mature, and routinely achieves NOx levels less than 1.5 g/BHP-hr, or about 85% less than a typical uncontrolled diesel engine. An alternative to spark ignition is *dual-fuel* operation, in which a small amount of diesel fuel is injected instead of a spark to ignite the natural gas charge. Recent developments in dual-fuel engine technology have resulted in emissions capabilities similar to those of spark-ignition engines. A third technology, still under development, is *direct injection* of natural gas, in the same way that diesel fuel is injected in a diesel engine. Although this approach has advantages in fuel-efficiency and power output, the NOx emissions from these engines are likely to be higher than from optimized spark-ignition or dual-fuel engines.

Existing diesel engines can be modified to operate as dual-fuel engines, thus offering the potential for cost-effective emission reductions from the existing vehicle population. Although many existing dual-fuel engine modifications are crude, and exhibit high CO and HC emissions, technology to achieve very low emissions in dual-fuel operation has been demonstrated. Dual-fuel engines also offer important advantages in fuel flexibility, as they can retain the capability to operate on 100% diesel fuel if gas is not available. This would allow the engine to operate normally on low-cost and low-polluting natural gas, while retaining the ability to operate on diesel alone if necessary.

Dual-fuel diesel/natural gas engines use natural gas as the primary fuel. Most dual-fuel engines induct the gas already mixed in the intake air, but gas can also be injected directly into the cylinder. Instead of a spark plug, a small injection of diesel fuel is used. The diesel fuel undergoes compression-ignition, just as in an ordinary diesel engine, and the burning diesel fuel ignites the natural gas. Compared to a spark-ignition engine, the widespread, high-energy combustion of the diesel fuel gives more reliable ignition and faster combustion of the natural gas charge (a particular advantage with very lean air-fuel ratios). More rapid and widespread combustion in the cylinder also reduces the time that the unburned gases are exposed to high temperatures and pressures, and thus reduces the tendency to knock. It is for this reason that many diesel engines can be converted to dual-fuel operation without reducing the compression ratio, when a spark-ignition engine at the same compression ratio would suffer destructive knock.

Dual-fuel engine performance and emissions vary depending on operating conditions and the sophistication of the control system. Dual-fuel engines perform best under moderate to high load, and can often equal or better the fuel-efficiency of a pure diesel under these conditions. Operating with a lean air-fuel ratio, they can also achieve much lower emissions (especially of NOx and particulate matter) than a pure diesel. Existing dual-fuel conversions suffer from major increases in CO and HC emissions and loss of fuel efficiency at light loads. This is because they operate unthrottled, so that the air-fuel mixture becomes leaner as the load is reduced. As the mixture becomes leaner, combustion eventually degrades, leaving large amounts of partial reaction products in the exhaust. Possible solutions to this problem include throttling the intake air at light loads, use of electronically-controllable turbochargers to reduce light-load airflow, or the use of skip-firing. Skip-firing means that a certain number of engine cylinders are shut off on a rotating basis. By supplying more gas to some cylinders in rotation, and none to others, it would be possible to ensure a combustible mixture in each cylinder. With this arrangement, enough cylinders could be fired to maintain the engine output while reducing fuel consumption and emissions. A skip-fire system can only be employed easily with a sequential multi-point injection system, since it requires the ability to shut the fuel supply off to a particular cylinder. At least one equipment manufacturer is developing such a system for dual-fuel engines.

In addition to light load emissions and fuel economy, dual-fuel engines may be limited by knock at high loads. Experience with dual-fuel engines on natural gas indicates that knock may be a limiting factor above about 190 PSI BMEP (in four-stroke engines, the limit in two-strokes is lower). The Cooper-Bessemer "Cleanburn" dual-fuel engine has 200 PSI BMEP at rated power (Blizzard, et al, 1991). Most diesel engines have BMEP levels in this region or lower, but some highly-rated truck, marine and locomotive engines have BMEP levels significantly above this (exceeding 300 PSI for GE locomotive engines and some recent truck engines, see Table 4). For these highly-rated diesel engines, conventional dual-fuel operation would require either derating or reduced substitution of gas for diesel under high-load conditions.

A third area of development for dual-fuel engines is in the diesel fuel injection system. To minimize pollutant emissions and diesel fuel use, it is desirable to reduce the pilot fuel quantity as much as possible, consistent with getting good injection and combustion characteristics. Flexible control of fuel injection timing is also important to optimize dual-fuel emissions and performance. The minimum pilot fuel quantity is presently limited by the injection system characteristics to about 5% of the quantity at full load on 100% diesel. Below this level, the fuel injection characteristics deteriorate. By using a separate "micro" injector, pilot fuel quantities less than 1% of full load fuel consumption are possible, and this also allows independent control of timing. The addition of a separate system would increase the hardware costs somewhat, of course. Since the pilot injectors would be small, however, it would be possible to use inexpensive electronic fuel injection systems developed for automotive diesels.

Table 4: BMEP for heavy-duty diesel and natural gas engines.

	Application	Fuel	RPM	Power (hp)	BMEP (psi)
	Two-stroke engines				
EMD 16-567	Rail, marine	Diesel	900	1,500	**73**
EMD 12-645E3	Rail, marine	Diesel	905	2,415	**137**
DDC 8V-149TI	Gen Set, Marine	Diesel	1900	800	**140**
EMD 16-645E3	Rail	Dual-fuel	900	3345	**143**
EMD 16-710G	Rail	Diesel	900	3950	**153**
	Four-stroke engines				
CAT G3516-TA	Gas Gen Set	SI NG	1200	1,085	**170**
Pielstick PA5 DF	Multi-purpose	Dual-fuel	1000	3,153	**185**
Pielstick PA4 185DF	Gen Set	Dual-fuel	1500	1,973	**189**
Waukesha AT25GL	Gen Set	SI NG	1000	2,587	**190**
CAT 3408BTA	Marine	Diesel	2100	585	**201**
CAT 3516 TA	Rail	Diesel	1800	2,075	**217**
Pielstick PA4 185	Rail	Diesel	1500	2,545	**244**
Wartsila GD32	Cogeneration	DING[1]	720	7,902	**281**
GE 16-7FDL	Rail	Diesel	1050	4,000	**282**
Pielstick PA5 DF	Research Eng.	Dual-fuel	1000	1,200	**282**

[1] Direct Injection Natural Gas.

An example of dual-fuel engine design with excellent emissions performance is the Cooper Cleanburn™ dual-fuel engine described (Blizzard et al., 1991). This is a modified 4-stroke diesel of approximately 8300 HP. The original engine was satisfactory in performance except for emissions: NOx was 11.5 g/BHP-hr, not much better than straight diesel. Smoke was also poor, with an opacity rating of 20%. Cooper engineers modified the combustion chamber and cylinder airflow characteristics to improve combustion, and added a separate prechamber (called a "torch cell" by the authors) for the diesel pilot injection. This prechamber is a self-contained unit mounted in the cylinder head, with its own injector. The diesel pilot fuel is injected into this prechamber, where it burns and shoots out into the main combustion chamber in a flaming jet - providing thorough ignition for the lean natural gas charge. The use of this torch cell made it possible to reduce the diesel pilot charge substantially, and greatly improve emissions. Originally, the dual-fuel engine used 6% diesel fuel and 94% natural gas at full power; the CleanBurn™ research engine burns only 0.9% diesel fuel at 200 BMEP. Smoke was virtually eliminated and the NOx was reduced 92% to 0.9 g/BHP-hr at rated speed and load. The engine also retained its original diesel injection equipment, giving it the ability to operate on 100% diesel if required.

Although the Cooper LSVB engine itself is too large, Cooper's technology or a similar one could be applied to locomotive engines, marine, and even truck engines. This would require modifications to the cylinder heads to incorporate the prechamber and its injector. A natural gas metering system and mixing system would also have to be supplied, and modifications to the existing injection system could be needed in order to prevent fuel in the injector tips from "cooking" and forming deposits with prolonged exposure to high temperatures.

NGVS AND GREENHOUSE WARMING

Concern has been growing over the possibility of a global temperature increase due to increased concentrations of "greenhouse" gases in the atmosphere. Two of the most important such gases are CO_2 and methane, which are projected to account for roughly 46% and 9% of the total anticipated warming, respectively (most of the remaining warming effect is due to chlorofluorocarbons (29%), increased tropospheric ozone (4%) and N_2O (6%)) (Ramanathan et al., 1985). One pound of methane in the atmosphere is estimated by some scientists to contribute about 65-70 times as much to global warming as one pound of CO_2. Unlike CO_2, however, methane in the atmosphere slowly oxidizes through photochemical processes, with an estimated lifetime of 5-10 years. Thus, the relative effects of methane emissions are less than would be anticipated from the effects of methane concentrations alone. The U.S. Office of Technology Assessment, in a recent study, has estimated the global warming potential of a kilogram of methane *emissions* at about 4.6 to 12 times that of a kilogram of CO_2 (U.S. OTA, 1991). Other estimates have ranged from 10 to 30 times.

One concern which is sometimes expressed about NGVs is that their methane emissions might exacerbate global warming. This argument is incorrect for a number of reasons. While additional methane emissions from NGVs would slightly increase global methane concentrations, this contribution would be negligible compared to other, natural sources. Total global methane emissions are estimated at 400 to 800 million tons per year, mostly from naturally-occurring sources such as swamps, wetlands, termites, and ruminant animals (Golomb et al., 1989). Fossil fuels account for less than 10% of global methane emissions (mostly from venting of associated gas from oil wells). In the most extreme conceivable case, the whole world vehicle population of 500 million vehicles, each traveling 10,000 miles per year and emitting 1.5 grams of methane per mile would result in about 7.5 million additional tons of methane per year, or 1 to 2 percent of total global methane emissions. This represents the maximum conceivable methane emissions from NGVs. To take a more realistic case, two million NGVs would emit only about 24,000 tons of methane per year at 1.2 g/mi, or less than 1/10,000th of total global methane emissions.

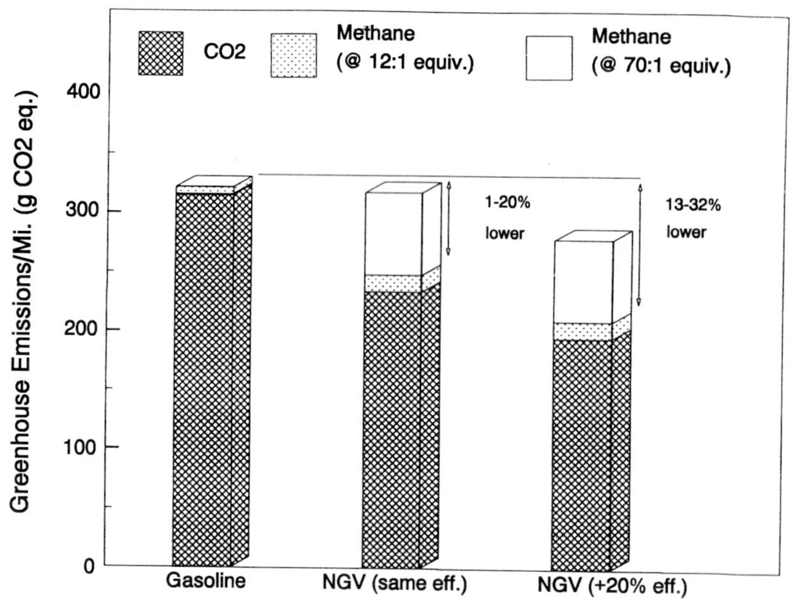

Figure 7: Global warming impacts of NGVs vs. gasoline vehicles.

Any slight increase in methane emissions from NGVs would be more than offset by a substantial reduction in CO_2 emissions compared to other fuels. Per unit of energy, natural gas contains less carbon than any other fossil fuel, and thus produces lower CO_2 emissions per mile, even if the energy efficiency is the same. The improved energy efficiency possible in an optimized NGV would further increase this advantage. Figure 15 compares the greenhouse impacts (expressed in grams of CO_2 equivalent) of a light-duty passenger car powered by gasoline with those of a similar NGV, with and without allowance for increased energy efficiency. The CO_2 equivalence of the methane emissions is calculated using a range of estimates: the OTA ratio of 12:1 and an ultraconservative 70:1 ratio. Since 70:1 is the ratio of the effects of methane concentration in the atmosphere to CO_2 concentration, using this latter ratio essentially ignores the well-established fact that methane remains in the atmosphere for a much shorter time. As Figure 15 shows, even using this ultraconservative approach, and with no credit for increased energy efficiency, the greenhouse impacts of an NGV emitting 1.2 g/mi of methane are still slightly lower than those of a gasoline vehicle. Taking the 12:1 ratio, which is probably more realistic, the greenhouse impact of the NGV would be 20% lower. The greater energy efficiency possible with natural gas vehicles would result in a further benefit - in the range of 13 to 32%.

A further reduction in global warming impacts would result from the lower CO emissions typical of NGVs. Although CO is not itself a greenhouse gas, it plays an important role in the atmospheric chemistry that determines concentrations of methane, ozone, and other greenhouse gases. In the atmosphere, CO undergoes photochemical oxidation, producing tropospheric ozone (which is a greenhouse gas) and competes with methane and other species for OH radicals. These radicals play a key role in breaking down methane and some other greenhouse gases. The reduction in CO emissions due to NGV use would increase OH radical concentrations in the atmosphere, and thus help to increase the global rate of methane decomposition. This more rapid breakdown could more than offset the small increase in direct methane emissions from NGVs.

REFERENCES

Blizzard, Donald T., Schaub, Frederick S., & Smith, Jesse G. (1991). Development Of The Cooper-Bessemer Cleanburn™ Gas-Diesel (Dual-Fuel) Engine, ASME ICE-Vol. 15, Fuels, Controls, and Aftertreatment For Low Emissions Engines, American Society of Mechanical Engineers, New York.

California Air Resources Board, 1991. "Proposed Reactivity Adjustment Factors for Transitional Low Emission Vehicles - Technical Support Document", Mobile Source Division, CARB.

California Air Resources Board, 1992. Mail-out # 92-03.

Carter, W.P.L., 1990. "Development of Ozone Reactivity Scales for Volatile Organic Compounds", U.C. Riverside.

Gabele, P., K.T. Knapp, W.D. Ray, R. Snow, W. Crews, N. Perry, and J. Lanning, 1990. "Ambient Temperature and Driving Cycle Effects on CNG Motor Vehicle Emission", SAE Paper No. 902069, SAE International, Warrendale, PA.

Golomb, D. and J.A. Fay (1989). "The Role of Methane in Tropospheric Chemistry", Energy Laboratory, Cambridge, MA, 1989.

Ramanathan, V. R.J. Cicerone, H.B. Singh, and J.T. Kiehl (1985). "Trace Gas Trends and Their Potential Role in Climate Change", J. Geophys. Res., v. 90, no D3, pp 5547-5566, June 20, 1985.

U.S. Congress, Office of Technology Assessment (1991). Changing by Degrees: Steps to Reduce Greenhouse Gases, U.S. Government Printing Office, Washington, DC.

Natural Gas Vehicles: Politics, Technology & Marketing

Jeffrey Seisler
Executive Director
The Natural Gas Vehicle Coalition
Arlington, Virginia, U.S.A.

Background:

Dr. Seisler has been the Executive Director of The Natural Gas Vehicle Coalition since its inception in 1988. In addition, he is currently a Vice President of The International Association of Natural Gas Vehicles.

NGV DEVELOPMENTS IN THE U.S. FOR 1992: POLITICS, TECHNOLOGY & MARKETING

Jeffrey Seisler
The Natural Gas Vehicle Coalition
1515 Wilson Boulevard, Suite 1030
Arlington, VA 22209

Commercialization activities for Natural Gas Vehicles (NGVs) in the United States are advancing rapidly on all fronts: politically, technologically; and in the market place. 1992 will be a significant year for NGV legislation (national and state), technology advancement, particularly from original equipment manufacturers (OEMs), and for the development of a national NGV fueling infrastructure. The NGV Coalition is playing a strong role on all fronts, as it implements a national 'NGV Campaign', designed to create a favorable national legislative environment, while obtaining new and higher visibility for NGVs through public relations and advertising.

NGV POLITICS

The Energy Policy Act of 1992

The U.S. Congress passed the Energy Policy Act of 1992 on October 8, creating new market opportunities for natural gas vehicles (NGVs) and alternative fuels. The bill comprises a relatively well-rounded approach to a fuels policy that includes a combination of mandates, incentives, funding for research and development, public information dissemination, as well as important provisions that, for the first time, demonstrate critical government leadership in purchasing alternative fuel vehicles (AFVs) for the nation's largest fleet. Note that funding for a number of the programs summarized below has been authorized to be budgeted, but will have to be **appropriated** later by Congress, and released to the various federal agencies to be spent.

- **Federal Fleet provisions:** The legislation requires the federal government to help create a market for AFVs by purchasing 5,000 AFV light-duty cars and light trucks in 1993; 7,500 in 1994; 10,000 in 1995; and phased in by 1999, of new vehicles purchased, as follows: 25% in 1996; 33% in 1997; 50% in 1998;and 75% in 1999 and thereafter. Since there are approximately 375,000 light-duty vehicles in the federal fleet, with about 20% annual turnover, the new law should result in about 125,000 to 160,000 alternative fuel vehicles by the end of the decade. The federal government is required to diversify their AFV purchases among a number of alternatives; however,cooperative arrangements with the NGV industry, federal fleet administrators and operators should allow natural gas to achieve over 50% penetration, representing some 60,000 to 80,000 **light-duty**, federal government NGVs to the end of the decade.

- **State Governments:** Provisions require centrally fueled state government fleets in metropolitan areas of populations of 250,000 or more to buy AFVs on the following schedule: 10% of their new vehicles in model year (MY) 1996, 15% in 1997; 25% in 1998; 50% in 1999; and 75% in MY 2000. Diversification among AFVs may not be as high a priority for states, so opportunities for NGV penetration among some 2.3 million qualifying state vehicles could result in as many as 805,000 NGVs in the state fleet market by the year 2000.

- **Private and Municipal Fleets:** Private and municipal fleets of 50 or more vehicles, with at least 20 vehicles in one location, and in metropolitan areas in excess of 250,000 population are required to phase-in **light-duty** AFVs, beginning with 20% of new vehicles purchased in MY 1999. The program is subject to the U.S. Department of Energy (DOE) determination in a rule making process (by December 1996) as to whether a private fleet program, in conjunction with other federal and state programs, will achieve a goal of replacing 30% of imported motor fuels with domestic fuels (or some lesser percentage set by DOE) by the year 2010. Up to seven million private, centrally fueled fleet vehicles would be included in this program.

- **Fuel Providers:** Private sector providers of alternative fuels (including gas and electric companies) are required to phase-in **light-duty** AFVS, beginning with 30% of their **new** fleet vehicles starting MY 1996; 50% in MY 1997; 70% in MY 1998; and increasing to 90% for MY 1999 and thereafter. All public utilities (including gas, electric, water and telephone) account for some

1.2 million vehicles, with the natural gas industry component estimated about 200,000. Although exemptions exist for fuel provider fleets -- vehicle and fuel availability requirement -- this statute could result in, within the gas industry alone, conversion (or new purchases) of some 96,000 NGVs by MY 1999. Approximately 28% of these are cars and about 60% are trucks, most in the weight classification affected by the legislation. We expect that the Big Three automakers will produce the trucks covered by the bill.

- **Tax Incentives:** Important financial incentives include federal tax deductions for businesses or individuals on the **incremental cost** of AFVs (over gasoline and diesel vehicles) at the following rate: $2,000 for vehicles up to 10,000 pounds gross vehicle weight (gvw); $5,000 for vehicles between 10,000 and 26,000 gvw; and $50,000 for vehicles above 26,000 gvw. Qualifying tax deductions for businesses installing fueling stations are up to a $100,000 ceiling. Taking into consideration depreciation already allowed for business deductions, the vehicle and station, in terms of **real dollar** value, the business deductions equate to the following: $111 for light-duty vehicles; $263 for medium duty vehicles; $2,630 for heavy-duty vehicles; and $5,539 for stations (at the $100,000 deduction level). For individuals claiming a tax deduction on a light-duty vehicle, the cash value of a $2,000 deduction (assuming 30% tax rate) is $600. The tax deductions begin June 30, 1993 (there is no "grandfather" clause for conversions or fueling stations installed before then), and begin to phase out in 2001, ending in 2004.

- **State Incentives:** States are required to consider a variety of incentive programs to help promote AFVs, including: use of AFVs in state fleets; exemptions from high occupancy vehicle (HOV) lanes; special parking in public buildings, at airports, and other transportation facilities; public education; and exemptions from certain state transportation taxes. DOE is required to issue regulations establishing guidelines for comprehensive state plans within one year of enactment of the bill. Although there are exemption provisions (and states can adopt alternative programs that achieve the desired emissions reductions), federal funding of state programs is tied to the states' level of activity and adoption of an AFV program.

- **Low Interest Loan Program:** A low interest loan program by the federal government to fund the incremental costs of vehicle conversions and purchases (including non-road vehicles) is authorized. **Importantly for NGVs, preference will be given to repayment schedules that enable the loans to be repaid by the borrower from the cost differential between gasoline and the alternative fuel.** $25 million for each fiscal year 1993, 1994, and 1995 is authorized. The program will begin in late 1993.

- **Exemptions from Regulation of Vehicular Natural Gas:** The legislation allows non-utilities (such as gasoline retailers) to sell natural gas at the fuel pump without being subject to regulation as a utility. The legislation codifies a FERC rule providing an exemption from the Natural Gas Act for non-utilities selling natural gas as a vehicle fuel; provides an exemption from the Public Utility Holding Company Act (PUHCA) for the sale of VNG; and preempts state laws in effect prior to January 1, 1989, with respect to regulation of VNG. These provisions overcome substantial regulatory hurdles that have blocked fuller development of the natural gas fueling infrastructure.

- **Urban Buses and School Buses:** DOE is authorized to enter into cooperative agreements and joint ventures with municipal, county, or regional transit authorities to demonstrate the commercial application of alternative-fueled mass transit buses in urban areas with populations of 100,000 or more. Private firms willing to provide cash or in-kind assistance may also participate. Federal funding will be provided for 80% of the project costs, and private or municipal sector funding is required for the balance.

  Funding for **dedicated** alternative fuel school buses will be made available for the incremental costs of new purchases, conversions (that meet warranty requirements), and for fueling stations.

  Authorized funding for the bus programs (but yet-to-be-appropriated) is $30 million for each of the fiscal years 1993, 1994, and 1995.

- **Non-road Vehicles:** Legislation requires a comprehensive study by DOE of whether the use of alternative fuels in off-road vehicles would reduce reliance on imported energy sources. The study is to include specific recommendations on increasing the use of alternative fuels in non-road vehicles.

- **Public Information:** DOE is required to disseminate information regarding the costs and benefits of AFVs for the purpose of promoting the use of alternative fuels.

State Legislative Initiatives

Developments in state-level NGV legislation are moving rapidly, as many local officials begin to look at alternative fuels, and NGVs in particular, as a partial but important solution to air pollution problems around the country. Legislation for alternative fuel vehicles and specifically natural gas currently exists in 26 states.

Regulatory Activities

One of the major objectives of the NGV Coalition since its inception in August 1988 has been to remove regulatory barriers that impede and/or prevent natural gas vehicles from competing in the open market with other fuels. Specifically, the Coalition has been working toward the development of an emission standard by EPA that recognizes the reduced ozone-forming hydrocarbon emissions -- as much as 85% -- from NGVs. The importance of the emission standard is twofold: 1) it will provide a target standard for original equipment manufacturers such as Ford, GM and Chrysler to build NGVs; and 2) it allows NGVs to qualify for substantial CAFE credits under the Alternative Motor Fuels Act (AMA) of 1988. NGVs can qualify for 130 miles per gallon under the AMA formula and, as such, will help auto manufacturers achieve fuel economy goals.

Additionally, the Federal Energy Regulatory Commission announced its decision to remove regulatory barriers relative to the sale-for-resale of natural gas, allowing non-utility companies to sell compressed natural gas at a public station without having to qualify as a gas utility.

NGV TECHNOLOGY

OEMs -- the 'Big Three' automakers in Detroit as well as engine manufacturers and vehicle body-builders -- are taking a more active interest in NGV developments. Legislation in the CAA and NES, as well as growing interest in the states is helping create a market opportunity for NGVs to which vehicle manufacturers are beginning to respond. Ford, General Motors, and Chrysler all have NGV development programs and will begin selling and field-testing larger quantities of dedicated NGVs this year than ever before. Engine, chassis, and vehicle body manufacturers all will be appearing with new market entry products in 1992.

- GMC Truck has been among the most aggressive with their early-market- entry, dedicated-natural-gas Sierra pickup truck. They will sell more than 2,000 vehicles in 1992 to utilities, their customers, and the federal government. These early-entry vehicles do not include, however, fully optimized engines, but the overall vehicle design is ready for a full scale launch into improved, dedicated vehicles that will meet the tougher emission standards well in advance of the CAA schedule.

- Ford is firmly committed to a multi-fuel alternative fuel program, looking at methanol (with a strong fuel-flexible vehicle (FFV) alcohol program), and an NGV van/pickup truck program. Market entry targets are not generally publicized, however, Ford is field-testing its first 50 dedicated Crown Victorias in 1992 using gas utilities as their partners.

- Chrysler was first-to-market with their medium-duty passenger van, two of which were sold to the General Services Administration (GSA) as part of the Federal Fleet Program developed from the 1988 Alternative Motor Fuels Act. Chrysler also is moving forward with the development of a natural gas mini-van, which is the most popular of their vehicle offerings.

Engine manufacturers such as Hercules, Cummins, Caterpillar and Detroit Diesel Corporation are developing and, in some cases demonstrating natural gas engines in a variety of applications, including school buses (in California and West Virginia), urban transit buses (there are more than 50 today on the road), and delivery vehicles. Chassis manufacturers such as Crane Carrier are forming alliances with these engine manufacturers, who also are using Thomas Built and Blue Bird school buses as well as Flxible and Bus Industries of America (BIA) transit buses operating on natural gas. Much of this work, but not all, has been funded by the Gas Research Institute (GRI).

The NGV Coalition also is moving forward with an aggressive five year NGV standards development program that includes creating standards for CNG cylinders (completed in March 1991); standards for compressors, compressor stations, fueling connectors (using a Sherex profile); fuel dispensers, and fuel composition. Fuel composition, in particular, will be an issue of major importance not only to the gas industry, but also to the auto and engine manufacturers who will be warranting their equipment with an eye to the quality of fuel that will be used in these vehicles and engines nationwide.

NGV MARKET DEVELOPMENTS

Also significant in 1992 is the increased development of the NGV fueling infrastructure. Not only are there more natural gas utilities installing compressor stations for customers, but petroleum companies now are taking an active interest in participating in the public sale of compressed natural gas. Amoco, Shell, Phillips, Texaco, Fina, Diamond Shamrock and Unocal all are working with utilities in cooperative arrangements to bring CNG to the public and fleets. With about three fueling stations being installed each week (there are about 550), the fueling infrastructure should expand significantly in 1992.

Other notable marketing activities include the following:

- Pacific Gas & Electric is planning to convert 25% of their local metropolitan area vehicles -- about 250,000 vehicles total -- to natural gas by the year 2000.

- Conversion and/or purchase of as many as 750,000 NGVs in Texas by the year 2002 as a result of the aggressive alternative fuels law. This is being accompanied by the opening of fueling facilities in Texas, including by Atmos Energy, Southern Union Gas, and Lone Star.

- Installation of as many as 50 stations by SoCal Gas in their service territory in the next several years.

- Expansion of the UPS NGV fleet program, to as many as 2,`000 vehicles.

- Expansion of the U.S. Postal Service fleet, the oldest of the federal government fleet programs.

- Significant expansion of the service station network in the New York area, by Brooklyn Union Gas (installing as many as 18 stations) and Consolidated Edison (opened recently a public fueling station in Manhattan).

NGVs -- INTO THE FUTURE

In the United States the proverbial 'chicken and egg' syndrome is becoming a metaphor of the past. With OEMs now moving toward production model NGVs and utilities and petroleum suppliers accelerating their development of fueling stations across the country, the prospects of successful NGV penetration grows brighter. Coupled with an aggressive Clean Air Act and the potential of the National Energy Strategy, as well as a more active state alternative fuels program, the prospects of an improving legislative and regulatory environment also looms likely. Federal and state governments will purchase NGVs, which will result in more production. More NGV production will motivate the development of the fueling network. And so the chicken gains enough strength and momentum to crack his shell, pop out, and walk about in the fresh air and sunlight.

But continued vigilance, perseverance, commitment, and creativity on the part of the NGV industry participants is critical. For the U.S. gas industry and NGV industry, we are attempting not only to commercialize a 'new' technology, but also are attempting to create a new market, with all the associated codes, standards, and regulations that demonstrate the economic viability, safety, and environmental benefits of natural gas for the transportation market.

But competition from reformulated gasoline, and 'cleaner' diesel with particulate traps will challenge the U.S. NGV industry. It is unclear, however, whether the liquid fuels will meet the most stringent U.S. and California emissions standards, and how much the new 'clean' liquid fuels will cost. Only an unrelenting push toward full commercialization will provide the impetus to capitalize on the opportunity created by organizations such as the NGV Coalition, the American Gas Association, the Gas Research Institute, and the host of aggressive, forward thinking utilities who have already devoted much to marketing NGVs.

The challenge for commercialization still looms ahead; the timing is right; the issues -- environmentalism, domestic fuel use vs. an imbalance of payments, and improved economics for fleet operators (not withstanding the requirement for up-front capital investments) -- will help make natural gas the premium fuel for vehicles in the very near term and far into the future.

CHAPTER 11 -- ENERGY STRATEGIES

Biomass Energy: A Summary of Environmental Benefits

Keith J. Fritsky
Project Engineer
Environmental Protection Agency
Triangle Park, North Carolina, U.S.A.

Background:

Having studied Mechanical Engineering at Drexel University, Mr. Fritsky has been working primarily on biomass-to-energy projects for the last year and a half in his position with the Environmental Protection Agency.

THE ROLE OF BIOMASS ENERGY IN STABILIZING GLOBAL CLIMATE CHANGE

Keith J. Fritsky
U. S. Environmental Protection Agency
Air and Energy Engineering Research Laboratory
Mail Drop 63
Research Triangle Park, NC 27711, USA

ABSTRACT

The causes, effects, and options for stabilizing global climate change are discussed with an emphasis on the global use of biomass energy as a feasible stabilization option. The mechanism by which biomass energy reduces emissions of carbon dioxide is discussed along with characteristics which make it an attractive energy supply option, particularly when biomass fuel is used to produce electricity. The role of biomass energy both in the present and future is discussed as are the potential barriers to its more widespread use.

GLOBAL CLIMATE CHANGE: CAUSES, EFFECTS, AND STABILIZATION OPTIONS

As a result of human activity, the atmospheric concentrations of greenhouse gases are increasing. For example, over the period from the industrial revolution to the present day, the concentration of carbon dioxide (CO_2), a major greenhouse gas, has risen from 280 ppm to 355 ppm [1]. Greenhouse gases are believed to create a warming effect by resisting the outward flow of infrared radiation more effectively than the inward flow of solar radiation. About 57% of the projected warming over the 1980-2050 timeframe is attributed to CO_2, making it the largest contributor [1]. The Intergovernmental Panel on Climate Change (IPCC) in their 1992 report estimates that, since the late 1800s, the global mean surface temperature has increased about 0.3°C [2]. The IPCC also estimates the potential global warming effects: a 0.3°C per decade increase in temperature and a 6 cm per decade rise in global mean sea level during the next century for a "business as usual" scenario [2].

In the case of CO_2, the most important source globally is fossil fuel combustion. Combustion of fossil fuels and the associated greenhouse gas emissions will increase globally with the growth in energy demand, which is estimated to be 2.4% per year between 1988 and 2005 [3]. Broad strategies for stabilizing global climate change brought on by an increase in greenhouse gas emissions from combustion sources include: (1) Increasing energy conversion efficiency and conservation associated with fossil fuels; (2) Utilizing non-fossil energy sources such as wind, solar, hydro, and biomass; (3) Utilizing fuel substitution such as replacing coal with natural gas; and (4) Sequestering carbon via reforestation.

BIOMASS ENERGY

Benefits

The use of biomass to produce energy in the form of heat, steam, electricity, or liquid transportation fuels reduces CO_2 emissions when used as a substitute for fossil fuels. If biomass is grown or harvested for energy at an amount equal to that burned for a given period, there would be no buildup of CO_2, because the CO_2 released in combustion is compensated for by that absorbed during photosynthesis. For example, when biomass is used renewably as a substitute for coal, 200 lb CO_2/ million Btu (86 kg/GJ) is prevented from entering the atmosphere.

Biomass energy has some important benefits in addition to being a viable global climate change stabilization option. Cultivation of biomass "energy plantations" would complement reforestation efforts by utilizing existing forested and agricultural lands which are "closed" to reforestation. As a result of revenue from the sale of electricity and liquid transportation fuels such as ethanol and methanol, the net cost of offsetting CO_2 emissions by substituting biomass for fossil fuels could be near zero or even negative, making biomass easier to implement than fuel substitution or energy efficiency improvements [4]. Unlike other renewable energy technologies, biomass energy needs no storage capability which tends to increase the cost of implementation. In addition to mitigating global warming, biomass energy has other environmental benefits including the reduction of biomass wastes and residues going to landfills; reduction of criteria pollutants like nitrogen and sulfur oxides since biomass contains negligible sulfur and has a low fuel-bound nitrogen content; improvement of air quality through reductions in ground level ozone and carbon monoxide (CO) when transportation fuels derived from biomass are utilized; and reduction of volatile organic compounds, CO, and carcinogens through prevention of the open burning of biomass residues.

Current and Future Role

Currently, only biomass residues in the form of wood wastes (mill and forestry residues) and agricultural wastes (rice hulls, sugar cane fiber, etc.), which are available at near-zero or negative cost, are used to produce energy. When energy in the form of electric power is generated from these low-cost biomass fuels, traditional steam-electric power technology is utilized, which tends to be inefficient and capital-intensive at the modest scales characteristic of biomass power plants [5]. These modest scales (less than 100 MW_e) are required as a result of the dispersed, limited supply of biomass fuel, which must be harvested from the countryside and transported to the plant [5]. Therefore, as a matter of economic sense, it is necessary to construct small, decentralized plants near the fuel supply to avoid the high cost of transporting fuel long distances.

The future role of biomass in the total energy supply mix is likely to grow with the development of energy plantations and improvements in energy conversion technologies. The development of energy plantations consisting of short-rotation, high productivity species may provide long-term, renewable supplies of biomass fuel and feedstocks. It is believed that, with further research, future productivities of 20 t/ha/yr can be achieved which would reduce the cost of delivering dry biomass from \$33-50 to \$24-32 per m^3 [6]. These higher cost "energy crops" and harder-to-recover residues are likely to be used for energy as low-cost biomass residues are exhausted. For electric power applications, in order to make the use of higher cost fuels/feedstocks economically attractive at modest scales, technologies are needed that offer high efficiency and low unit capital cost [5]. These requirements are likely to be met by adapting integrated gasification/combined cycle (IGCC) technology to biomass. Net efficiencies for biomass IGCC plants could exceed 50% which would surpass the 20-25% efficiencies of similar scale, biomass steam-electric plants [7]. The unit capital cost for the gas turbine in IGCC plants is relatively insensitive to scale unlike the steam turbine in steam-electric plants [5]. Biomass is a good candidate for gasification, relative to coal, since it contains a high volatiles content and gasifies rapidly. Also, unlike coal gasification plants, biomass IGCC plants would not require a process step to remove sulfur from the fuel gas.

Overcoming Hurdles

Before the above developments are fully implemented, technical, economic, and environmental "hurdles" must be overcome. The largest technical hurdle involves the removal of alkali compounds, formed primarily from sodium and potassium in the biomass, and particulates from fuel gas exiting a biomass gasifier. Such "cleanup" of the gas is required to prevent alkali compounds from entering the gas turbine and condensing on turbine blades, thereby fouling the turbine. These alkali compounds may either condense on particulate or be in the vapor phase prior to entering the gas turbine. Demonstrations of a cleanup system and the biomass IGCC plant as a whole are needed at a small scale and at increasingly larger scales in order to evaluate cost and performance. This offers an opportunity for public/private partnerships to accelerate the development, demonstration, and commercialization of these technologies. Plant owners/operators need to secure long-term supplies of biomass in order to guarantee stable fuel/feedstock costs and availability. Environmental implications of cultivating large-scale energy plantations also need to be addressed; these include maintaining adequate nutrient and carbon contents in the soil, controlling the use of fertilizers, pesticides, and herbicides, maintaining biological diversity, conserving agricultural lands, and controlling erosion.

REFERENCES

1. Princiotta, F. T., "Greenhouse Warming: The Mitigation Challenge," Presented at the 1992 Greenhouse Gas Emissions and Mitigation Research Symposium, August 18-20, 1992, Washington, DC, Session I, 1-24.

2. Intergovernmental Panel on Climate Change (IPCC), "Climate Change 1992: The Supplementary Report to the IPCC Scientific Assessment," 1992.

3. Rosillo-Calle, F. and Hall, D. O., "Biomass energy, forests, and global warming," *Energy Policy*, **20**, 2, 1992, 124-136.

4. Hall, D. O. , Mynick, H. E., and Williams, R. H., "Cooling the greenhouse with bioenergy," *Nature*, **353**, 1991.

5. Williams, R. H. and Larson, E. D., "Advanced Biomass Power Generation: The Biomass-Integrated Gasifier/Gas Turbine and Beyond," Proceedings from the Conference on Technologies for a Greenhouse-Constrained Society, June 11-13, 1991, Oak Ridge, TN, Session I: Technologies - Biomass, 105-158.

6. Hall, D. O. and Woods, J., "Biomass: Past Present, and Future," Proceedings from the Conference on Technologies for a Greenhouse-Constrained Society, June 11-13, 1991, Oak Ridge, TN, Session I: Technologies - Biomass, 83-104.

7. Hall, D. O. , Mynick, H. E., and Williams, R. H., "Alternative Roles for Biomass in Coping with Greenhouse Warming," *Science and Global Security*, **2**, 1991, 1-39.

Catalysis and Clean Environment What Do These Two Have in Common?

Devinder Mahajan
Scientist/Group Leader
Brookhaven National Laboratory
Upton, New York, U.S.A.

Background:

Presently a Group Leader (Catalysis) in the Department of Applied Science, Dr. Mahajan has been at BNL for the last twelve years. His six U.S. patents and over twenty publicaitons in the area of natural gas and coal utilization through catalysis reveal his prime interest in developing novel technologies. Dr. Mahajan also serves as a Technology Transfer Consultant with the Electric Power Research Institute.

CATALYSIS AND CLEAN ENVIRONMENT: WHAT DO THESE TWO HAVE IN COMMON?

Devinder Mahajan,* Nandita Gupta, and Atsushi Kobayashi
Department of Applied Science
Brookhaven National Laboratory
Building 815
Upton, NY 11973, USA

ABSTRACT

The theme of the 1990s, "Continual Expansion through Sustainable Development," is the outcome of interdisciplinary studies undertaken in the preceding two decades to arrest ongoing global environmental deterioration. A critical component of various international efforts to formulate and implement environmentally-sound technologies is the planned involvement of developing countries and countries with economies in-transition which are poised for explosive growth well into the twenty-first century but do not have policies or financial resources to simultaneously manage their environment. Though CO_2-producing, technologies utilizing relatively abundant and inexpensive carbon-based feedstocks will still play a major role on the scene. Related to this, the aim of the ongoing effort at Brookhaven National Laboratory is to develop technologies for enhancing natural gas and coal usage through liquid phase catalysis. A successful application of this concept in the synthesis of oxygenates is described. The potential impact of this activity on methanol-energy, non-petroleum feedstocks development, H_2-energy, CO_2-recycle, and several other areas in relation to CO_2-mitigation, is also discussed.

INTRODUCTION

Since petroleum became a major energy source in the 1900s, the world economy has been dictated by the price of oil. To gain independence from this energy source, a concerted effort was dedicated in the US after the first oil embargo in 1973. In the following years, several potential feedstocks were recognized. But the overall economics of these alternatives were tied to the price of oil which is dictated by politics. In early 1980s, with a glut of oil in the world market, any incentive to utilize feedstocks other than petroleum faded, and several promising technology developments were abandoned by major corporations. It is only in the past few years that revival of this activity is being noticed. The thrust behind this effort is based on **environmental concerns**. In the US, severe pollution problems in several major cities (Los Angeles, Denver, to name a few) prompted the US Congress to pass, in 1990, Amendment to the Clean Air Act. This Act specifically mandates increased utilization of oxygenated fuels and natural gas and decreased use of petroleum.

The trend towards clean environment is growing world-wide. To meet this goal, the utilization of natural gas, a clean-burning gaseous fuel, is being expanded. For example, in Japan where raw materials are limited and the people/land ratio is relatively high, a constant effort is being made to stay at the forefront of technology development to meet the energy needs with minimum environmental impact. An impressive lead taken by the Japanese government is to implement methanol/oxygenates (recent projects funded by MITI through Mitsubishi and Mitsui) and natural gas (one of the biggest LNG projects undertaken in Australia a few years ago). These examples demonstrate Japan's national commitment to introduce state-of-the-art and environmentally-sound technologies to meet energy demands in the twenty-first century. To this effect, several end uses of desired raw materials are being developed.

It is important to mention that utilities, transportation, steel manufacturing, and other industries together emit > 70% of the total CO_2 generated. Therefore, a wise selection and implementation of technologies affecting these targeted areas is of prime interest.

ENERGY STRATEGIES--ROLE OF DEVELOPED COUNTRIES

Though adverse effects of industrialization on environment are now well documented, it is hard to imagine giving up all modern day comforts. It is clear that energy usage will expand as world population continues to grow [1]. In this scenario, it is critical that CO_2-mitigation strategies must involve developing countries which are poised for explosive growth well into the next century but do not have policies or resources to effectively manage their environment. With a theme, **"Continual Expansion through Sustainable Development,"** several international efforts are under way. Discussed below are basic elements of two such efforts.

US Initiative

Due to availability of large scientific talent and resources, a large percentage of novel technologies is still developed in developed countries with the US taking a lead. Earlier this year, in response to management of global environment, the US initiated the ADEPT (Assisting Deployment of Energy Practices and Technologies) program through International Technology Cooperation Branch of the Department of Energy. This program comprises six components (Need Studies, Technical Information, Institution Building, Training, Technology and Practices Adaptation, and Technology and Practices Demonstration) and targets developing countries and countries with economies in-transition to promote wise selection and implementation of energy technologies.

Japanese Initiative

To integrate energy policy with global environmental issues of the 1990s, the Industrial Technology Inquiry Commission recently issued an Interim Report [2]. The following goals are set forth:

- Fundamentals of developing energy and technology that will make continual expansion possible.
- Implement "Greenhouse Effect Prevention Action Plan."
- Provide support to neighboring developing countries in energy/environmental areas.
- Implement "Earth Regeneration Plan."

The outcome of this report is formulation of "The New Sunshine Project." A critical aspect of this project addresses "Development of Broad Area Energy Utilization Network System--Basic Concept of Eco-energy City" [2,3]. The ultimate aim is to reduce overall energy needs by 6% and CO_2 by 9% in Japan. The heart of this topic is storage/transport/utilization of low grade industrial waste heat (< 200°C). To implement this program, two innovative technologies are sought:

- Low temperature liquid phase methanol synthesis/decomposition.
- A workable metal hydride system for reversible H_2 storage.

PRIMARY ENERGY OPTIONS

There are several non-petroleum (non-nuclear) energy sources which are of interest:

- Solar
- Wind
- Geothermal

- Natural Gas
- Coal
- Biomass
- H_2O/H_2

Solar and wind are the most favored but at present, these cannot be utilized on a major scale. Water (from oceans) is abundant but an economical H_2-production process needs to be developed. Natural gas and coal are the most abundant and inexpensive available feedstocks. **A sensible exploitation of these C-based resources is desirable.**

CHEMICAL SYSTEMS FOR SPECIFIC APPLICATIONS--A CASE STUDY

Since Japan meets most of its energy needs with imported feedstocks, there is a pressing need to address environmental issues. A recent study [4] was dedicated to consider six coupled chemical systems for transportation of clean fuels.

1. $H_2O \rightarrow H_2 + \frac{1}{2}O_2$
   $H_2 + \frac{1}{2}O_2 \rightarrow H_2O$

2. $3H_2 + CO_2 \rightarrow CH_3OH + H_2O$
   $CH_3OH + H_2O \rightarrow 3H_2 + CO_2$

3. $2H_2 + CO \rightarrow CH_3OH$
   $CH_3OH \rightarrow 2H_2 + CO$

4. $2H_2 + HCOOCH_3 \rightarrow 2CH_3OH$
   $2CH_3OH \rightarrow 2H_2 + HCOOCH_3$

5. $3H_2 + N_2 \rightarrow 2NH_3$
   $2NH_3 \rightarrow N_2 + 3H_2$

6. $3H_2 + C_6H_6 \rightarrow C_6H_{12}$
   $C_6H_{12} \rightarrow 3H_2 + C_6H_6$

For Case 1, H_2 was assumed to be generated via water electrolysis. For electric power generation, the cost calculation followed the order: $H_2O/H_2 > CH_3OH/CO_2 > CH_3OH/HCOOCH_3 > C_6H_{12}/C_6H_6 > NH_3/H_2 > CH_3OH/CO$, with the lowest and the highest cost differing by a factor of two (assuming 4,000 MW power plant and fuel transportation distance of 5,000 km). Since systems containing large quantities of ammonia (Case 5) and benzene (Case 6) may have a problem gaining environmental acceptance, a preference for methanol-energy (Systems 2, 3, and 4) is clearly demonstrated.

It is important to review general criteria used to favor non-petroleum feedstocks for environmental applications. Table 1 lists energy densities [5] and CO_2 emission (per fuel heating value) [4] for selective fuels. The emissions decrease in the order: coal > gasoline > methanol > CH_4 > H_2. H_2 is assumed to be generated through water electrolysis [6,7] and methane is in the form of LNG. The picture changes when H_2 is produced from fossil fuels and the total process (mining → combustion) is taken into account [4]: under this scenario, CO_2 emissions follow: H_2 ($\geq$ 1.0) > methanol (0.8) $\geq$ LNG (0.65 - 0.85). From the energy density point of view, H_2 and CH_4 appear favorites but technical problems associated with storage need to be resolved [8]. Therefore, liquid methanol is of immense interest.

ROLE OF CATALYSIS

Considering that "Catalysis" accounts for > $2 trillion worldwide product stream, it is not surprising that this topic is a subject of several recent reviews [9,10,11,12]. But of interest here is application of catalysis to environmental problems. It is envisioned that CO_2-mitigation could be achieved through more energy-efficient processes. For example, in a recent catalysis meeting, of the 10 critical areas identified for future development, **3 involve methanol and other oxygenates** [13]. These are:

1. Direct methane conversion to methanol using low temperature liquid phase catalysts.

$$CH_4 \rightarrow CH_3OH \tag{1}$$

2. Since > 40% of all manufactured organic chemicals utilize ethylene feedstock, oxidative coupling of methanol to ethylene is highly desirable.

$$2CH_3OH \rightarrow C_2H_4 \tag{2}$$

3. The Anti-Markovnikov addition of H_2 and NH_3 to unsymmetrical olefins to selectively produce primary alcohols and primary amines respectively.

$$C = C \xrightarrow{H_2O/NH_3} ROH/RNH_2 \tag{3}$$

Obviously, catalysis is poised to play an important role in development of energy technologies.

ONGOING BNL CATALYSIS EFFORT

Given the criteria set forth in diverse applications, it is convincing that methanol and other oxygenates will play an important role in the application of clean-energy technologies. It is, therefore, critical that more energy-efficient technologies to synthesize methanol and other oxygenates be developed. Coincidently, a need to develop such catalysts was realized several years ago at BNL. Since Natural Gas and Coal are the most abundant and inexpensive carbon-based non-petroleum feedstocks available, a wise development of these resources is of interest.

Since economical processes for direct conversions of coal (direct liquefaction) and natural gas (direct methane conversion) are yet to be targeted for commercial applications, the BNL effort is dedicated to synthesize oxygenates via the "Indirect Route," i.e., through synthesis gas. The stoichiometry of synthesis gas, produced from these two sources, is of interest:

$$\text{Coal} \xrightarrow{\text{Gasification}} CO + xH_2 + CO_2 \ (x < 1) \tag{4}$$

$$CH_4 \xrightarrow{\text{POX}} CO + 2H_2 \tag{5}$$

$$CH_4 \xrightarrow{\text{Steam Reforming}} CO + 3H_2 \tag{6}$$

Thus, the H_2/CO ratio varies from < 1 for coal-derived syngas to 3 for syngas from steam-reforming of natural gas. In order to maximize C utilization, the BNL effort is dedicated to a One-Step synthesis of identified products of commercial interest without complications of other possible secondary reactions that normally accompany during synthesis of these products. Thus, the Catalyst-By-Design

(CBD) approach for synthesis of methanol and higher oxygenates is ongoing under the "BNL Catalyst Development" program. The central theme of this effort is to achieve:

- ideal syngas source/product match
- favorable thermodynamic parameters
- high activity
- enhanced product selectivity
- liquid phase performance
- one-step synthesis
- low temperature and pressure operation
- better process economics

It is noteworthy that this "Homogeneous Catalysis" approach is finding a niche in the commercial domain of heterogeneous catalysis [14,15,16].

<u>LLTMeOH Technology</u>

The methanol synthesis reaction is commercially catalyzed by a Cu/ZnO heterogeneous catalyst [17].

$$CO + 2H_2 \overset{\text{Cat.}}{\leftrightarrow} CH_3OH \qquad \Delta H^\circ_{298} = -21.6 \text{ kcal.mol}^{-1} \tag{7}$$

The following disadvantages:

- large capital investment
- < 20% conversion per pass requiring gas recycle

render the present technology unsuitable for several desirable applications which could potentially expand the role of methanol.

A successful application of the aforementioned BNL concept utilized thermodynamically favorable parameters to design a series of catalysts which synthesize methanol in liquid phase with > 95% selectivity and > 90% per pass conversion with balanced gas. This process requires <u>no recycle</u> and the catalyst is <u>inert to CH_4 and N_2</u>. Partial oxidation of natural gas with air is envisioned making the overall process of methanol synthesis from natural gas economically attractive. Summarized below are salient features of the BNL LLTMeOH technology:

- The liquid phase catalyst employed is inexpensive and easy to prepare.
- Liquid medium provides better heat control.
- Methanol (product) is the main solvent used in the process.
- The system operates efficiently between 50° to 150°C at pressure as low as 0.5 MPa.
- A > 90% per pass conversion is routinely achieved.
- The process is highly selective (> 95%) to methanol with methyl formate generated as a byproduct.
- At low levels (< 0.1%), the catalyst is tolerant to sulfur.
- Since the catalyst is inert to N_2, syngas manufacture with air (instead of O_2) is envisioned which eliminates the need for an O_2-separation plant.
- An EPRI-sponsored study indicates that at $1/MMBTU natural gas, methanol could be produced at ~ $0.20/gallon.

The LLTMeOH technology has been patented [18,19] and is the subject of several publications and presentations [20,21,22].

A comparison of operating conditions of methanol process (BNL LLTMeOH versus commercial) is shown in Table 2. Initially, the importance of BNL technology for the following applications was realized:

- Coupling with the Integrated Gasification Combined Cycle (IGCC) System which requires a Once-Through-Methanol (OTM) process.
- Remote natural gas transport where a skid-mounted unit will be economical.

Table 3 compares energy delivery efficiency of BNL LLTMeOH with transported LNG. Given the advantages associated with the BNL approach, several potential environmental applications of methanol synthesis via this technology are being identified.

<u>LLTMF Technology</u>

As emphasized earlier, one of the goals of BNL approach is to maximize energy efficiency. Realizing that,

- Syngas derived from natural gas is ideal for methanol synthesis ($H_2/CO = 2$ via POX).
- Syngas from coal is always CO-rich ($H_2/CO < 1$).
- Alcohols, higher than methanol, normally produce H_2O as a byproduct.

Methyl formate (b.pt. = 31.5°C; m.pt. = -100°C; F.pt. = -19°C) was identified as a product of interest due to the following reasons:

- Storable liquid
- Miscible with alcohols
- Potential as a fuel <u>if</u> produced cheaply. Presently, this product is a specialty chemical.
- Higher CO utilization ($CO/H_2 = 1$)
- Can be easily decomposed to methanol
- IGCC application

The commercial synthesis of methyl formate is carried out in two steps via methanol (equations (7) and (8):

$$CH_3OH + CO \rightarrow CH_3O_2CH \qquad (8)$$

Recognizing that methyl formate was produced as a byproduct during methanol synthesis via the LLTMeOH process, the BNL methanol catalyst was successfully modified to develop a new liquid phase catalyst for synthesis of this C2 oxygenated product [23].

$$2CO + 2H_2 \rightarrow CH_3O_2CH \qquad (9)$$

This patented process [24] has these advantages:

- One-step synthesis
- Low temperature (50° - 200°C) and low pressure (0.5 - 5 MPa) operation.
- High yields with methanol as a byproduct
- Initial estimate indicates that methyl formate is produced at about the price of methanol.

This potentially inexpensive route is expected to promote utilization of methyl formate as a desirable feedstock for several downstream processes [25]. This BNL LLTMF technology is expected to find distinct applications as environmental concerns continue to grow.

SUMMARY

It is beyond doubt that the 1990s will be the "Decade of the Environment" and several state-of-the-art technologies will be implemented during this period. Due to economics and their sheer abundance, natural gas (near term) and coal (long term) will still play a major role in energy supply. But applications of these carbon-based materials will be dictated by their relevance to environmentally-sound technologies. In this respect, catalysis will play a key role.

This paper describes advantages of new technologies, LLTMeOH for methanol synthesis and LLTMF for methyl formate synthesis, based on the liquid-phase concept and developed at BNL. Several diverse emerging applications of these materials are reviewed. A few noteworthy are:

<u>Methanol</u>

- Fuel (M100/M85)
- Fuel-additive
- Feedstock (MTBE, and other downstream chemical processes) coupled with IGCC system
- Waste heat utilization
- Liquid energy carrier for transport of H_2 (12.5 wt% H_2).

<u>Methyl Formate</u>

- Several applications are very similar to methanol.
- A C2 feedstock

Obviously, new applications are emerging for which these technologies were not originally intended. Now, the challenge is to customize technologies for specific applications. To this effect, participation in planned international efforts to implement CO_2-mitigation technologies will ensure that the ultimate goal of a "Clean Earth" is realized.

THE FUTURE

Looking into the future, it is not unrealistic to assume that if the presently targeted global CO_2-stabilization is not achieved, even further CO_2-emission reduction may be implemented through a Carbon Tax proposed in the US (replaced with a BTU Tax which is replaced with a 4.3¢/gallon Gasoline Tax, as of today!) and elsewhere [26]. This scenario may provide incentive to reduce CO_2 at the emitting source. At BNL, CO_2-recycle through catalytic conversion is being considered:

$$CO_2 \xrightarrow{\text{catalyst}} \text{Oxygenates} \qquad (10)$$

This reaction is endothermic and coupling with H_2 is an option to make the overall process exothermic. This and several other activities, ongoing at BNL, are:

- The CO_2-recycle has specific applications to stationary CO_2-emitting sources. But transport of cheap H_2 to site is required for oxygenates synthesis.
- Technologies related to methanol (MeOH/H_2 cycle) as a liquid energy carrier for H_2 storage are desirable.
- Improvement in kinetics of the Mg/MgH_2 system (7.6 wt% H_2) for development as an energy carrier for H_2.
- A vision of <u>flexible</u> oxygenates synthesis (for specific environmental applications) skid-mounted (for economical transport of feedstocks located in remote areas) plant is to be realized. The

key to this strategy is development of oxygenates synthesis technologies with closely-related chemistry that will allow such integration.
- Incorporation of waste utilization technologies in the overall scheme.

These and other related technologies, being developed at BNL, will ensure that BNL plays a role in addressing global environmental problems now and in the future.

NOMENCLATURE

ADEPT	Assisting Development of Energy Practices and Technologies
BNL	Brookhaven National Laboratory
CBD	Catalyst-By-Design
EPRI	Electric Power Research Institute
IGCC	Integrated Gasification Combined Cycle
LLTMeOH	Liquid Low Temperature Methanol
LLTMF	Liquid Low Temperature Methyl Formate
LNG	Liquefied Natural Gas
MITI	Ministry of International Trade and Industry (Japan)
OTM	Once-Through-Methanol
POX	Partial Oxidation
US DOE	United States Department of Energy

ACKNOWLEDGEMENTS

The authors thank the Electric Power Research Institute (EPRI RP 3214-01) and the Department of Applied Science, Brookhaven National Laboratory (US Department of Energy Contract No. DE-AC02-76CH00016) for financial support. Thanks are also due to Ms. Margaret Bogosian, Deputy Manager, and Ms. Doris Tooker, Coordinator, both of the BNL Office of Technology Transfer for their continued commitment to promote BNL participation in energy-related activities.

REFERENCES

1. World Development Report 1987. Oxford University Press. New York. 1987.
2. Overall Evolution of "The New Sunshine Project." Interim Report issued by the Industrial Technology Inquiry Commission, MITI, Japan. December 1992.
3. Development of Broad Area Energy Utilization Network System Technology-Basic Concept of Eco-energy City. Report prepared by the Agency of Industrial Science and Technology, MITI, Japan. December 1992.
4. Outline of the Activities on the Global Warming. Mitsubishi Heavy Industries, Ltd. Report for MITI. January 1992.
5. B. Bogdanović, A. Ritter, and B. Spliethoff. Angew. Chem. Int. Eng. Ed. 29(3) 223-234 (1990).
6. Proceedings of the Symposium on Industrial Water Electrolysis. S. Srinivasan, F. Salzano, and A. Landgrebe, Eds. The Electrochemical Society, Vol. 78-4 (1978).
7. Solar-H_2 Energy System. T. Ohta, Ed. Pergamon Press. New York (1979).
8. For example: J. Wegrzyn, H. Weismann, and T. Lee. Low Pressure Storage of Natural Gas on Activated Carbon. Annual Automotive Technology Development Contractors' Coordination Meeting. Dearborn, MI. November 1992.
9. G. C. Pimentel. Opportunities in Chemistry. National Academic Press. Washington, DC (1985).
10. National Research Council Panel on "Catalysis Looks to the Future." National Academy Press, Washington, DC, 1992.

11. J. A. Cusumano. Chemistry for the 21st Century. J. M. Thomas and K. I. Zamaraev, Eds. Blackwell Scientific Publications, 1992.
12. J. A. Rabo. Catalysis: Past, Present, and Future. 10th International Congress on Catalysis. Budapest, Hungary. 1992.
13. J. Roth. 13th North American Catalysis Society Meeting. Pittsburgh, PA. May 1993.
14. F. J. Waller. Recent Achievements, Trends, and Prospects in Homogeneous Catalysis. J. Molec. Catal. 31, 123-136 (1985).
15. Symposium on Industrial Applications of Organometallic Chemistry and Catalysis. J. Chem. Educat. 63(3) 188-225 (1986).
16. G. W. Parshall. Trends and Opportunities for Organometallic Chemistry in Industry. Organometallics, 6(4) 687-692 (1987).
17. S. Lee. Methanol Synthesis Technology. CRC Press, Inc. Boca Raton, FL (1990).
18. R. S. Sapienza, W. A. Slegeir, T. E. O'Hare, and D. Mahajan. Low Temperature Catalysts for Methanol Production. U.S. Patent #4,614,749 (1986); 4,619,946 (1986); 4,623,634 (1986).
19. D. Mahajan, R. S. Sapienza, W. Slegeir, and T. E. O'Hare. Homogeneous Catalyst Formulations for Methanol Production. U.S. Patent #4,935,395 (1990); 4,992,480 (1991).
20. T. E. O'Hare, R. S. Sapienza, D. Mahajan, and G. T. Skaperdas. Liquefaction of Natural Gas to Methanol for Shipping and Storage. Intern. Gas Research Conf. Canada. September 1986.
21. D. Mahajan, R. S. Sapienza, and T. E. O'Hare. Bench-Scale/Laboratory Result-Novel Methanol Catalyst. EPRI Proc. of 13th Annual Conf. on Fuel Science and Conversion. Palo Alto, CA. May 1988.
22. D. Mahajan and N. Gupta. Catalytic Reduction of Carbon Monoxide--Selective Synthesis of C1 and Higher Oxygenates. Manuscript submitted to the AICHE Microfiche Center, New York, for the 1993 AICHE Spring National Meeting, Houston, TX. March 28-April 1, 1993.
23. D. Mahajan and R. S. Sapienza. Potential Improvements in the Synthesis of Methyl Formate. Florida Conf. on Catalysis, Palm Coast, FL. May 1988.
24. D. Mahajan, R. S. Sapienza, W. Slegeir, and T. E. O'Hare. Low Temperature Synthesis of Methyl Formate. U.S. Patent #4,613,623 (1986).
25. J. S. Lee, J. C. Kim, Y. G. Kim. Methyl Formate as a New Building Block in C1 Chemistry. Appl. Catal. 57, 1-30 (1990).
26. Chem. Eng. October 1992. p. 27.

TABLE 1. ENERGY DENSITY AND RELATIVE CO_2-EMISSION OF SELECTIVE FUELS[1]

Fuel	Energy Density $\times 10^{-3}$ kJ.Kg^{-1}	CO_2 Emission from Combustion[2]
Coal	29.9	1.0
Gasoline	44.5	0.7
Methanol	21.6	0.65
CH_4	48	0.55[3]
H_2	125	0.0

[1]Data Sources: References [4] and [5].
[2]Relative Ratio
[3]LNG

TABLE 2. A COMPARISON OF METHANOL SYNTHESIS PROCESS CONDITIONS [20]

	BNL LLTMeOH	Conventional[1]
Reactor temperature, °C	110	265
Reactor pressure, psia	150	750
Equilibrium CO conversion, %	94	61
Operating CO conversion, %	90	16
Volume of gas recycle, mols CO/mol prod.	0.11	5.25
Reactor feed, mols CO/mol prod.	1.11	6.26
Overhead gas cooling duty, Btu/mol prod.	4,100	71,000
Separator temperature, for 95% product recovery, °F	163	77

[1]Supp. E. Hydrocarbon Processing, March 1984 (pp. 71-75); July 1984 (pp. 340-345)

TABLE 3. ENERGY DELIVERY EFFICIENCY [20]

	Billion Btu/Day (Higher Heating Value)		
	BNL LLTMeOH	Methanol Conventional	LNG Mixed Refrigerant
Energy Delivered			
Fuel	323	323	323
Electricity	27.5	----	----
Total	350.5	323	323
Energy Consumed			
Process Feed	476.6	487.5	397.3
Process Fuel	16.3	32.8	included
Total	492.9	520.3	397.3
Delivery Efficiency %	**71**	**62**	**81**

Energy and the Global Economy

Geoffrey K.C. Pardoe
Chairman
General Technology Systems, Ltd.
Middlesex, U.K.

Background:

Dr. Pardoe was responsible for some of the first British guided weapons and later started UK space activity. Currently he is Chairman of the international consultancy, General Technology Systems, which he created 20 years ago, specializing in a wide range of disciplines. He is Chairman of the UK Watt Committee on Energy and has a major involvement in global energy affairs.

Dr. G.K.C. Pardoe
Chairman
General Technology Systems, Ltd.
Brunel Science Park, Kingston Lane,
Uxbridge, Middlesex, UB8 3PQ, U.K.

ENERGY AND THE GLOBAL ECONOMY

As the last speaker at the conference, I have the benefit of hindsight of all the excellent papers we have enjoyed already - but with a challenge to say something not already covered! I chose the title of this paper partly to give plenty of freedom of movement in this respect, but mainly because I feel particularly strongly that whilst we must use the best technologies to meet our energy needs, we must have a clear understanding of how we can best apply them to the benefit of the economies of the nations of the world. The conditions in these countries of course vary dramatically - some with prodigious quantities of indigenous resources, but at the other extreme, the unfortunates with virtually no conventional energy resources - but perhaps they have more energy options than they may realise, with the advent of new technologies!

It is the purpose of this address to take an overview of these global issues and needs.

Where have we come from, and where should we go? The industrial countries have a quarter of the world's population, but consume three quarters of the energy. The 'hi-tech man' of today uses a hundred times the energy of his forbears. (fig. 1) World population is projected to be about 10 billion by 2040 - equivalent to adding a new United Kingdom every six months. These are macro figures we all toss around - but at micro-level there are some interesting facts, for example in India a woman uses five times as much energy to cook the evening meal as does a European housewife. We, in the advantaged countries, have more efficient equipment, and we know how to use it.

Which brings me to the vital triad (fig. 2) of Energy, Information and Materials. These are the <u>only</u> resources available to mankind and in the olden days when we needed to make or grow something, we did not care really how much energy was used as long as the job was done and so the triangle was roughly balanced. However, we have begun to think more carefully about economising on materials and energy, and with the great increase of 'information' available, the more intelligent use enables us to economise on our resources and yet finish up with the same product. Of course there are other driving factors such as environmental issues, cost, availability etc. but there is the undisputable trend that "information" equals "savings" when properly applied. So we have to arrange in the future that the Indian women (and a few billion like her) can get an <u>appropriate</u> form of energy using utensil - and know how to use it. This, in conjunction with family planning imperatives in order to keep the numbers down is not easy - but necessary.

There are clearly several influence factors within the whole subject of energy and the global economy, and this diagram reminds us of the main ones (fig 3). I have already touched on one or two - now let us take the problem of population.

The largest national population is China with about 1 billion; I was impressed to find when I visited Beijing for the first time in 1984, that already they had over 600 anerobic digesters working with biomass, in that city, producing energy for domestic and industrial use, and had most impressive surveys - well written in perfect English - of nationwide activity in this sector of renewables. Their surveys covered detailed measured improvements in health and prosperity in villages, once digester units were introduced to replace disease - ridden urine pits (as in India).

I found one interesting policy feature in China - first noted in the space sector it so happens, in connection with the development of remote sensing technologies from spacecraft platforms in orbit. After 10 years or so of public sector investment in their many relevant satellites and analytical techniques, the central Government then said to the appropriate agencies around the country "we have paid you to develop these techniques and equipment - now use them - and demonstrate their benefit if you want any more money". This sharpens the mind, and it was revealing to me to find the number of people approaching me for advice on the ways to do this, to the benefit of the economy. This lesson (appropriately applied) could be used with good effect around the world in the energy and other sectors, where the take-up of ideas is disturbingly low and left largely to the private sector, which could be better. Some top-down pressure is effective but rarely exerted.

What about the supply side? There is the obvious list and I will comment on a few factors, as we have covered the full spectrum in this conference in technical detail. I remind ourselves that, of world-wide energy sources, currently, fossil fuels provide 77% (17% gas, 33% oil and 27% coal) renewables (hydro, wood etc) account for nearly 20% and nuclear nearly 4%. We are told we have 57 years remaining of natural gas, 46 of conventional crude oil and 310 of bituminous coal. Oil could be extended from harder-to-win sources like shales, but at a price. In this, an American consumes 60 barrels of 'oil - equivalent' per annum, a Western European about 25, and an inhabitant of a less developed country (LDC) about 5 barrels. The inequalities persist in different ways.

I do not want to persist in these comparisons, but globally we have no absolute shortage of energy - but a massive problem of even availability. It is not always reflected by numbers either. The world is not always as it seems. In South Africa, ESKOM, the national electricity generation company is the fifth largest in the world. Sitting here in the heart of high-energy USA, that may seem extraordinary, but ESKOM provide most of the power to all the Southern countries of Africa. A few switches flicked off indiscreetly in their central control room and the lights go out in half of a continent! Not generally known, is that that half of Africa has depended on trade and such infra-structure availability from S.A. for many decades.

So we now have another ingredient to contend with in the energy pie - politics. The elements now appear not as a tidy ring, but with many cross-influences. Various of the factors act in conflict, and taken overall, the global situation is unstable, and must inevitably shift with time regarding at least the balance between different sources of energy. Clearly, the more dependence certain nations of the world have on others, the more potential for stress. This was manifestly the case in the oil crisis of the 70's. On the other hand, the more a nation can depend on its own resources - particularly renewables - the less likely the problem of external pressures. Towards that end, we are sitting here today in one of the meccas of hydrogen expertise, worldwide, and one of the obvious benefits as this technology develops, is the ease and availability of water. Even here, with hydrogen, the perceptions vary; some 5 years ago, the Joint Research Centre of the European Community concentrated much work on hydrogen within the context of the total energy cycle. By this I mean the impact of an energy resource from its first availability, to the time beyond its use until any after-effects are finally disposed of environmentally. This diagram (fig. 4) demonstrates the benefit of hydrogen with no after-use costs, compared with, say, any fossil fuel. Massive industrial interest from different European countries was deployed at this time, and Canada and other countries sought to join in this JRC initiative. The British Government however chose to maintain its position that hydrogen was not likely to be important until the year 2030 in spite of this new evidence which suggested to the other participants that 2000 was now a more likely date to watch.

This lack of willingness to consider possible strategies adequately far ahead is a weakness of the British Government (any various others). An obsession that market forces will establish priorities for a country, is a dangerous premise - certainly for medium and long term needs. A British Minister giving the opening speech at a consultation on energy I recently attended in London, stated that the Government had a firm policy on energy but said (with some pride) that they had no plan for implementation and so had avoided falling into the trap of getting a plan wrong! Therefore they would leave planning to market forces. I presumed to say, that if one drew an analogy with industry, that was like saying that a company had a firm policy to make money - but they had no plan as to how to achieve this - except to hope that the marketing department might think of something to sell, and might mention it to the engineers in case they could make it! In such circumstances, a Board of Directors would be thrown out!

I mention this anecdote because I do not equate with inevitable intervention an intelligent and professional forward analysis by Governments and other public sector organisations of options and possible priorities. I come from industry and believe strongly in the power of the market. But in many areas, it requires dialogue and collaboration between public and private sectors to achieve success - particularly in high-capital cost, hi-tech projects where front-end funding is necessary. Such an example is the in-orbit solar power station (SSPS). This, conceived in USA in 1968, has been worked on by some of us around the world as an option for providing major energy inputs. With the roller-coaster history of oil prices the pressure came off such work in the late 70's. However, the lead time to do a 10GW Satellite Solar Power station is about 20 years (15 x 5Km of solar cells, transmitting via a microwave beam from the 22,300 miles orbit via a 1Km dia spacecraft antenna to a 15Km diameter receiver on earth and thence into the grid). 20 years covers four British parliament life-times, and so it is very difficult to get political support for any early commitment even though our studies revealed there are no scientific, technological or economic programme stoppers for such a project. It is a classic example however, of the importance of being aware - just in case! Incidentally, the Canadians have been flying a small aircraft for the last few years, able to be driven by transmitted electrical power from the ground, and the Japanese put up their own aircraft using the same technique in February of this year. We have been able to transmit power at quite reasonably high kilo-watt levels for about 20 years, with about 80% receiver efficiency.

This is really leading up to one main point I wish to make, which is that the energy options worldwide are very fragmented, and whilst they come together occasionally for discussion (as in this excellent conference), on a macro-level I believe there is insufficient overall systems type analysis and synthesis. It should be possible to develop scenarios which take the economic needs of the nations of the world as a stronger driver, and indicate optimum routes to the provision of appropriate delivered energy. I say indicate not impose, as we have no global mechanism for that, and may feel it not appropriate in any case. But there is no excuse not to be informed, with all the aid programmes for developing countries, and the structures available within industrialised ones, options can then be pursued by the more wide-awake countries.

It is worth spending a few minutes therefore on the global structures which are relevant to energy;

We have no cohesive global structure for energy planning. We have special bodies such as the International Atomic Energy Agency which provides an excellent monitoring and regulatory function, but not in an executive role so far as deployment of nuclear power generators is concerned - this is left to individual nations. Some have scared their populations to the extent they have opted not to use nuclear energy, others have continued relentlessly with a nuclear policy, and France for example has 80% of its electric power provided by nuclear plants. Incidents such as Chernobyl and Three Mile Island have caused disproportionate reaction around the world in terms of perceived health hazard to the population as a result of the various global nuclear levels of industrial activity. The facts speak otherwise but with such emotive reactions, it is difficult for administrations to decide and implement a programme of supply which is based on the facts of safety etc. rather than the perception. Britain is reviewing its nuclear policy at the end of this year.

If one looks at the manner in which the world uses expendable oil resources pumped out of the ground, it is arguable that we should not use such a high proportion of this declining resource by burning it off to create heat, or refining it to petrol for transportation etc - but leave that to other resources, and so conserve such mineral feedstock for the petro-chemical industry. This provides at least the opportunity to make permanent, or recycled, materials in the polyurethane, polyethelyne categories etc. Deciding what is the right balance between different energy sources would be difficult enough - but even if one did (and got it right) there is no mechanism to implement this on an international scale. There are developments afoot which in any case can distort the need; one prime example is a British invention of which I am aware, in which excellent quality polyurethane products - competitive in price and quality with mineral derived products - are being made from polyol which is produced by small $3 million plants using feedstock comprising either vegetable, fish or animal oil! These plants have been set up in many countries, and the growth of this process will have an increasingly significant effect on the need for mineral-based polyol - with consequent energy and environmental effects via global applications.

It probably is inappropriate to have global planning - but it cannot be wrong to have global awareness of options, and this is a theme I shall return to later.

The "carrying capacity" of the Earth measured in food supply, evenly distributed, has been computed as 6.5 billion people based on a vegetarian diet - much less on diets containing meat. We are at 5.5 billion at the moment. This sharpens the mind. The growth of population obviously has a direct effect on global energy demand. However, the efficiency with which we then convert and deliver energy, has a massive bearing on the problem and offers one of the most cost-effective solutions to the future demands. The rational use of energy is a prime theme of the UK Watt Committee on Energy which I chair. I am told that the world average conversion efficiency (i.e. available energy converted to kinetic energy or light) is 3%. In Europe it is 5%, but in the USA only 2.5%. In the old Eastern Europe it was and is appallingly bad, and that is a region of the world with massive need for improvement, particularly with regard to ecological and environmental issues, and their reverse effects on energy needs. It is hard work to get across to so many energy users, that a professional and informed approach to meeting environmental standards, need not cost money, but indeed save money when energy-conserving measures are used.

There are some dramatic developments around in terms of reduced energy demand from better engineering. An Australian approach to a new way to energise fluorescent tubes for lighting, saves over half the energy and triples the tube life. The same rationale applied by the same inventor to other electrical appliances also shows massive savings. From another UK inventor, there is now also available a new wire-less, (and therefore mobile) digital communications system about to become operational. This links by radio an intelligent module, capable of monitoring and/or control, into a micro-local or standard-local area network, regional network or global. The power of this new ability to operate group facilities efficiently, is startling in various beneficial applications such as energy, operational performance and so on. When then coupled to new energy saving techniques and devices, the reduced energy demand when integrated on a national scale to cover all situations, is quite dramatic. When so integrated, the numbers are interesting - on the above lighting system alone, if installed throughout say Singapore, it would reduce their national needs by 1½ power stations!

This then is the point. Knowledge is power, and I believe there is need for the world to be better informed - not of just individual opportunities which conferences such as this do so well, but of the total system economic and other implications of the many facets of this fascinating and essential matter of energy. I believe there is much scope for more integrated thinking and rational approach. Access to such information may well encourage users and suppliers to adopt an improved stance, without the presumption of control. When control is appropriate, then the better availability of synthesised data must help considerably.

What are the mechanisms for this? There is no better way to catch the real attention of individuals or organisations than to offer them money or power - and then follows a descending line of ways of appealing to people which can be based on such things as improved environment, improved health and so on - including simply suggesting something should be done because it is right and good for the Community. Supporters appear and gather to support all these reasons - but the most powerful one regretably, is personal gain. I once prepared a document for a Swiss Government Minister, who asked me to help with his problem of dealing with the eternal triangle of Politicians, Industrialists and Scientists, who were in confrontation and not harmony - and giving him a rough time! I developed an approach by starting with a blunt postulation of what each type of individual in his (or her) black heart really wanted if being honest. This could be summarised by saying - a politician wants power and image without needing to persuade people to vote for him, the industrialist wants maximum profit, minimum tax, with least hassle with workers conditions - and the typical scientist wants unlimited funds and facilities for blue sky research, with automatic publishing worldwide of all his works - and no timescales to meet! Each views the other categories with suspicion and so on but the analysis of the thought process is revealing.

If you set down these crude motivations and explore then how each of the trio views each other, it is possible to build up a very effective approach as to how to persuade the other party to meet ones own interests by apparently asking for something different, but really going for what you want yourself.

What do we want ourselves in the energy scene? In realistic terms, I think we would settle for the optimum use of each energy source, involving the very conservative use of hydro-carbons and fossils, so that we stretch out these irreplaceable resources for many hundreds of years, if not indefinitely. This would mean by implication, the maximum use of renewables, with a healthy, balanced use of nuclear. Hydrogen is in a specially favoured category, as whilst not exactly renewable, could be effectively unlimited in supply so depending on how it is derived.

So what stops us meeting these ideals? Well we cannot seem to overcome the problem of cost-effective conversion of renewable energy sources on a major scale to compete with coal, oil and gas, with its (currently) plentiful availability at a price which we have come to accept. I came back to my Swiss story - do the oil and gas companies really want to see renewable technology succeed? Some make gestures, but overall I find it strange that we have in the last few decades, failed to harness renewables in a world which has succeeded in so many other more technologically challenging fields (space etc.). Does the problem arise from the disparate interests and diffused industrial scene or is it lacking some central direction? To what extent are vested interests at fault - if at all? Do we have such an established traditional energy infra-structure that we are locked in too firmly to break into a new pattern? It would be interesting to postulate from where the 11 billion citizens of the world will derive their energy in say 2050 - and even more intriguing to do the same (for the what - 18-20 billion?) in the year 2100? I am a great believer in establishing some idea of such long-term scenarios, which even if inevitably wrong numerically, they can be very revealing about where the short-term situation may lead.

One thing is clear, and that is, by that time in the future, it will all be very different from what it is now but an area ripe for speculation - and reward!

Well ladies and gentleman, I did not set out to wave a wand and offer magic solutions, but rather to provoke some thoughts. I hope I have done that, and thank you for listening.

Human daily per capita energy consumption

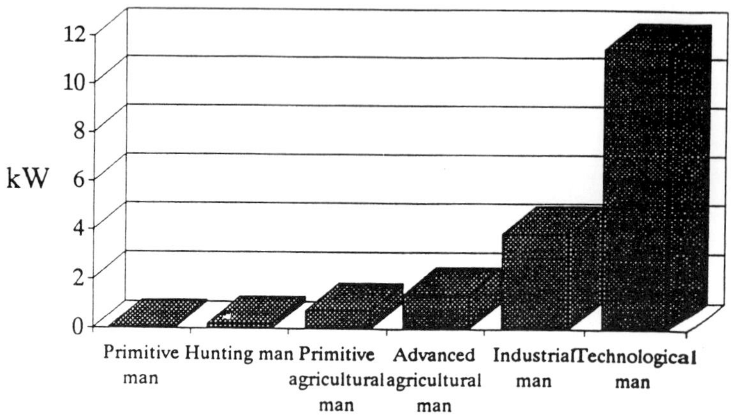

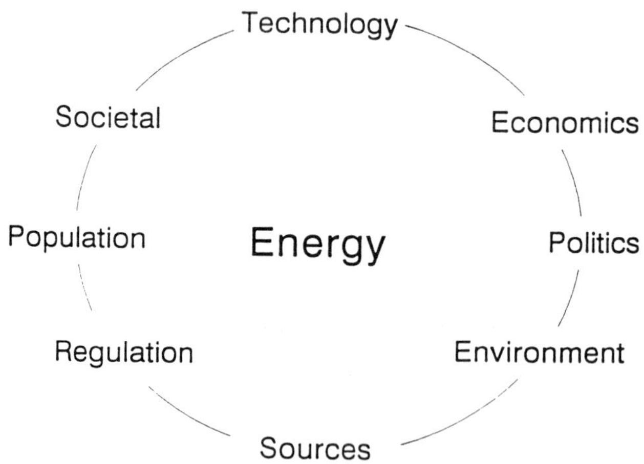

General Technology Systems

CANDIDATE SOURCES

* HYDROCARBONS
* FOSSILS
* NUCLEAR
* RENEWABLES
* SPACE SYSTEMS
* EFFICIENCY

General Technology Systems

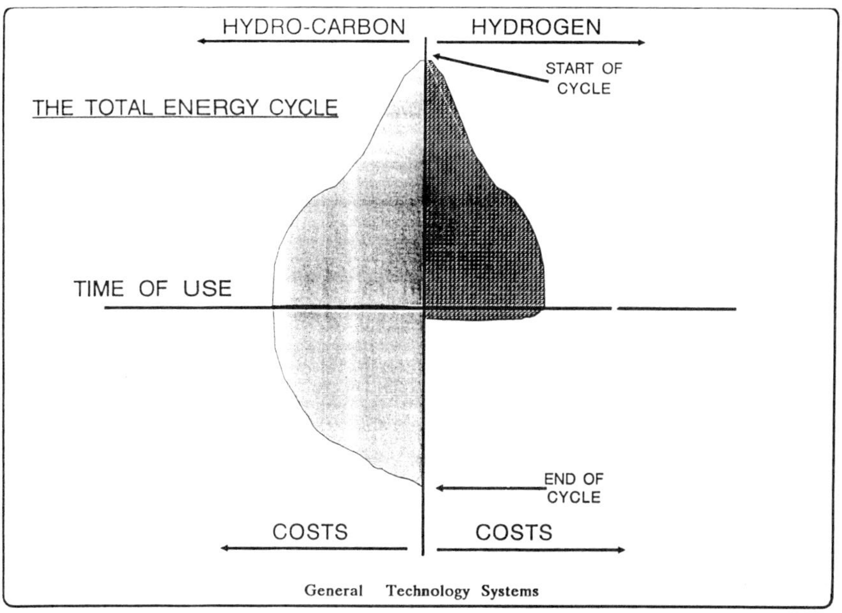

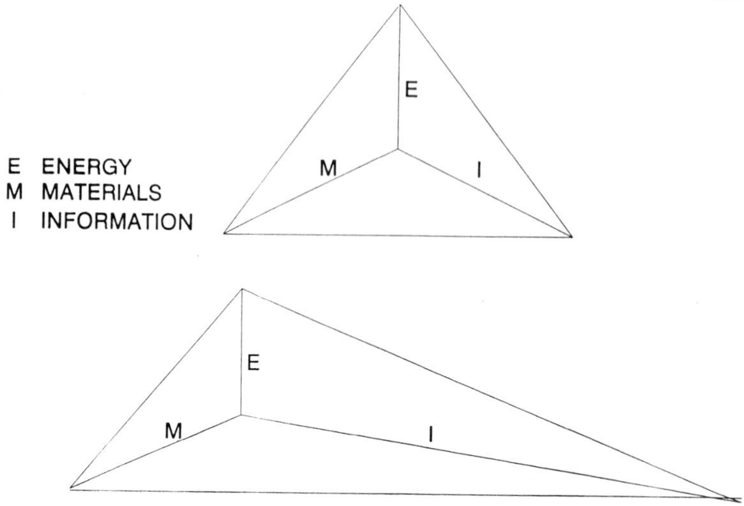

THE TRIAD OF RESOURCES

General Technology Systems

SIMPLE GLOBAL STRATEGY

Optimum use of each resource:-

- o Conservative use of hydro-carbon and fossil,
- o maximum use of renewables,
- o extended but balanced use of nuclear,
- o major use of hydrogen,
- o adequate research on future possible sources.

Ensuring all have minimum adverse environmental effects.

General Technology Systems

CHAPTER 12 -- KEYNOTE ADDRESS

Implementing Our Sustainable Energy Future

Willis Hawkins
Senior Advisor
Lockheed Corporation
Calabasas, California, U.S.A.

Background:

Dr. Hawkins has been Senior Vice-President of Lockheed Corporation and President of Lockheed California, and has played a major role in the design and development of airplanes, missile systems, and space vehicles. He has held managerial responsibility for extensive space and aircraft programs at Lockheed for over fifty years, and was awarded the National Medal of Science in 1988 by President Ronald Reagan.

IMPLEMENTING OUR SUSTAINABLE ENERGY FUTURE

Dr. Willis M. Hawkins
Senior Advisor
Lockheed Corporation
4500 Park Granada Blvd.
Calabasas, CA 91399

After a lovely introduction like that I must conclude I am dead. I am really quite alive and I have enjoyed this superb conference. The international character of it, the quality of the people that are here is impressive beyond belief. I am a little humble about trying to make any kind of a contribution myself. However, when I take a look at it, I have had a long history worrying about how to use hydrogen in aircraft. I would like to talk to you about that. I am impressed by a new public emphasis on the environment that I believe is quite strong compared to what it used to be. I would like to talk about that because I think some of it is not quite rational, but it is providing a window for something new to be done. I have a few suggestions on how we might be able to create a hydrogen economy, and this conference has shown that we are pretty close to being ready.

Let me go back over some of my hydrogen history at Lockheed. It is not very well known, but in the late fifties ---- nearly forty years ago, we had a rather interesting project at Lockheed. We had built the U2 which, bless our hearts, performed admirably for our nation. We, at the time, were able to fly over belligerent nations without anybody worrying about us very much. We were reasonably sure, however, that if those nations thought about it a while they would figure a way to counter the U2, which they did. To avoid this we started a supersonic U2. It was supposed to do the same job but much faster, which would make it a little tougher for anybody to knock us down.

The supersonic U2 was not easy because it had to go a long, long way. The high altitude was not all that bad, but the required range was not easy. We finally discovered that the only way we were ever going to accomplish the mission was to use liquid hydrogen as a fuel. This was forty years ago at Lockheed.

In great secrecy we started the program, and we had it pretty well along. As a matter of fact, the Pratt and Whitney Co. had the engine running in their Florida Laboratories. We had a full scale tank and fuel system developed with all of the functioning cryogenic elements , including all of the insulation on all the pipes and tanks, and we had the fuel system running: the pumps were working. We built a kind of bomb shelter to do this in with a very weak roof, the idea being that if we did not really know what we were doing and had an accident it would at least go up instead of sideways.

We began to learn. The fuel system was running quite well. We did have adventures. We thought that one of the things we really had to figure out was how we kept this cryogenic fuel at low temperature while we were going at a speed of mach 2 or over, which meant that the outside of the airplane was pretty warm. We did not have all that much money and so we built ourselves a wood shell around the fuel tank and all of the pumps, in order to heat up the surrounding air. We got a little careless. The heating system that was supposed to heat up the space between the tank and this wooden shell wound up setting fire to the shell. We had all sorts of fire extinguishers around and we tried to put it out. It did not work. So we had to call in the local fire department. The fire department at the airport came storming in and sprayed everything. Of course strange things began to happen at this point. They put the fire out, but the water hitting the cold tank walls caused a fog. You should have seen it. The room was about half the size of this auditorium, and the fog rose up to about five feet. There were nothing but heads around the room. Nothing blew up, and we shooed out the fire people because they were not cleared. They never did find out what we were doing. They thought we were making some sort of system for freezing food, and they finally went away. When we finally put things back in order, our fuel system was still running. Everything was going fine. At that point we became a little more relaxed about how to handle liquid hydrogen.

As the development proceeded we realized, along with the Air Force, that we had not thought the total system through. We were not going to build very many airplanes, maybe a dozen. The Air Force had decided that there should be fifty-seven bases from which this airplane was to be operated. Adding up all of the dollars for liquefaction facilities at fifty-seven bases around the world, and some of them were in pretty remote places, the dollars were staggering. The whole program died on the dollar vine. But we did learn about liquid hydrogen nearly forty years ago.

Then came the fuel crisis of the seventies. Everybody worried about what to use for fuel. We dragged out all of our information and suggested to NASA that some studies should be done on how to make liquid hydrogen fueled aircraft of a different variety than a Supersonic U2. We obtained several study contracts, and the results surprised us. One study concerned subsonic transports and we evolved a beauty. We also studied supersonic transports, but hydrogen did not solve the problem of the sonic "boom". So that was an exercise in futility. We even had a hypersonic transport which we thought would fly high enough so that the "boom" would not give the populated areas of the world trouble. This design had a lot of the elements of what eventually became the aerospace plane that you read about in the newspapers today. This has not gone very far and probably will not until the program is reoriented to focus on the power plant itself.

The subsonic transport had so much promise that we decided to tackle the challenge of how to certify such an airplane through the FAA. We concluded that we could create an experimental cargo airline that would fly internationally with a converted current airplane, that used liquid hydrogen. Of course we selected for conversion a Lockheed L1011 transport. We were going to fly from Pittsburgh to somewhere in Canada -- we did not quite get that sorted out -- then Birmingham and Frankfurt and Riyadh in Saudi Arabia. The reason we picked these terminals was to obtain a source for hydrogen. We were going to have coal to hydrogen conversion in Pittsburgh, coal to hydrogen in Birmingham and Frankfurt, and in Riyadh we were going to use natural gas which they traditionally burned off. In Canada we were going to use hydraulic power sources in Nova Scotia and deliver the liquid hydrogen from there. The concept was wonderful and almost every nation agreed that this was a good idea. Canada, Great Britain, Germany, and Saudi Arabia literally committed the money to do their part to create facilities. We fell apart in the United States. The plan was for the United States to support modification and certification of four of these airplanes. With such an operating cargo airline we could have proven that hydrogen as a fuel would be safe and operationally suitable. We never managed to sell that to the United States.

The L1011 had a conclusive set of characteristics. This was not an optimized airplane for liquid hydrogen. We used as a design base the long-range L1011, which with its full payload range capability weighed 515,000 pounds. We modified it (on paper) to fly on liquid hydrogen. We had to increase the length of the fuselage 65 feet. This made a pretty long airplane for the amount of wing. Even in spite of that the airplane was capable of carrying the same payload for the same range. But, instead of weighing 515,000 pounds, its takeoff weight was only 390,000 pounds -- over a 100,000 pounds lighter for the same payload and range. That was not an optimum airplane. It would have been a fine program and we would know a lot more today about how to design and operate liquid hydrogen aircraft.

While we were at it we tried some other things on the drawing board. I say "on the drawing board". I am a very old man but I have a computer. It has got a software on one end and it's got a printer on the other end and in between it has the computer body. It is called a pencil. So back to the drawing board. We designed a cargo sea plane. The Lockheed Georgia division had done studies of how to make a bigger airplane than the C-5. The C-5 was pretty big in those days, and still is. We discovered a very interesting fact. You can make a bigger airplane, but it would fall through the pavement on nearly all of the airports. The runways and the taxiways particularly, the dimensions of the airplane compared to the gates and all of the service facilities precluded making something that was much bigger than a 747 or a C-5. We decided to try a sea plane. We eventually evolved a two hull sea plane, liquid hydrogen powered. It was a beaut. This transport carried 700,000 pounds of payload. It carried it, not 5,500 or 6,000 miles like the C-5, it carried its payload over 12,000 nautical miles. The idea was to carry a very heavy payload of cargo and to wherever you wanted to go (which was usually 5,500 - 6,000 miles), unload the payload, reload the airplane and fly back home without refueling. This minimized the number of facilities for liquid hydrogen. It was a fine, fine airplane -- a big, two-hull sea plane. We put all of the technology into it we could think of. I think it was a feasible concept and it had a cost per ton mile about half of what you could do with anything else.

Another design exercise involved our old U2 problem. By this time we had the SR71 flying and it was proving itself to be nearly invulnerable, but we decided to see if we could do better. The idea was to make an airplane that flew high enough that it could reconnaissance over somebody's land, but did not really have to violate the nation's borders. It was also hoped that such an airplane could stay up for a long, long time. We found that such a design, done properly with liquid hydrogen, makes an airplane that will hover around 85,000 feet and stay in the air about a week and a half.

That poses a certain amount of difficulty for the crew. In fact one does not need a crew. Such an airplane can be operated very much like a satellite. Send it up and tell it what it is supposed to do, or change the rules while it's there, and let it transmit back to you what it is seeing or listening to.

This concept is intriguing, but there is a social problem that has to be meet. The base from which such and airplane operates is probably close to some concentrated civilization. About the time that the population discovers that this very large airplane -- it has a very long wing span -- has no one on board, (an unpiloted flying overcast), it is predictable that social problems will emerge. To cope with this, we created a cockpit for a backup pilot in order to keep people happy on the ground. We equipped this cockpit with all sorts of amenities, everything but a bathtub and lots of mystery stories, because the pilot did not have a darn thing to do except to be there when it landed. This is a useful, practical project and one of these days maybe somebody will do something about it.

The upshot is that a liquid hydrogen airplane has a lot of benefits. It is really quieter than most, because for any given payload and range it does not take nearly as much power. Therefore it is quieter on the airport. It has absolutely no pollution. It has some perception problems. It has a perception problem on cost, because some kind of infrastructure is necessary to deliver the liquid hydrogen at a reasonable cost. It has another problem and that is the perception of safety. This, unlike the cost problem, is a perception that is erroneous, but we will need some program of demonstration before this perception is erased.

NASA did not really trust us to evaluate safety because we were pretty enthusiastic about hydrogen. They did give us a contract to evaluate safety in potential crash situations for a liquid hydrogen airplane, but they also gave a study to Arthur D. Littles, to get two views on the subject. Both of us came to the same conclusion, that liquid hydrogen is probably the safest fuel for a transport airplane. I could go on into great detail, but the fact is that it is safe.

This is supposed to be a joke. I can predict one problem with the environment. As soon as we start flying a hydrogen airplane all of this lovely exhaust which is nothing but water vapor is going to turn into trails in the sky. I can guarantee that we will have an anti-straight cloud organization in the United States of America.

This meeting went way, way beyond airplanes. It went into a very vital subject, the atmosphere, and what we are breathing. It went into the dependence on non-renewable energy, which one of these days we will run out of.

One subject that has not been talked about much, but is very important, is that space can contribute substantially more knowledge about our atmosphere than we have obtained so far. It can contribute substantially more knowledge of our atmosphere than we have really planned to find out. It can see and explore our environment in substantial detail. Our instrumentation state of the art is unbelievably full.

We have a scientist at Lockheed who works in our research laboratories, Dr. Loran Acton. He is also an astronaut. He became an astronaut because his focus and his talent is to figure out how to look at the sun in a way that no one else has accomplished. He has developed some fascinating cameras to do this trick. He has flown on the shuttle and he has taken these unique pictures and lots of others. The most impressive picture, which he will hit you over the head with if you will listen to him, is a picture of the earth where you see the whole horizon nearly North to South -- a gorgeous picture. There is a thin white line that goes all the way along the horizon. Dr. Acton points out to you that this thin white line is all the atmosphere there is. We need to be extremely careful of it.

Unfortunately there is another side to the atmospheric question. Knowledgeable people have been impressed by the ability of great computers and programs to model and predict complex interactive phenomena. We also have in this country what I call "promotional scientists". They take untested programs concerning what happens to the atmosphere and they predict all sorts of strange and disastrous things that are going to happen to that atmosphere if you blow your nose. These programs are intensely important. We ought to work on them. But, we ought to seek the facts these programs need -- facts which we can get from space. We must not guess, because if we start guessing and feed these guesses into these programs, we can scare ourselves to death. If we scare ourselves to death we will not only be frightened, but we will probably destroy our economy in the process because of the things we erroneously predict we have to do. Every one of us who works in this field should dedicate their efforts to rationally plot a future with space instrumentation as a core. We really need to know what is truly important.

This meeting has underlined one thing that we know how to do that will create a great deal of good, and nobody is going to quarrel with us -- the sooner we get to using hydrogen the better.

That leads me to conjecture a little about how do introduce hydrogen gently? This won't be easy, but I think it is possible. Then I want to make what I think is an important statement. We do not need the United States government in the act. It will cost us three to five times as much via government support if we are going in the right direction. The government, bless its heart -- in all good will -- starts out with prejudice, based on the defence of programs already established and politically oriented.

Because I am supposed to be an expert, I believe that the way to start is not with an airplane. Once upon a time I was intrigued with what my pencil type computer had created. I thought it would be a good idea to start with an airplane to introduce hydrogen as a fuel, but it is not. The investment cost is fierce for an industry that is just barely staying alive at best. It creates a singular infrastructure that does not help anybody else. It is ultimately a sound idea, but it is not where we should start.

My candidate is the automobile. Everybody is working on an automobile. It is the most obvious polluter we have in the country, in the world. The automobile reaches every voter, particularly in the United States. We are obsessed with our automobiles. We need political support to do something right. Political support from the ground, not from the top. Furthermore, if we ever get the automobile converted it will produce the broadest infrastructure of any other kinds of hydrogen-based efforts.

If we can figure out how to start a little at a time on commercial, not government, dollars I think we will have made a pretty good start. I heard in this meeting a tiny touch of what my pet approach is. In a speech a couple years ago I suggested that if I were going to try to design an automobile to start the use of hydrogen I would start on the luxury end of the line, not the kiddy car end. I would have a twin engine automobile. Maria Sanchez mentioned that this afternoon in one of her talks. The reason that I would like to suggest a luxury vehicle is that one should strive to make it absolutely as glamorous as possible. Two engines may be needed early on for reliability. The car will need full electrics, the drivers at the wheels, regenerative brakes, the whole works. My suggestion for a storage mechanism Roy McAlister talked about a little today, mainly the high pressure carbon adsorption system. That is looking pretty good. What I am suggesting is what I call an "envy" market. This "envy" market is a market that can pay the price. This is the kind of thing that a Saudi Arabian prince would buy and somebody that is as rich as heck in the United

States would buy. Out-mercedes Mercedes! What we have to do is be sure that Mercedes does not out-mercedes us. This kind of an automobile let's you do the whole system at once. That may be kind of tough to start, but once you have it done you have shown the true advantages of hydrogen as a future fuel for nearly everything.

This market can pay the price. Lawrence O'Connell would not really say what the price of the Chrysler T-Van was. I read an article in the Los Angeles Times that said it was going to be $120,000 dollars. If I am going to pay that much for an automobile, I do not think I want to buy a van. I want to buy something a little bit more comfortable. It has a range of about 80 miles I think. Our sights must be set higher.

We also need to know how to build the infrastructure. It seems like a backwards step, but I am really worried about how we can build the infrastructure a little at a time. It is sort of like shooting a cannon off a little at a time. You know it is not easy. I think that what we ought to do perhaps is to use natural gas -- it is already piped nearly wherever we want it -- reforming it to hydrogen at the filling station. That gets us started and we can do it a lot sooner than we could create any other kind of infrastructure. Later when we have some renewable sources and when the market is a little larger, we can start piping the hydrogen itself around.

I worry about the Department of Energy and I am sorry for you folks in the audience who are from the Department of Energy. I detect a strong prejudice. This ethanol-methanol effort seems to me to be a dead end. We are going to spend a great deal of money and we won't have any place to go afterwards. I hope we take another hard look at that. We must not go down dead ends. We can't afford it.

Renewables are all around us. It is obvious that there is only one energy carrier that fits them all: solar farms, Otec, the hydro and wind sources, and that is hydrogen. We have not really done a convincing job of defining how much the atmospheric damage going on now is costing us. T. Nejat Veziroglu has done a credible beginning analysis. We have got to recognize those costs. I want to use commercial dollars. There may be a powerful incentive for the commercial world. It fits the politics of today, and has been suggested by the others. What we ought to do is tax pollution. The commercial world is absolutely expert at avoiding taxes. This will get them going. They will be willing to spend money to avoid such a tax, we will get money to start with. I am very serious. I think we ought to create a major pollution tax.

What have I said? I hope I have said in an articulate and convincing way that we must all be on guard and not let political scientists panic the nation. We ought to use space as soon as we can to get the facts on what is going on in our atmosphere. We should try to do whatever we are going to do in introducing hydrogen by commercial dollars, and see if we can figure out a way to start small and grow. We ought to have in mind where we are going so that it is a good systematic start. We ought to try if at all possible to avoid government bureaucracy. They are trying to do their best, but they are forced into short term political goals and biases. These biases turn out to be technical prejudices, and we cannot afford to have our energy future depend on bureaucratic planning and research directives.

I suggest that, in doing this within industry, we take a hard look at a kind of a development process that has taken place in some defense programs, just to spread the talent base and risks. This process involves substantial industry partnerships -- industry partner agreements that even include universities. This is not quite a consortium. It is different, because a partnership like this has a leader. One entity is in charge. Such an organization does not have to have a day by day consensus, as in a consortium, and progress will be much faster.

We ought to start an "envy" market. If that market can start driving beautiful automobiles that do not make a sound when they go by, we will have something that all people will want. These voters will see to it that the government doesn't get in the way and it won't be too long before we'll find out how to make it affordable for everyone.

Lastly, I think that the time is very close technically, politically, and actually for us to start. I hope we do it soon. Once started, the airplane will come. It will attach itself to whatever infrastructure is developed. A hydrogen transport airplane can be a major step forward. Let's move fast; I'm not young and I would like to stay around and watch for the airplane.

CHAPTER 13: STUDENT ENERGY AWARENESS COMPETITION

Winning Essays

The Student Energy Awareness Competition, a science essay contest involving high school and college students from all over the world, was held in conjuction with Project Energy '93 for the purpose of encouraging young people to become actively involved in solving the environmental/energy crisis which affects the world community.

Nearly 200 students entered the contest, writing to the topic "An Energy Solution We Can Live With". The Grand Prize-winning entries follow.

An Energy Solution We Can Live With

Stefan Griebel
Dolores, Colorado, U.S.A.

An Energy Solution We Can Live With

Stefan Griebel
P.O. Box 240
Dolores, Colorado 81323
U.S.A.

Architecture integrated with passive solar heating technology is becoming more and more popular due to concern about the earth's fossil fuel depletion, global warming, and economic strain. Utilizing solar energy for space heating is a relatively simple idea first employed by early man who situated his house in such a way as to take advantage of warmth provided by the morning sun. Since this first use, the idea of using passive solar energy for heating has been expanded upon a multitude of times through analytical studies culminating in books, data, and charts that provide informative patterns for modern building applications.

The following essay describes and explicates the designing of a sample home (Figure 1) that incorporates passive solar heating systems into its design to maximize the energy efficiency of the house. The house design was developed only to the extent of fulfilling these heating and cooling requirements and should not be considered a professional construction design.

This home was conceived incorporating passive solar energy systems into the original house plan to better the efficiency and reduce cost in the long run. The site for the house to be located is approximately 40° NL at an elevation of 8000'. This setting is an ideal Colorado mountainous setting that has fairly, but not extremely cold winters, and mild summers. A passively heated home was chosen over one heated by active means because passive systems are virtually silent and require less upkeep and maintenance.

The overall shape of the building is elongated about the east/west axis to expose more surface area to the south sun during the winter. It is also tall (three levels) to maximize vertical surfaces which receive the most winter sun. The north side is partially underground on a south-facing hillside to reduce heat loss. This type of construction also allows for easier landscaping such as evergreen trees above and to the north of the house, and deciduous (ex. Aspen) trees below and to the south of the house. These trees would help shelter the house from wind and other climatic conditions.

Part of this home, mainly the garage and other buffer spaces, is located underground. The constant underground temperature is approximately 58°, and is very ideal in winter when the average outdoor temperature in January for this location is below freezing. This automatically raises the temperature of interior spaces in winter and keeps them cooler in summer.

A major problem associated with passive solar heating is one of heat loss at night, mostly through direct gain solar windows so movable insulation, operable windows, and a back up wood stove will make up for this heat loss. Since a manually-controlled window insulation system requires operation by the building occupant at least twice daily, an automatic timer system will be employed. This timer will be programmable and can he turned on and off at will if window views are desired at night. The insulation around the windows, clerestories, and skylights will be made of 1-1/2" thick folding polystyrene panels with an R-3 insulation value rating. A chain track system will provide the mobility for this insulation; kind of like track lighting, but track insulation. Infiltration is also a major source of heat loss. To reduce this heat loss, all entries are recessed and airlocked to minimize infiltration.

In this house, the main type of system used is direct gain. For example, vertical thermal masses are used to heat northern interior spaces such as Bedroom #1 and Bedroom #2. All of the masonry in the house is either brick or concrete. Each has a magnesium additive to increase the rate of conduction. This magnesium additive can make the masonry conduct heat more efficiently. The floor of the entire first level is made of one-foot thick concrete to serve as thermal mass and as the foundation of the house. Other concrete structures include the entire garage, being one-foot thick on the floor, eight inches thick on the south side, and six inches thick for the underground walls and ceiling. An underground building should be built out of concrete and metal for strength, stability, and ease of waterproofing. The floor of Bedroom #3 and the Master Bedroom are made of the same type concrete with Magnesium additive, except only four inches thick. Four inches is the minimum thickness for thermal gain and a thickness of more than four inches constitutes too much weight for a non-ground level. A floor thickness of more than 4 inches is also very hard to support and build.

The ratio of Solar Window square footage to space floor area in this house is approximately 0.4 square feet of south-facing direct glazing (solar windows) to one square foot of floor area in every room with a Direct Gain System. This allows for adequate heat absorption by the masonry floor without excess heat flow. Some thermal walls are simply vertical thermal masses constructed of brick with Magnesium additive and are located on the north side of the Sunroom, Den, Bedroom #2 and the Master Bedroom. They are also located around the wood stove and on the south wall of Bedroom #2. These interior vertical masses also serve as support for second- and third-level floors and are six inches thick, except for the one on the north side of the sunroom. The sunroom's masonry wall is only four and one-half inches thick because it receives nearly twice as much sun as the other masses, and the room is supposed to maintain the highest temperature of any room in the house during the day.

This house is designed to have 0.2 - 0.6 square feet of thermal wall for each one square foot of floor space and approximately 0.5 - 1.0 square feet of direct sunlight for each one square foot of vertical mass. The area of the floor was calculated and then an approximation of the size of the vertical mass was determined, taking into account the amount of light that could be admitted to that space. Because clerestories and skylights allow for maximum freedom when placing a thermal mass, the vertical masses in Bedroom #1, the Master Bedroom, and the Den, are all heated by clerestories, and the mass in Bedroom #2 is heated by a skylight and has the wood stove pipe running through it for additional heat.

The masonry around the wood stove is constructed, again, of brick with a masonry additive and is six inches thick. It is located so that when the stove operates, heat is stored next to it and radiated into the Living Space and Washroom, thus decreasing the amount of time needed for wood to be burned.

There is one true indirect-gain thermal wall in this structure, and it is formed in front of the kitchen. It is eight inches thick and set four inches behind a large 3' 6" x 7' 6" double-glazed, double-paned, solar window. Thermosiphoning vents are located at the bottom and top of the wall with dampers over each vent to control unprecedented heat gain or loss. This type of wall is also known as a Trombe Wall.

The south-facing exterior wall is eight inches thick, has a total R-value of 20, and is composed of the following: one inch of outside stucco, one-half inch insulating board sheathing followed by six inches batt-blanket fiberglass insulation supported by two-by-six's and studs. Finally, the inner layer is one-half inch gypsum board or paneling. The R-value is slightly over 20, but for simplicity is just 20.

Northern exterior and underground walls are six inches thick and are composed of one and one-half inches expanded extruded waterproofing polystyrene followed by four inches of solid concrete and one-half inch gypsum board for interior aesthetics.

The roof has an R-value of 24 and is composed of an outer skin of reflective metal for snow removal and summer cooling purposes. Beneath this is one inch insulating board sheathing, six and one-half inches batt/blanket fiberglass insulation and interior gypsum board, supported by a roof truss system.

Interior Walls are simply one-half inch gypsum board on each side supported by two-by-four's (two by three and one-half inches) and studs with no insulation.

Shading overhangs are located above every south-facing window/wall. These overhangs are one-third the length of the wall being shaded and so in all cases are three feet long at an angle of 30" to the horizontal. This will block out nearly all direct summer sun rays, while letting nearly all direct sun rays in winter to penetrate the glazing.

A trellised overhang is located along the outside of the living space. This overhang will have green perennial plants interwoven about it to provide adequate summer shading because of their leaves, and yet even more direct sunlight penetration in winter because of their autumn leaf loss.

The garage is located on the east side of the building so that morning sun can help melt off the snow in winter, and because the majority of Colorado winter storms hit from the east. The roof and floor of the garage are sloped to the east at approximately a four- to five-inch angle to allow for run-off drainage in spring and late winter. A gutter will carry the water away from the driveway to the south.

Passive solar energy is an excellent alternative heat source for any home. In northern climates, a properly designed house can be very efficient, using all or most of the natural energy available to it. It would be possible to construct a home that is 100% efficient in the winter, but the same house in the summer would possess an unbearable overheating problem. This is what makes the integration of solar technology into architecture so difficult. There is a delicate balance between the amount of consideration given to heating in winter, and to cooling in summer.

Direct Gain is by far the easiest and most efficient system to use, and therefore has the largest amount of heat-gain contribution to this home. Other systems work well also, but take up more space and time on the owner's part. Incorporating a Direct Gain system must be done before the home is built to be truly efficient. Other systems can be retro-fitted to already existing homes, but without prior planning, solar efficiency is minimal.

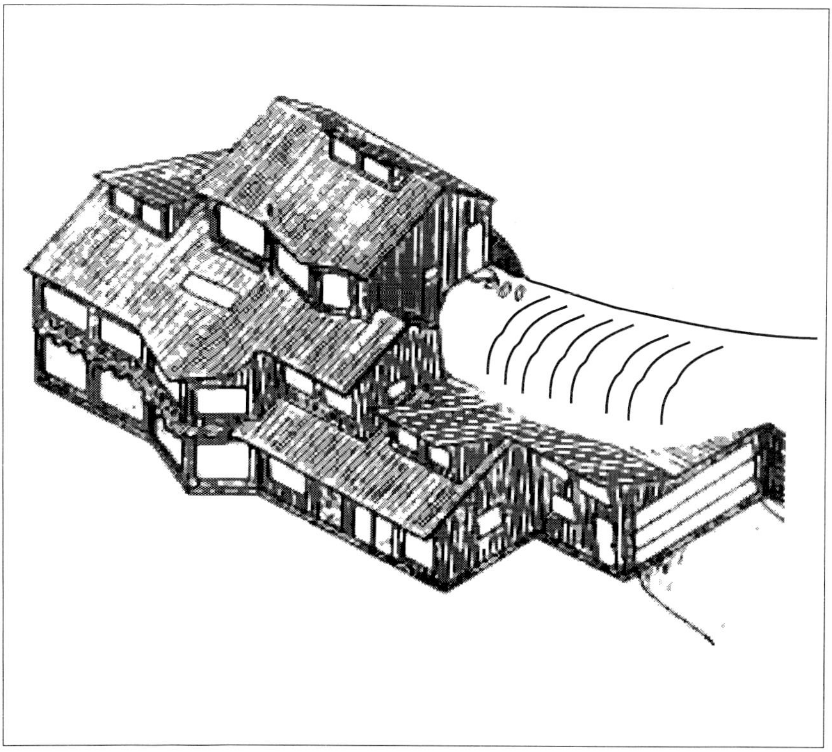

Figure 1

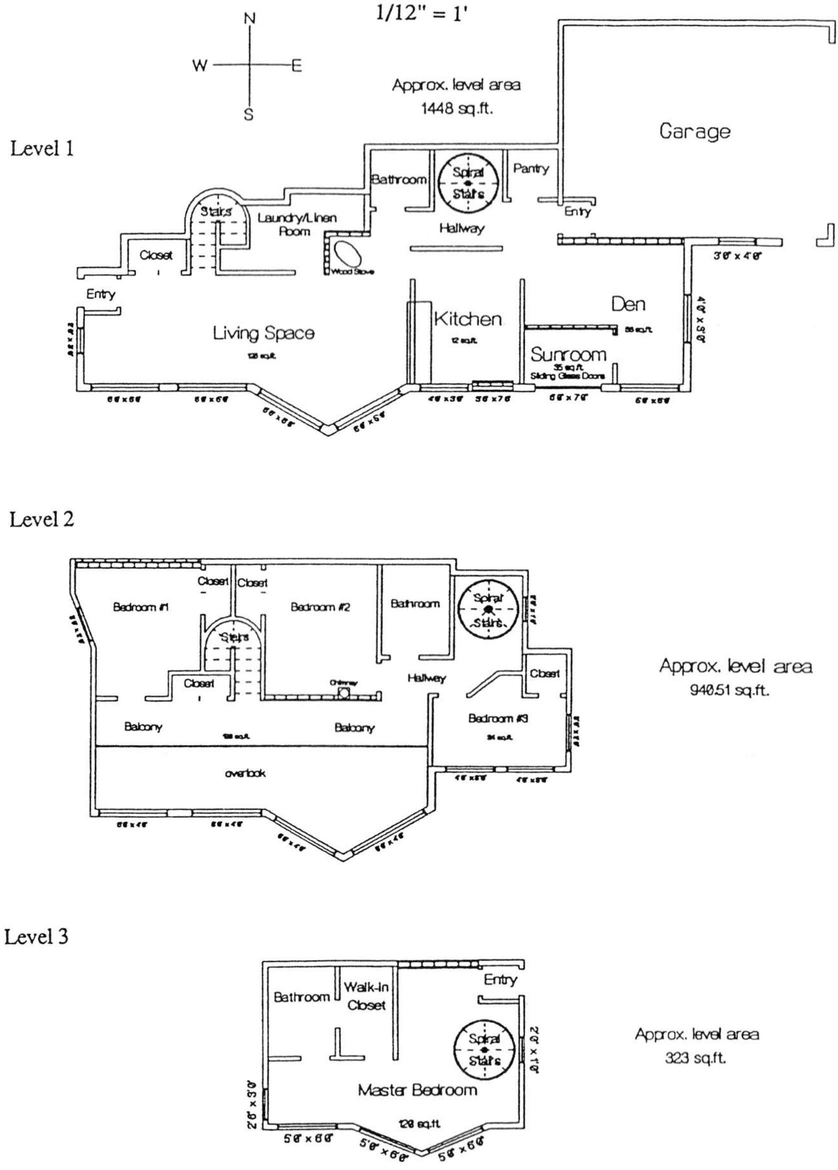
1/12" = 1'
N
W
E
S
Approx. level area
1448 sq.ft.
Level 1
Garage
Bathroom
Spiral Stairs
Pantry
Stairs
Laundry/Linen Room
Entry
Hallway
Closet
Wood Stove
3'0" x 4'0"
Entry
Living Space
Kitchen
Den
Sunroom
Sliding Glass Doors
Level 2
Closet
Closet
Bedroom #1
Bedroom #2
Bathroom
Spiral Stairs
Stairs
Closet
Chimney
Hallway
Closet
Balcony
Balcony
Bedroom #3
overlook
Approx. level area
940.51 sq.ft.
Level 3
Bathroom
Walk-In Closet
Entry
Spiral Stairs
Master Bedroom
120 sq.ft.
2'0" x 1'0"
2'6" x 3'0"
5'0" x 6'0"
5'0" x 6'0"
5'0" x 6'0"
Approx. level area
323 sq.ft.

Figure 2

An Energy Solution We Can Live With

Richard Miller
London, U.K.

An Energy Solution We Can Live With

Richard Miller
32 Aldridge Road Villas
London W11 1BW
England

On June 14th 1992, negotiators from 172 governments pushed, cajoled and pulled each other towards the formal close of history's largest international conference, the Earth Summit in Rio de Janeiro, (which I attended as Special Adviser to the UN Youth and Environment Ambassador, Robert Swan). The denouement was centered on the two unfinished chapters - finance and the protection of the atmosphere - of Agenda 21, the blueprint for survival in the 21st Century. Whilst the gravity of these issues and the threats to the planet's existence were recognized, possible solutions could be agreed. Richard Bendick, veteran diplomat, energy expert and senior adviser to Earth Summit Secretary General, Maurice Strong, had to conclude, "The most important single thing . . . is energy, because that's going to save money and it's going to save resources. This is an aspect of the conference that so far has been ignored, and the world ought to know about it." Let us hope that Project Energy '93, the International Academy of Science and all those gathered here in Kansas will help overcome the ignorance of these issues.

The way we produce and consume energy touches every part of our existence and is at the heart of the problems we face. We have now reached the point where the destructive side-effects of our short term attitude to energy are outweighing the benefits. Some side effects are immediate and local, while others are subtle, indirect, long-term and far more insidious. We have to confront life-threatening pollution, global warming, acid rain and ozone depletion, energy security and geopolitical struggles, to name but a few. Nothing that has been agreed at climate control conventions, such as CO_2 limits, gives much comfort, when the reality is that even the most conservative governmental projections predict dependence on nuclear and fossil fuels and thereby the increasing emission of greenhouse gasses and pollutants. It is clear that we are now in a survival situation and that we cannot rely on governments or private sector industry alone to promote or produce a real energy solution. We have to respond not with despair, but with innovation.

Science and technology are blamed all too often for the scale of our problem, but in truth, these are neutral tools whose effects are defined by the wisdom of their controllers. We must bend the great power conferred on human purpose by science and technology to environmentally sound and sustainable pathways of development, which lead to reduced consumption of resources, in both energy and materials, and to new resources to produce the required quality of life and opportunities. We need to examine our attitude towards energy. Do we really want energy or do we want energy services? Is it coal, nuclear power, oil, electricity that we want or the facilities to provide warmth, illumination, cooked food, clean water, manufactured products and mobility? The focus should be on satisfying the services required and not purely on volume energy production.

The path to the real energy solution is not just an external technological one. To begin with, we need positive environments and structures that allow the space for inner creative energy to innovate solutions. How do we go about this? The key is the process of collaboration, as expressed in the Rocky Mountain Institute philosophy, "People can solve problems through collective action and their own common sense, and by understanding the interconnections between resources often many problems can be solved at once." It is estimated that Nobel Prize winners spend 70% of their time collaborating with their peers around the world and there is no doubt that the energy of working together has been the crucial factor leading to major advances and innovation; so it will be in the future, particularly if the means are enhanced. At Rio, it was this prime opportunity for the beginning of a truly global collaboration on energy, which was lost. Kansas City offers another chance.

However, the Earth Summit did illustrate that, in the world at large, there is a discernible will to collaborate, driven by the dawning realization that nations and peoples cannot survive independently from each other, that our local problems are shared ones. Environmental threats are no respecters of national boarders and manmade divisions. Therefore, we require new bridges of energy consciousness.

Now, more than ever, we need to establish fora to stimulate international collaboration and coordination of research and development of better processes and for their promotion, implementation and introduction into our societies throughout the world. The process and methodology of these fora can be mapped very simply as follows: identify focus/problem; analyze problem; generate ideas/solutions; organize and evaluate ideas; form policy/strategy;

evaluate strategy; implement strategy. This process can feed back into itself continuously.

Many would say that this formula is not only obvious, but already exists and can name and enumerate institutes and organizations that perform collaborative roles. Of course I will agree. However, the collaboration that exists is not efficient and effective enough to produce processes that will match the complexity of the threats we are facing day to day. The problem is how to facilitate the required level and spread of collaboration. The good news is that the tools are now becoming available. It is only recently that we have begun to turn our minds to the real potential that computers, software, networks, databases and satellites have for enabling groups to maintain beneficial relationships and therefore to solve problems and reach the right decisions. Modern business now recognizes that inspiring and managing people effectively is the route to sustainable success and that means supporting the interaction of the knowledge workers rather than just enhancing production methods or ever increasing random data distribution. "Groupware", the next generation of networking or communication software that supports the networking of minds, can accelerate decision making and problem solving in an efficient and effective manner in a way which is easily scaleable and not restricted by the constraints of face-to-face meetings. Users of these new collaborative tools have achieved remarkable efficiencies (by factors of 10 and more) in man hours to complete tasks and by conserving time, our most valuable non-renewable resource, we will perhaps have the opportunity to innovate and survive.

Once an effective global energy organization that can respond to shifting requirements and is supported and enhanced by "Groupware" has been established, the process of finding, developing, implementing and promoting appropriate solutions can begin. What are they likely to be?

There are two parts to the energy equation, the production of energy carriers and their use. It is clear that if energy is used more efficiently, less needs to be produced to satisfy demand. Improved energy hardware, insulation, building and heating systems, coupled with the use of ambient energy (passive solar heating) can dramatically reduce the need for services (current efficiencies can be as low as 1% of potential) and costs. The behavioral education of users will make a considerable contribution. Some enlightened authorities are beginning to make significant inroads in this area, but more awareness raising and lobbying for positive legislation as well as financial support is required. Energy efficiency will save money and resources and give us time to tackle the problems of energy production.

Turning to the supply side, let us consider the ideal attributes of an energy carrier. It should be renewable, non-pollutant, safe to use and capable of supplying future requirements. Beyond that it is desirable for energy carriers to be generated virtually at the point of use to save the expense and inefficiencies of transporting them about. The primary renewable energy source is the sun. Energy from the sun drives the climate and biosystems on the planet from which we can derive energy. While wind, wave, tidal and conventional solar technologies can contribute to the overall picture where geography permits, the majority of these alternatives are either not sufficiently concentrated to provide adequate levels of service on their own or are too concentrated to control (i.e. hurricanes). However, developments in other potent power sources including geothermal, both natural and hot dry rock, biomass, photovoltaic and hydrogen hold out the most significant hope for sustainable energy solutions.

Perhaps the most cogent argument could be made for a solar hydrogen combination, where photovoltaic could be used to produce electricity with surpluses stored as hydrogen through the medium of fuel cells. The extraordinary versatility of hydrogen and the range of uses that fuel cells allow is providing the more innovative of our energy scientists with a wealth of options for clean and sustainable energy. Hydrogen, safely stored with metal hydride technology, provides a powerful, clean-burning fuel capable of servicing home, office and industrial needs and will probably be the solution to our seemingly intractable transport problems. Research also indicates that hydrogen could be collected from blue-green algae. All these are very real possibilities and combinations which need to be explored.

It is certain however, that none of these possibilities will come to light without the social and political pressure to provide the funds, both public, private and philanthropic, or without the room to manoeuvre created by collaborative alliances, enhanced by communication technology, and available energy efficiencies. Amulya Reddy, in "Energy for a Sustainable World", made the poignant observation that although it is extremely difficult to implement new ideas, it is less difficult than continuing with the present impossible ones."

The future is indivisible with the present, a seamless continuity with the past. The quality of our future to the extent of our survival depends on nothing more, but certainly nothing less, than the fully expressed belief in a cooperative stewardship of the great gifts with which we toy. The analysis is simple: the choice of an energy solution we can live with is ours and at hand.

INDEX OF AUTHORS

INDEX OF AUTHORS